Fundamentals of Aviation Law

Fundamentals of Aviation Law

Raymond C. Speciale, Esq., C.P.A.
Mount St. Mary's University

McGraw-Hill
New York Chicago San Francisco Lisbon London Madrid Mexico City
Milan New Delhi San Juan Seoul Singapore Sydney Toronto

The McGraw·Hill Companies

Cataloging-in-Publication Data is on file with the Library of Congress

Copyright © 2006 by The McGraw-Hill Companies, Inc. All rights reserved. Printed in the United States of America. Except as permitted under the United States Copyright Act of 1976, no part of this publication may be reproduced or distributed in any form or by any means, or stored in a data base or retrieval system, without the prior written permission of the publisher.

4 5 6 7 8 9   DOH   15 14 13 12

ISBN 0-07-145867-0

The sponsoring editor for this book was Stephen S. Chapman and the production supervisor was Pamela A. Pelton. It was set in Times by International Typesetting and Composition. The art director for the cover was Anthony Landi.

Printed and bound by RR Donnelley.

This book was printed on acid-free paper.

McGraw-Hill books are available at special quantity discounts to use as premiums and sales promotions, or for use in corporate training programs. For more information, please write to the Director of Special Sales, McGraw-Hill Professional, Two Penn Plaza, New York, NY 10121-2298. Or contact your local bookstore.

Information contained in this work has been obtained by The McGraw-Hill Companies, Inc. ("McGraw-Hill") from sources believed to be reliable. However, neither McGraw-Hill nor its authors guarantee the accuracy or completeness of any information published herein, and neither McGraw-Hill nor its authors shall be responsible for any errors, omissions, or damages arising out of use of this information. This work is published with the understanding that McGraw-Hill and its authors are supplying information but are not attempting to render engineering or other professional services. If such services are required, the assistance of an appropriate professional should be sought.

To Xiaoping

Contents

Table of Figures xi
Table of Cases xiii
Reviewers xv
About the Author xvii
Preface xix
Acknowledgments xxi

1 Fundamentals of the U.S. Legal System 1

Introduction to the Law 2
Functions of the law 2
Sources of law 3
Classifications of law 6
Law, ethics, and morals 8
Effective legal systems 9
Dispute Resolution 11
Court system 11
Jurisdictional matters 13
Litigation process 17
Alternative methods of dispute resolution 25
Basics of Legal Research 26
Case law 26
Statutes and regulations 27
Law review articles 27
Cases and Commentary 27
Discussion Cases 31
Endnotes 36

2 The U.S. Constitution and Aviation 37

Overview of the U.S. Constitution 38
Structure and organization of the federal government 39
Distribution of authority between the federal and state governments 41

Basic rights of individuals and businesses 49
Cases and Commentary 52
Discussion Cases 59
Endnotes 64

3 Impact of Criminal Law on Airmen and Air Carriers 67

Basics of Criminal Law 68
Classifications of criminal law 68
Elements 69
Criminal procedure 69
Constitutional protections for defendants 71
Criminal Laws Affecting Aviation Activities 74
Federal provisions 74
State provisions 76
Cases and Commentary 76
Discussion Cases 83
Endnotes 89

4 Tort Liability and Air Commerce 91

Intentional Torts 93
Intentional torts against persons 93
Intentional torts against property 95
Negligence 97
Elements of negligence 97
Defenses to negligence 102
Wrongful Death 104
Strict Product Liability 104
Elements 104
Defenses to strict product liability 105
Cases and Commentary 108
Discussion Cases 114
Endnotes 117

5 Administrative Agencies and Aviation 119

Administrative Agencies 121
Department of Transportation 122
Federal Aviation Administration 122
National Transportation Safety Board 123
Functions of Administrative Agencies 123
Rulemaking 124
Enforcement 126
Adjudication 131
Special considerations in FAA matters 133
Checks on Administrative Agencies 138
Oversight by traditional branches of government 138
Equal Access to Justice Act 140
Freedom of Information Act 141
Privacy Act 141
Sunshine Act 142
Cases and Commentary 142
Discussion Cases 153
Endnotes 156

6 Commercial Law Applications to Aviation-Related Transactions 159

Contracts 160
Mutual agreement 160
Consideration 161
Capacity 162
Lawfulness 162
Defenses to contracts 164
Rights, duties, and remedies for breach 167
Sales Law 170
Sales law versus contract law 171
Warranties 171
Transfer of title and risk of loss 173
Debtor-Creditor Legal Issues 174
Credit—unsecured and secured 175
Bankruptcy 177
Cases and Commentary 179
Discussion Cases 187
Endnotes 189

7 Entity Choice for Aviation Enterprises 191

The Sole Proprietorship 192
Establishing a sole proprietorship 192
Liability issues for sole proprietors 193
Taxation of sole proprietorships 193
Partnerships 194
Establishing a partnership 194
Operation of a partnership 194
Liability issues for partnerships 195
Ownership of assets 196
Taxation of partnerships 196
Termination of a partnership 197
Limited Partnerships 197
Establishing a limited partnership 198
Liability of general and limited partners 198
Taxation of limited partnerships 200
Limited Liability Company 200
LLC formation 200
LLC operating agreement 200
Liability of LLC members 201
LLC taxation 202
Corporations 202
Formation of a corporation 203
Operating the corporation 204
Duties of directors and officers 204
Liability issues for corporations and shareholders 205
Taxation of corporations 206
Cases and Commentary 208
Discussion Cases 212

8 Property Law Issues for Aircraft Owners and Airport Operators 217

Real Property 218
Ownership rights in real property 219
Multiple ownership 219
Transfers of real property 220
Easements 221
Local zoning issues 221
Leasing 222
Airports 223

Personal Property   224
Tangible property   224
Intangible property   230
Insurance   230
Hull insurance and liability insurance   231
The insurance contract   232
Other types of insurance   234
Cases and Commentary   236
Discussion Cases   247
Endnotes   251

9 Employment Law and the Aviation Industry   253

Agency Law   254
Defining agency   254
Employee versus independent contractor   254
Creating and terminating agency relationships   255
Duties of parties   257
Contract liability   258
Tort liability   259
Labor Unions and Employment   261
Major federal labor union laws   261
Basics of collective bargaining   261
What if collective bargaining fails?   262
Strikes and related issues   263
Employee Protection   263
Employment discrimination   263
Employee safety and security   265
Cases and Commentary   269
Discussion Cases   276
Endnotes   278

10 International Aviation Law   279

Fundamentals of International Law   280
Origins of international law   280
Public International Aviation Law   281
The beginnings of public international aviation law   281
The Chicago Convention   282
Open skies agreements   283
Private International Aviation Law   283
The Warsaw Convention   284
The Cape Town Convention   286
International Civil Aviation Organization   287
Cases and Commentary   287
Discussion Cases   295
Endnotes   297

Appendices

A   Case Briefs   299
B   The Constitution of the United States of America   303
C   NAS ASRS Form 277B   321

Selected Bibliography   325
Index   327

Table of Figures

Figure No.	Caption
1-1	Features of common and code law
1-2	Sources of the law
1-3	Comparison of civil and criminal law
1-4	Federal and state court systems
1-5	Complaint
1-6	Summons
1-7	Answer
1-8	Case citations
2-1	Constitutional checks and balances
3-1	Levels of fault in criminal cases
3-2	Summary of constitutional protections in criminal cases
4-1	Determining proximate cause
5-1	FAA Letter of Investigation
5-2	FAA Notice of Proposed Certificate Action
5-3	Appeal of FAA order
5-4	Flowchart of FAA enforcement actions (nonemergency cases)
5-5	Flowchart of FAA enforcement actions (emergency cases)
6-1	Illustration of contract assignment
6-2	Two-party secured transaction
6-3	Three-party secured transaction
7-1	Sole proprietorship liability
7-2	Sole proprietorship taxation
7-3	Partnership liability
7-4	Partnership taxation
7-5	Limited partnership liability
7-6	LLC liability
7-7	LLC taxation
7-8	Corporate management structure
7-9	Corporate taxation
8-1	FAA bill of sale
8-2	FAA aircraft registration application
8-3	Initiation of an aircraft escrow transaction
8-4	Closing an aircraft escrow transaction
9-1	Summary of employee protection provisions

Table of Cases

Case No.	Case Name
1-1	*Lucia v. Teledyne*
1-2	*Helicopteras Nacionales De Columbia, S.A. v. Hall et al.*
2-1	*United States v. Causby et ux.*
2-2	*City of Burbank et al. v. Lockheed Air Terminal, Inc. et al.*
2-3	*Braniff Airways, Inc. v. Nebraska State Board of Equalization and Assessments et al.*
3-1	*Robert David Ward v. State of Maryland*
3-2	*United States of America v. SabreTech*
3-3	*United States v. Evinger*
3-4	*California v. Ciraolo*
4-1	*Brockelsby v. United States of America and Jeppesen*
4-2	*Altseimer v. Bell*
5-1	*Air Transport Association of America v. DOT and FAA*
5-2	*Garvey v. NTSB and Merrell*
5-3	NTSB Identification: NYC05FA001
6-1	*Dallas Aerospace, Inc. v. CIS Air Corporation*
6-2	*Edward Miles, Richard V. Keenan and Kenneth L. "Dusty" Barrow, Appellants, v. John F. Kavanaugh, Appellee*
6-3	*In re: UAL CORPORATION, et al., Debtors*
7-1	*Klinicki v. Lundgren*
7-2	*Nelsen v. Morris*
8-1	*Philko Aviation v. Shacket et ux.*
8-2	*James Bowman v. American Home Assurance Company*
8-3	*Western Food Products Company, Inc. v. United States Fire Insurance Company*
8-4	*South Carolina Insurance Company v. Lois S. Collins*
9-1	*David E. Hollins v. Delta Airlines*
9-2	*Professional Pilots Federation, et al. v. FAA*
10-1	*Air France v. Saks*
10-2	*El Al Israel Airlines, Ltd. v. Tsui Yuan Tseng*
10-3	*Resham Jeet Singh, Gursharan Jeet Kaur, Individually and as Guardians of Gurpreet Kaur v. Tarom Romanian Air Transport*

Reviewers

William V. Cheek, Esq.

Professor Bill Cheek is an associate professor and grants coordinator at Embry-Riddle Aeronautical University (ERAU) in Prescott, Arizona. He earned his Juris Doctor degree at the University of New Mexico. Professor Cheek has presented more than 30 papers on law and aviation economics and published several papers on airline economics and law. He is the managing partner of an aviation consulting firm, and has served in executive capacities in several airlines, including general counsel of Alaska Airlines.

Judge John E. Faulk

In 1968, Judge Faulk was appointed as an administrative law judge with the Civil Aeronautics Board and served in that position for approximately 6 years. On April 1, 1974, he joined the National Transportation Safety Board as an administrative law judge where he served until his retirement in November 1990. From 1983 to June of 1992, Judge Faulk was a member of the faculty of the College of Aeronautics, Florida Tech, where he served as a professor of aviation law and regulations. He is currently an adjunct professor of aviation law at Florida Tech. Judge Faulk is associated with the law firm of Jerry H. Trachtman, P.A., in Melbourne, Florida.

Ronald D. Golden, Esq.

Ron Golden graduated from West Point and Catholic University School of Law. He is an active pilot and aircraft owner with commercial and instrument ratings. For nearly 30 years he has practiced law with Yodice Associates and served as associate general counsel to the Aircraft Owners and Pilots Association (AOPA). Through his work with AOPA, Mr. Golden has handled scores of airport-related cases throughout the United States, and hundreds of FAA enforcement cases involving general aviation, corporate, and air transport pilots. He has also assisted hundreds of aircraft owners in aircraft title and taxation issues.

Joel K. Lyon, Esq.

Joel K. Lyon is currently working as a senior claims representative for AIG Aviation, Inc., in New York. He is a commercial pilot, flight instructor, CFII, and MEI. He earned his Juris Doctor and B.S. in aeronautical science from the University of North Dakota. He served as research editor on the North Dakota Law Review Board. Mr. Lyon served as an associate for Yodice Associates (AOPA counsel). He also worked for a summer with Jackson, Wade & Blanck, L.L.C., in Kansas City where he gained experience in aircraft transactions as well as regulatory and tax matters. He has completed a flight operations internship with Delta Air Lines and is a former flight instructor for the University of North Dakota.

Richard Theokas, Esq.

Professor Richard Theokas is an associate professor of aeronautical science at Embry-Riddle Aeronautical University (ERAU) at Daytona Beach Florida. Professor Theokas earned his Juris Doctor from Mercer University School of Law after a nearly 30-year career as an officer and pilot in the U.S. Air Force. During his Air Force career, Prof. Theokas flew C-130 tactical airlift and KC-135 aerial refueling aircraft. He currently teaches aviation and business law classes at ERAU.

Brett D. Venhuizen, Esq.

Professor Brett Venhuizen received his bachelor of science and Juris Doctor degrees from the University of South Dakota. He has been an active pilot and certified flight instructor since 1990. Currently, Prof. Venhuizen is an assistant professor teaching aviation law at the University of North Dakota's John D. Odegard School of Aerospace Sciences in Grand Forks, North Dakota. He also has a law practice representing pilots in FAA enforcement actions.

About the Author

Raymond C. Speciale is a practicing attorney with Yodice Associates, counsel to the Aircraft Owners and Pilots Association (AOPA) for over 40 years. During more than 15 years as an aviation attorney, he has provided legal services to hundreds of aircraft owners and pilots. Mr. Speciale is an active pilot and flight instructor (CFII). Also a certified public accountant, he has written several booklets and articles for the AOPA related to aircraft ownership and taxation issues. He teaches law and accounting classes at Mount St. Mary's University, where he is an assistant professor. Mr. Speciale is a member of the Lawyers-Pilots Bar Association and the National Transportation Safety Board Bar Association. He lives in Frederick, Maryland.

Preface

This book was primarily designed to serve students and instructors in aviation law classes at the university, college, and community college levels. The Council on Aviation Accreditation (CAA) *Accreditation Standards Manual* requires that most undergraduate and graduate programs in aviation cover national and international aviation law and regulations. The content of this book was specifically selected to meet CAA's accreditation standards for the typical aviation law course.

In certain respects, the title of this book is misleading. There is no universal recognition of "aviation law" as a distinct legal subject matter. Often, legal matters concerning airlines, aircraft, airports, and/or airmen involve issues that are more directly addressed in traditional legal subject matter such as commercial law, tort law, employment law, and property law. Therefore, when a reference is made to aviation law, it is often referring to the legal environment of aviation.

It is important for all aviation professionals including pilots, executives, air traffic controllers, and mechanics to have a fundamental understanding of the legal environment in which they operate. The United States is a nation with a foundation built on the rule of law. The world of aviation is no exception. Almost every facet of the aviation industry is impacted by some body of law or another.

The purpose of this book is not to turn students into lawyers. However, it is intended to bring legal context to many of the issues that aviation professionals will come in contact with during their careers.

Each chapter begins with an overview of the subject matter to be discussed. For some students who have already taken other law courses, these overviews will serve as a review. For students who have never been exposed to a law class, the overviews are meant to provide the legal basics of the subject addressed. An understanding of these basics is necessary to get students to the point where they can apply the law to aviation-related situations.

After the fundamentals are addressed, a "Cases and Commentary" section is available for deeper study of how the law affects aviation-related activities and situations. The cases are edited with most footnotes, citations, and some text omitted to allow students to focus on the primary issue(s) of the cases. At first, students may find reading the cases challenging. With a bit of time and practice, the hope is that students will come to a greater appreciation of the complex issues faced by lawyers and judges when arguing and deciding cases. Students are encouraged to read the full text of the cases to evaluate dissenting opinions and other cases cited within the opinions. Based on time available, instructors may wish to pick and choose the cases reviewed in class.

After the "Cases and Commentary" section students get an opportunity to tackle the issues under "Discussion Cases." The discussion cases are meant to prompt written responses that can be reviewed during class time. Many of the discussion cases are based on actual litigated cases. Students should not be frustrated by the fact that there may not be clear yes or no responses to the questions put by the discussion cases. In many respects, the cases selected were meant to prompt questions and debate—that is what the law is all about. Beyond exposing students to the legal environment of aviation, these cases are also meant to develop students' critical thinking and writing skills.

As is true with any text, there are bound to be some errors or omissions—especially in a first edition. Any comments, suggestions, or corrections are welcome. Students and instructors are invited to write to the author directly at speciale@msmary.edu.

Acknowledgments

Putting a book together can sometimes be difficult and lonely work. I needed lots of help to get this project completed.

First and foremost I want to thank my wife, Xiaoping. She sacrificed much to allow me the time to write this book. Xiaoping also provided expert technical assistance with manuscript preparation and the associated charts and figures in the book. She is my greatest blessing.

My work was made a lot less lonely because I had such a great group of reviewers. They provided much needed technical assistance and writing suggestions throughout the year that I was preparing the manuscript for this book. Having these experienced and knowledgeable colleagues available gave me confidence that I was headed in the right direction. The mix of professors, a former judge, practicing attorneys, aircraft owners, and pilots in the reviewer group led to balanced and useful feedback.

I also want to thank all my supportive colleagues and friends at Mount St. Mary's University where I teach and at Yodice Associates (AOPA counsel) where I serve of counsel. I am very fortunate to be engaged in the professional worlds of teaching and practice. Sometimes it is difficult to switch back and forth between these two worlds, but it regularly provides me a fresh perspective on both endeavors.

Last but not least, I want to thank my publisher, McGraw-Hill. I am very grateful for its confidence in this book. I am particularly indebted to Steve Chapman, my editor, for his patient and encouraging efforts to see this book through from beginning to end.

Fundamentals of Aviation Law

1 Fundamentals of the U.S. Legal System

INTRODUCTION TO THE LAW 2
Functions of the law 2
Sources of law 3
Classifications of law 6
Law, ethics, and morals 8
Effective legal systems 9

DISPUTE RESOLUTION 11
Court system 11
Jurisdictional matters 13
Litigation process 17
Alternative methods of dispute resolution 25

BASICS OF LEGAL RESEARCH 26
Case law 26
Statutes and regulations 27
Law review articles 27

CASES AND COMMENTARY 27

DISCUSSION CASES 34

ENDNOTES 36

We live in a world where virtually every aspect of our personal and work lives is touched by the legal system. Aviation activities are no exception. From the moment a pilot embarks on a flight, her every action is guided by a complex set of laws and regulations designed to enhance safety. Whenever aircraft or aviation equipment is sold, there is a set of laws in place governing the rights and duties of the parties to the transaction. When airports are constructed or expanded, local, state, and federal codes must be considered. When things go wrong and people and/or property is harmed due to an aviation accident, there is a well-established body of law in place to identify those who might be at fault and compensate the victims.

The purpose of this book is to provide you with a fundamental working knowledge of how the law impacts aviation activities. The purpose of this chapter is to introduce you to the law and the basics of the U.S. legal system.

INTRODUCTION TO THE LAW

The law has been defined as "that which is laid down, ordained, or established. A rule or method according to which phenomena or actions co-exist or follow each other. Law, in its generic sense, is a body of rules of action or conduct prescribed by controlling authority, and having binding legal force...."[1] This widely accepted definition of the law expresses the often prevailing view that the law is a series of related rules that govern the conduct of human behavior.

For centuries, philosophers have wrestled with a definition of the law. One fitting summary of the diverse perspectives on the law states:

> We have been told by Plato that law is a form of social control, an instrument of the good life, the way to discovery of reality, the true reality of the social structure; by Aristotle that it is a rule of conduct, a contract, an ideal of reason, a rule of decision, a form of order; by Cicero that it is the agreement of reason and nature, the distinction between the just and the unjust, a command or prohibition; by Aquinas that it is an ordinance of reason for the common good, made by him who has care of the community, and promulgated [thereby]; by Bacon that certainty is the prime necessity of law; by Hobbes that law is the command of the sovereign; by Spinoza that it is a plan of life; by Leibniz that its character is determined by the structure of society; by Locke that it is a norm established by the commonwealth; by Hume that it is a body of precepts; by Kant that it is a harmonizing of wills by means of universal rules in the interests of freedom; by Fichte that it is a relation between human beings; by Hegel that it is an unfolding or realizing of the idea of right.[2]

Functions of the law

As you can see, even among some of the world's greatest thinkers, there is no universal agreement when it comes to defining the law. However, there are enough common threads among these varying definitions to argue that the purpose of the law is, at least in large part, to encourage or ensure certain standards of conduct in human behavior.

Despite the difficulty that people have had in pinning down a definition of the law, we can generalize that the law accomplishes its purpose (in large part) by (1) providing for orderly dispute resolution, (2) protecting property rights, and (3) preserving the structure and integrity of government.

As will be discussed a bit later in this chapter, it is inevitable that disputes will arise between parties. Sometimes the disputes will involve criminal behavior. Other times the disputes will involve noncriminal or civil matters, including aircraft accidents and safety violations. Disputes always carry a potential threat to order. That is why a legal system must have a well-developed and flexible dispute resolution system in place. For all practical purposes, the legal system and its courts take the place of less desirable remedies such as revenge or self-help.

Preservation of private property and rights to private property is another key function of any worthy legal system. For an economy and commerce to thrive, participants must be assured there are rules and procedures in place to protect their rights. For instance, aircraft owners must be assured that if they follow the rules, the aircraft they purchase will belong to them and cannot be taken by another. Contractors building new airport runways need to be certain that the contracts they entered into with airport owners to build the runways will be enforceable under the law. If basic property rights are unprotected, people will be unwilling to engage in commerce, thus diminishing the growth of the economy and the benefits that growth brings to individuals and communities.

A good legal system also protects the government put in place by the people. Our legal system puts protections in place to ensure that leadership changes and changes to our laws take place through a sanctioned (although often untidy) process. The U.S. legal system accomplishes this objective through elections, lawmaking, referenda, and court decisions. Less developed systems suffer change through revolution, rebellion, and sedition.

Sources of law

A well-established body of law has emerged in the United States over the last few centuries. To understand the U.S. legal system, it is important that you understand the primary sources of the law as we know it today. The following discussion provides a brief overview of those sources and the hierarchy of those sources.

Constitutions
The primary source of law in the U.S. legal system is the U.S. Constitution. The Constitution is often referred to as the supreme law of the land in the United States because any law—even an otherwise validly adopted federal, state, or local law—that conflicts with the Constitution is deemed to be illegal and unenforceable.

An essential function of the Constitution is the establishment of structure for the federal government. The Constitution created the three branches of government: the legislative (with the power to create and enact laws), the executive (with the power to enforce laws), and the judicial (with the power to interpret and determine the validity of the law).

Any powers not included in the Constitution are said to be reserved to the states. Each state has its own constitution. Most state constitutions are patterned after the federal Constitution but may be more detailed and provide for even broader protections of individual liberties (however, a state constitution cannot roll back federal Constitutional protections).

As indicated earlier, state constitutional provisions are valid unless they conflict with the federal Constitution.

The role of the federal Constitution in aviation matters is pervasive. Chapter 2 will discuss the interaction of the federal Constitution and aviation activities in greater detail.

Treaties

The federal Constitution authorizes the President, with the advice and consent of the U.S. Senate, to enter into treaties with foreign states. These treaties become part of the supreme law of the United States. In today's world with evergrowing international trade and economic relationships, the role of treaties will continue to grow. In the world of aviation, the ability to travel freely from one country to another necessitates strong treaties between nations, in order to provide for the safe and efficient transport of persons and property across international boundaries. A more detailed overview of how international treaties impact the aviation community can be found in Chap. 10.

Common law

Common law generally means law that has developed from adjudicated cases. It is sometimes called case law in the United States; this common law may relate back to case law originating in the ancient and unwritten laws of England. Essentially, common law is any law that was not derived from legislative bodies. You may have heard the term *legal precedent*. This means that a similar case was decided in a certain manner in the past. Although a court in a particular case may not be held to the precedent of a previous case, typically a court will feel obligated to follow the same rule in the case currently being decided. That approach to deciding cases by using past cases is called the doctrine of *stare decisis*.

In most cases, trial courts will be bound by the mandatory authority of appellate courts above them that have decided similar controversies in a certain manner. For example, a Maryland trial court judge will usually be bound by the decision of the highest appeals court in Maryland that decided a factually similar case. Sometimes the trial court judge may be forced to observe the precedent of a higher court even if she vehemently disagrees with the outcome it dictates in the case before her. On the other hand, if a judge in Maryland is faced with a case that has no clear precedence in Maryland, he may be able to turn to courts in other states for precedent. However, this type of precedent, from another state, will only rise to the level of persuasive, not mandatory, authority. As you will see, the doctrine of *stare decisis* can even influence the decisions of an administrative law judge from the National Transportation Safety Board (NTSB) dealing with Federal Aviation Administration (FAA) enforcement matters.

We'll explore the impact of common law on aviation-related matters in Chaps. 4 ("Tort Liability and Air Commerce"), 6 ("Commercial Law Applications to Aviation-Related Transactions"), and 8 ("Property Law Issues for Aircraft Owners and Airport Operators"). However, throughout your studies, you will be regularly exposed to cases where courts are faced with applying statutory law or treaties to specific circumstances.

Code law

If federal, state, or local lawmakers have enacted rules related to certain issues, those rules become code or statutory law. Today, almost all federal and state laws are the result of legislative enactments. Most aviation-related rules are an outgrowth of federal statutes. Rules related to corporations (and other business entities), commercial law, and criminal law are substantially all statutory in nature.

Legislative enactments often relate to the structure and day-to-day functions of government. For instance, it is imperative that legislators spell out the rules for operating aircraft. Leaving such matters to the courts for interpretation and guidance over time could prove unwise and certainly unsafe. Concerns such as aviation safety require fully developed and detailed laws and regulations in order to provide as much certainty as possible to aircraft and airport professionals. Discussions related to criminal law (see Chap. 3), business entities (see Chap. 7), and labor and employment (see Chap. 9) will deal largely with statutory law and its impact on the aviation environment. For an overview of the differences between common law and code law, see Fig. 1-1.

Administrative rules and regulations

The legislative and executive branches of federal and state governments are empowered to create administrative agencies. These administrative agencies are usually formed when the subject of legislation is so complex and/or enforcement of the laws is such a substantial task that a specialized government entity is required to support the legislature.

Administrative agencies are given the power to interpret the law. To do so, most administrative agencies adopt rules or regulations that serve as interpretations of the legislatively created law. Of course, the government agency with the biggest impact on aviation activities is the Federal Aviation Administration. A fuller discussion of administrative agencies and how they function can be found in Chap. 5 ("Government Regulation of Aviation").

Priority of laws

When you are sorting out the priority of laws in the U.S. legal system, the U.S. Constitution ranks first. The second highest priority goes to ratified treaties. Federal statutes rank above

Features	Common law	Code law
Creation	Created by judges ruling on particular cases	Drafted by elected legislators in formal lawmaking process
Form	Common patterns found by lawyers and courts in similar cases	Codes and statutes
Scope	Narrowly limited to case at hand	Throughout jurisdiction affected
Political and other outside forces	Depends on insulation of judges from societal and political pressures	Typically a high level of political and external pressure

Figure 1-1
Features of common and code law.

Figure 1-2
Sources of the law.

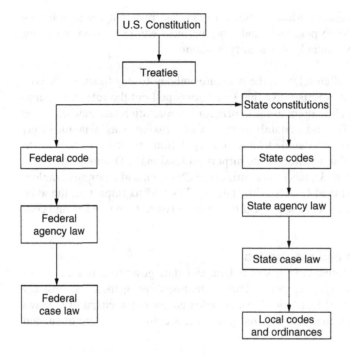

state statutes and take precedence over conflicting state statutes. State constitutions are the highest state laws, followed by state statutes and state regulations. State statutes take precedence over local laws. See Fig. 1-2 for an overview of the sources of law and priorities.

Classifications of law

There are several ways to classify the law. While an attempt to classify the law is not meant to ignore the interdependence of the various types of law, the classifications can serve to break the law down into component parts, thus making it easier to study, analyze, and compare. The most traditional classifications of the law are as follows:

- Subject matter
- Federal and state law
- Civil and criminal law
- Public and private law

Subject matter

One of the most common ways to classify the law is by subject matter. In fact, this text attempts to break down the study of the legal environment of aviation into various subject matter classifications. Most law schools and legal textbooks attempt to view the law as a collection of separate, but connected subjects including but not limited to

- Administrative law
- Agency
- Business entities
- Commercial law

- Contracts
- Criminal law and procedure
- Evidence
- Family law
- Negotiable instruments
- Real and personal property
- Sales
- Taxation
- Torts
- Wills and estates

Federal and state law

One other common approach to classifying the law is to look at where the law comes from. In some cases the law may come from the federal legislature. There are several important areas of federal law, not the least of which relates to aircraft and airport operations. However, there are far more areas of the law that are based on state statutes and/or common law. Although it may seem as though state laws generate 50 completely different sets of rules, this is not necessarily the case. In fact, most state laws parallel those existing in other states.

Civil and criminal law

Civil laws are laws that govern how individuals, businesses, and government agencies interact with one another. If someone breaches a contract or injures another due to negligence, civil laws may have been violated. If the parties involved in a civil action cannot settle their dispute, the person claiming to have been harmed becomes the plaintiff and the person who allegedly committed the civil wrong is the defendant.

The remedies sought in a civil case are typically monetary damages—with the idea that the monetary damages sought will compensate the victim of a civil wrong for his or her damages. This kind of case is common when an aircraft accident causes someone harm. The victim will sue for the damages suffered as a result of the accident. Sometimes the remedy sought may be an order called an injunction. An injunction is an order from the court requiring a party to do something or to refrain from doing something. One example of a plaintiff seeking an injunction is an aircraft owner who wants to stop an airport operator from imposing a curfew that the aircraft owner believes to be contrary to law. If the aircraft owner prevails on the injunction, the airport operator will be unable to enforce the curfew.

The usual standard of proof required in a civil case is "by a preponderance of the evidence." In most courts this means that the plaintiff will succeed if she or he proves that it is more likely than not that the defendant committed a civil wrong. For those who prefer to quantify this sort of thing, it might be fair to say that this standard implies greater than 50 percent probability that the plaintiff's charges are correct.

Criminal actions are actions brought by a federal or state government against persons or entities (corporations and other businesses) that have violated a criminal code. In most-cases, a criminal case involves a wrong so serious that it could require the imposition of large fines or imprisonment of the person or persons who violated the law.

In criminal cases, the usual standard of proof is "beyond a reasonable doubt." Essentially this means that a criminal defendant can be convicted only if the government has proved its case to a near certainty. Sometimes lawyers will say that this means a jury or a judge must be 99 percent sure that the defendant has committed the crime that he or she has been accused of committing.

Serious crimes are called felonies. Lesser crimes are generally referred to as misdemeanors. Typically, felonies carry penalties of jail time over 1 full year whereas misdemeanors carry smaller fines or shorter sentences.

You should note that sometimes a wrongful act can be a civil wrong and a criminal wrong at the same time. For instance, someone who strikes another person with a weapon can be sued in a civil action for monetary damages and at the same time can be prosecuted by government authorities for a crime. See Fig. 1-3 for a summary comparison of criminal and civil law.

Private and public law
Sometimes lawyers and legal scholars will refer to the distinctions between private and public law. Usually, references to private law reflect legal issues between nongovernmental parties. For instance, a lawsuit by an airline passenger against an air carrier for lost baggage would typically be classified as a private law issue. On the other hand, if the Federal Aviation Administration were attempting to fine an airline for substandard maintenance practices, it would be classified as a public law issue. As a general rule, criminal law and administrative law matters are often viewed as being within the realm of public law.

Law, ethics, and morals

To be more effective in your study of the law, you should begin considering the distinctions between law, ethics, and morals. As indicated earlier, the law deals with standards of conduct that society enforces to keep order. To some extent, the law can be looked

Figure 1-3
Comparison of civil and criminal law.

Key characteristics	Civil law	Criminal law
Initiation of action	Person alleging harm files a complaint with the courts	Federal or state government files a complaint
Function	Compensation Reimbursement Deterrence	Punishment Deterrence Rehabilitation Preservation of civil order
Burden of proof	Preponderance of the evidence—more likely than not	Beyond a reasonable doubt—virtual certainty
Sanctions	Monetary damages Injunctions Equitable relief	Capital punishment Incarceration Monetary fines Community service

upon as a sort of bare minimum standard of behavior. We obey the law to stay out of trouble.

The term *ethics* is often used to describe accepted professional standards of conduct. These are typically codes of conduct created by communities of professionals to guide them in their field of endeavor. For instance, although it might not be contrary to the law for a physician to accept a referral fee from a specialist or a surgeon, accepting such a referral fee might be a breach of ethical conduct. Therefore, ethics will often involve standards of conduct that exceed legal standards.

Perhaps the highest standards of conduct are based on an individual's determination of what is right and wrong. These standards are often referred to as morality. Moral standards are often based on an individual's family background, religion, community, and a whole host of other factors. It would be a mistake to confuse moral and legal standards. They often intersect in areas of criminal law; for instance, it is both morally and legally wrong to kill another person. However, these standards often part ways when it comes to more difficult questions. For instance, should a person be legally responsible to assist another who is in peril? As a general rule, our legal system does not require us to assist someone who is in peril.[3] However, many people would insist that it is a moral imperative to come to the aid of another person who is in danger.

Effective legal systems

As you study the law and how it affects aviation-related activities, you should also reflect on whether the current system of laws benefits or hurts the aviation community and the communities it serves. To fully understand the law, it is not enough to simply know the rules. You must be able to see the purpose of the law and determine whether the law meets its stated purposes. At times, even the purpose of the law may come into question. Part of the joy (and challenge) of studying law and legal systems lies in critically analyzing the law and its effectiveness.

Scholars who have studied different legal systems tend to believe that there are certain characteristics of effective legal systems. As a general rule, the most important requisites for a vibrant and resilient legal system appear to be (1) stability, (2) adaptability, (3) availability, and (4) fairness. Legal systems that embody these qualities tend to be effective and long-lived. A brief discussion of these attributes is found below. You should consider these qualities as you critically review the laws you encounter in your studies.

Stability

For a legal system to be effective, it must create and operate within an environment of stability. Many of our activities—aviation being just one—are based on the assumption that the guiding principles of law will remain unchanged. For instance, it would be hard to find an aircraft finance company willing to extend credit to airlines looking for aircraft if the company were not certain that the law would require the airlines to meet their obligations to pay the loans. The behavior of individuals and businesses is in large part guided by their faith in the stability of our legal system. If you take that stability out of the mix, the legal system will fail and ultimately the community served by that legal system will be harmed.

Adaptability

A legal system also needs a strong measure of adaptability to be effective. The emergence of aviation in the early 1900s demonstrated the flexibility of our legal system. Prior to the existence of aircraft, it was widely understood that a landowner owned rights to everything that existed below and above her or his property. This would become an obvious problem for aviators. Under the law as it existed, any aircraft flying above a landowner's property could be liable for trespassing. Our legal system had to adapt to the new technology of aircraft—and it did, with the legal system permitting reasonable use of airspace for the purpose of air navigation.

Of course, the evolution of aircraft technology, jet propulsion, satellite navigational systems, and countless other innovations forces the law to adapt constantly. Standards for aircraft certification have to be revised, and new rules for air navigation and sharing airspace must be developed. Our legal system needs the flexibility to adapt to these changes to maintain its credibility and effectiveness.

Availability

No legal system can be effective unless the people and entities using it are able to avail themselves of the law as needed. Practically speaking, this means that citizens must have reasonable access to the law and the legal system.

The U.S. legal system is one of the most thoroughly evolved systems of law ever. People seeking access to rules and regulations only need to go to the Internet to find many of the rules pertinent to their questions. It is routine for federal, state, and local agencies to create electronic web sites with links to code, cases, and regulations pertaining to their areas of regulation. For complex questions, it is always important for individuals and businesses to be able to consult with experts on the law, typically lawyers. Again, the U.S. legal system makes lawyers available—but some would argue at too high a cost—for advice and counsel.

When disputes arise, most courts and regulatory agencies must also be available to assist in the dispute resolution process. A bit later in this chapter, we will examine the basics of the U.S. dispute resolution system.

Fairness

Perhaps the most important characteristic of a viable legal system is fundamental fairness. For a legal system to function, citizens and participants must respect the system and abide by the rules—even if they don't necessarily agree with the rules.

For a legal system to be fair, it generally must comport with the same sense of right and wrong as those subject to the rule of law. If it does not, it may be viewed as irrelevant and ultimately disregarded or replaced.

In the history of the U.S. legal system, perhaps the most poignant example of a law out of touch with reality was the prohibition amendment which unequivocally outlawed the sale and distribution of liquor throughout the United States. This law was so harsh that it caused widespread and blatant noncompliance. The contempt and dissatisfaction generated by the law ultimately led to its repeal a few short years after its enactment.

DISPUTE RESOLUTION

For better or for worse, one of the primary functions of any legal system is dispute resolution. In the United States, there is a well-developed system of courts at the federal, state, and local levels designed to manage disputes among the government, individuals, and business entities. Increasingly, alternative means of dispute resolution such as arbitration are becoming more popular due to the sometimes inefficient and costly process of litigating cases in the public courts. This section will deal with the systems in place for dispute resolution. It will be important for you to become well acquainted with various dispute resolution systems now. Your subsequent study of aviation-related matters will include significant numbers of cases and rulings—most borne from disputes that were resolved in the court system.

In the United States, courts are designed as neutral or impartial tribunals (judges). Courts are established and funded by government bodies to serve the function of resolving disputes. The U.S. system is a dual court system. Each of the 50 states has its own court system, and the federal government has a separate court system.

Court system

Federal court system

Article III of the U.S. Constitution states that the judicial branch of the United States shall be vested in a Supreme Court with Congress having the power to establish inferior (often referred to as lower) courts. Since the inception of the U.S. Constitution, Congress has put in place a federal court system comprised of courts of appeal, district courts, and special courts. Judges must be appointed to these courts by the President and confirmed by the Senate. Once a judge is confirmed to sit on a federal court, her or his appointment to the court is for a lifetime. Presumably, this was designed to take the dynamics of politics out of the judicial equation to the extent possible.

District courts serve as the general trial courts in the federal court system. Nearly all federal cases commence in district courts. The district courts decide matters of fact and law. Usually only one judge presides over any given case at the district court level—in some special cases up to three judges may preside. Appeals from district courts typically go directly to circuit courts of appeal. In very rare instances, appeals from district courts will go directly to the U.S. Supreme Court. Located in every state is at least one federal district court, while some states, such as New York, have as many as four. To a large extent, the number of district courts assigned to a state is based on population and need.

Our federal system also houses 12 judicial circuits (11 numbered circuits plus the District of Columbia circuit). Each of these circuits has a court named the court of appeals (also known as U.S. Circuit Court). A court of appeals will often hear appeals from district courts located within its circuit. Additionally, a court of appeals will hear appeals from the decisions of administrative agencies including, but not limited to, the National Transportation Safety Board (see Chap. 5). Each court of appeals has nine judges. Usually, a court of appeals hears cases by using a panel of three judges. In unusual circumstances, it may agree to hear a case *en banc*, where all nine judges of the circuit participate in deciding a case. The function of these appellate courts is to review the record of a case on appeal and to decide whether the case was properly handled by a lower court. If the court of appeals determines that the lower court committed error, it may reverse or modify the lower court's decision. If error was committed, the court of appeals may also remand the case back to the lower court for further review in light of the court of appeals decision. If the court of appeals determines that the lower court did not commit error, it may simply affirm the decision of the lower court.

Our federal court system also includes some special courts (sometimes referred to as courts of special jurisdiction). The highest of these special courts is the U.S. Court of Appeals for the Federal Circuit. This special court of appeals hears appeals from the U.S. Court of Federal Claims (a trial court specializing in claims against the federal government), the U.S. Patent and Trademark Office, and the Court of International Trade. Other special courts include the U.S. bankruptcy courts (federal courts with exclusive jurisdiction over bankruptcy cases) and the U.S. tax court (federal courts with exclusive jurisdiction over certain federal tax cases). To a large extent, these special courts were put in place because of the need for special judicial expertise in certain areas of the law such as bankruptcy, taxes, intellectual property, and other highly technical subjects.

Of course, the highest federal court is the Supreme Court. The Supreme Court has nine justices (a Chief Justice and eight Associate Justices). These justices sit as a group at the Supreme Court in Washington, D.C. As indicated in the Constitution, there are some cases (e.g., cases involving admiralty matters) in which the Supreme Court has the right to hear a case first (often referred to as original jurisdiction). Nonetheless, the vast majority of cases reviewed by the Supreme Court involve appeals from U.S. Courts of Appeals and the highest state courts. Most of the time, the Supreme Court gets to pick the cases it wishes to review. To bring a case before the Supreme Court, a party seeking an appeal must file a petition for a *writ of certiori*. Only a small percentage of the cases presented to the Supreme Court for review are ever accepted. Typically, the Supreme Court will be most interested in reviewing cases involving matters of great national importance or cases that expose a rift in opinions between various courts of appeal.

State court systems

As indicated earlier, each of the 50 states (and the District of Columbia) has its own unique court system. In most of these systems, judges are either politically appointed or elected, or a combination of both is done. While it is impossible to generalize about all 50 systems, it is possible to outline the common threads that run through the structure of most state court systems.

Most state court systems employ courts of limited or inferior jurisdiction to handle small civil and/or criminal matters. Often, these courts are identified as small claims courts, traffic courts, or justice of the peace courts. Usually these courts do not keep a written record of proceedings. However, decisions of these lower courts can be appealed. The appeals are heard most likely by the next court up the ladder, the trial court of general jurisdiction.

Every state has trial courts of general jurisdiction. These courts are often referred to as superior courts, circuit courts, county courts, or courts of common pleas. There is no dollar limitation on the judgments in the jurisdiction of these courts. Trials are more formal at this level, and most often legal counsel is required. In addition to trial courts of general jurisdiction, most states have special jurisdiction courts for matters such as probate and estates, family law matters, and tax matters. Appeals from trial courts of general jurisdiction and these special jurisdiction courts are usually heard by appeal courts within the state system.

State appellate court systems typically include an intermediate court of appeals and the state's highest court, sometimes, but not always, known as the state's supreme court. It is important to note that a decision by a state's highest court may be reviewed by only the U.S. Supreme Court.

A flowchart illustrating the basic structure of the federal and state court systems can be found in Fig. 1-4.

Jurisdictional matters

For a court to hear a case and make a decision, it must have legal authority to hear a case. The special term used in the law for the authority to hear and decide a case is *jurisdiction*.

Courts need two types of jurisdiction before they can properly assert authority over a case. The first type of jurisdiction required is subject matter jurisdiction. In plain terms, the court must have the legal authority to deal with the subject matter of the case presented. If subject matter jurisdiction is lacking, the court's decision will have no legal effect.

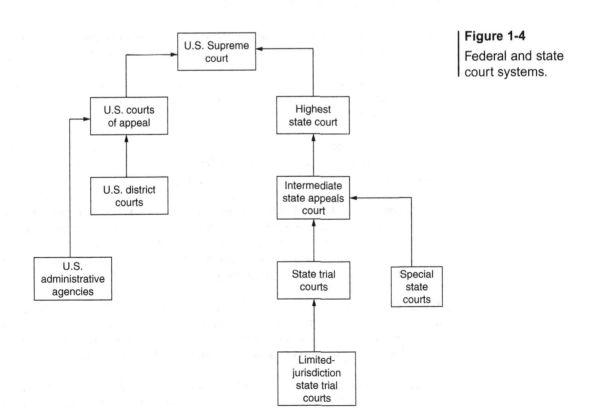

Figure 1-4
Federal and state court systems.

The second type of jurisdiction necessary is jurisdiction over the parties. This is often referred to as personal jurisdiction. Personal jurisdiction requires that the court be able to exercise legal authority over the persons or entities involved in a case. Just as with subject matter jurisdiction, a court lacking personal jurisdiction cannot render a legally effective or binding ruling.

Subject matter jurisdiction

The U.S. Constitution gives federal courts exclusive jurisdiction over a limited, but significant number of matters. All other cases are properly within the jurisdiction of state court systems.

Federal jurisdiction Congress has provided the federal court system with exclusive jurisdiction over the following types of cases:

- Admiralty
- Bankruptcy
- Federal criminal cases
- Antitrust
- Patents, trademarks, and copyrights
- Lawsuits against the United States
- Other cases specified by federal statute

As indicated above, any cases falling in these categories must be presented to the federal court system. They cannot be entertained by a state court under any circumstance.

However, under certain special circumstances both federal and state courts may properly exercise jurisdiction over cases. This shared jurisdiction is commonly referred to as concurrent jurisdiction.

The first type of concurrent jurisdiction arises when there is a federal question that does not give the federal courts exclusive jurisdiction. It is hard to pin down all cases in which a federal question may be present, but as a general rule, questions related to the U.S. Constitution, federal statutes, and federal treaties (that are not within the exclusive jurisdiction of the federal courts) can create a situation in which federal and state courts share jurisdiction concurrently. In essence, for a federal question to create concurrent jurisdiction, the plaintiff's complaint must contain substantial issues related to federal or constitutional law. There is no minimum dollar amount for a federal question claim to be considered for concurrent jurisdiction.

The second type of concurrent jurisdiction is exercised in cases in which the parties involved are from different states or foreign countries. If the case involves diversity of citizenship between the parties and the amount in controversy exceeds a good faith claim of $75,000, the case may be brought before a federal or state court.

The rules for gaining access to the federal courts in cases involving diversity of citizenship require that all plaintiffs be citizens from different states than the defendants.

Additionally, if either the plaintiffs or the defendants are U.S. citizens and the other party is a foreign citizen, diversity of citizenship jurisdiction may be obtained.

If a federal court hears a case based on diversity of citizenship jurisdiction and no federal question is presented, the federal court will be required to apply the appropriate state law in deciding the case. Sometimes this will create a question of which state law should be applied. This issue is discussed below under the heading "State Jurisdiction."

In any case involving concurrent jurisdiction, the plaintiff has the ability to bring the case to either a federal or a state court. However, if the plaintiff brings the case to a state court, a defendant may have the ability to have the case removed to a federal court within the district serving the state. It is hard to generalize about when a plaintiff or defendant might prefer to bring a case to state or federal court when there is a choice. Sometimes the decision may be based on perceived bias in a local or state court or a preference regarding certain judges or trial locations. Decisions regarding which court to use in a case of concurrent jurisdiction are typically left to the lawyers involved in a case, and the decisions typically are based on experience and intuition regarding the most favorable court system for a particular case.

State jurisdiction   In all cases where there is no exclusive federal jurisdiction or concurrent jurisdiction, parties wishing to resolve a matter before the courts must use state courts. Typically, state courts hear matters related to state criminal law, contracts, property matters, commercial transactions, and agency relationships.

Sometimes state courts may determine that it is appropriate to use the laws of a different state in deciding a case. For example, let's say that a Virginia resident sues a Maryland resident in Maryland related to a minor aircraft accident that occurred in Delaware. It is possible that the Maryland court will choose to apply the substantive law of Delaware in the case because the accident took place in Delaware. Each state has its own conflict of laws rules that it may apply in such a case. The rules vary from state to state.

Jurisdiction over the parties

The second test of jurisdiction that a court has to meet before it has authority over a case is jurisdiction over the parties. This is often referred to as personal jurisdiction.

As a general rule, courts may obtain personal jurisdiction over the parties in any one of the following three ways:

- *In personam* jurisdiction
- *In rem* jurisdiction
- Attachment jurisdiction

Each of these requisites for establishing jurisdiction over the parties is discussed below.

***In personam* jurisdiction**   *In personam* jurisdiction is another way of saying jurisdiction over the person. It is a way of distinguishing a court's jurisdiction over persons as opposed to its jurisdiction over property. A court will most commonly obtain *in personam* jurisdiction by serving process (court papers summoning an appearance) within the same state as it is located. Under this approach, a court may also obtain *in personam* jurisdiction over a nonresident who happens to be temporarily located within a state but is served while in the state. *In personam* jurisdiction may also be obtained by consent. This often happens when the parties to a contract agree within the contract that any disputes will be resolved by a particular state's courts.

Sometimes, a defendant resides, and must be served, outside the state. The question then becomes whether a state court can obtain jurisdiction over a nonresident defendant by service of process outside the court's state. In order to allow a court to take jurisdiction in this type of situation, most states have enacted "long-arm statutes" that allow a court to extend its reach beyond state borders. To comport with due process considerations in the U.S. Constitution (discussed in greater detail in Chap. 2), there are specific circumstances and limits attached to the exercise of long-arm jurisdiction. As a general matter, the law does not want to offend traditional notions of fair play and justice. Therefore, most long-arm statutes require that, to obtain jurisdiction over an out-of-state defendant, one or more of the following circumstances must exist:

- The defendant committed a tort (civil wrong) within the state.
- The defendant owns property within the state that is the subject of the lawsuit.
- The defendant entered into a contract within the state.
- The defendant transacted business that is the subject of the lawsuit within the state.

While these rules may seem cut and dry, they are sometimes difficult to apply. Examples of how different courts have applied long-arm jurisdictional tests can be found in the "Case Studies and Commentary" section of this chapter.

***In rem* jurisdiction**   If property is situated in a state, the courts in that state will generally have jurisdiction over any disputes or claims related to that property as long as the plaintiff provides reasonable notice to those parties who may have an interest in the property. *In rem* jurisdiction essentially gives a court power over a thing (property) and in turn allows the court to preside over a case involving the property. As an example, assume that Peters and Dean are involved in a dispute that relates to airport property that Dean owns in Delaware. Even if Peters were from Maryland and Dean were from New Jersey, a Delaware court would likely have *in rem* jurisdiction over their dispute as long as each party was given reasonable notice and an opportunity to be heard in litigation.

Attachment jurisdiction   Attachment jurisdiction is purely jurisdiction exercised over property, not persons. It is sometimes referred to in legal circles as *quasi in rem* jurisdiction. It allows a court to take jurisdiction over property that is not related to a lawsuit. As an example, let's assume that plaintiff Petry, a citizen of Virginia, has obtained a judgment against defendant Dent, a citizen of Pennsylvania, in a matter related to a computer sales contract. Petry would be allowed to seize Dent's aircraft located in Virginia to satisfy his court judgment against Dent.

Venue issues   Venue relates to the physical location of a court selected to hear a dispute, not to the type of court selected. Most state court systems have specific rules related to venue that serve to allow for a convenient forum for disputes. For instance, if an aircraft accident involving Maryland residents occurs within Maryland, proper jurisdiction in a negligence case against the pilot most likely resides in the trial court of general jurisdiction, the circuit court. However, there are more than 20 counties with circuit courts located in each Maryland county. Venue rules will help determine which county circuit court should be used to hear a case. In most situations, venue is appropriate where at least one defendant resides, where an incident or accident occurred, or where property subject to a dispute is located.

Litigation process

Our legal system is designed to permit citizens to settle disputes in a peaceful manner and is facilitated by the litigation process in the U.S. court system. Typically, when we discuss litigation, we are referring to dispute resolution in a court system funded by the government. Later in this chapter, we will take a look at alternative methods of dispute resolution that are often paid for by the parties to the dispute.

Civil procedure
Civil procedure is the process applied by court systems to resolve noncriminal disputes. The courts and legislators continuously refine civil procedure in efforts to make the process fairer and more efficient. Although there are distinct differences between the various court systems in the United States, it is fair to say that most systems follow the basic approach to civil procedure outlined in this section.

To enable you to better understand civil procedure, we will focus the discussion on an example. Let's assume that Doug Defendant, a resident of Delaware, is taxiing his aircraft to depart an airport where he stopped for fuel on his way to a business destination. While taxiing out of the ramp, he nearly collides with another aircraft being taxied by Pete Plaintiff, a resident of Maryland. Doug's aircraft suffers no noticeable damage, and Doug walks away without injury. However, Pete's aircraft is severely damaged when he takes evasive action and taxies his aircraft into the side of a maintenance shed. Pete suffers serious injuries requiring hospitalization for several weeks. It is estimated that Pete's medical expenses will be approximately $30,000, and damage to his aircraft is estimated at $50,000 to repair. Pete hires a lawyer who sends a letter to Doug, claiming monetary damages for Pete's property damage and hospital bills. Doug hires a lawyer and denies any fault in the matter. The parties attempt to settle the matter over the course of several months. However, all attempts to settle the matter fail.

Pete decides that he will sue Doug in court, to settle this dispute once and for all. Because of the diversity of citizenship and the amount of damages claimed, Pete decides to sue Doug in federal court. We'll follow the progress of this case through the civil procedure process in the discussion below.

Summons and complaint
The first step in litigation is typically the summons and complaint. The complaint in our example will be drafted by Pete's attorney. A well-drafted complaint lays out the relevant facts and the basis of the plaintiff's legal claim. The complaint must also specify the amount of damages that the plaintiff is seeking. An illustration of a complaint that might be filed by Pete's attorney in our example can be found in Fig. 1-5.

Figure 1-5
Complaint.

```
Pete Plaintiff,            *    IN THE
  Plaintiff                *    UNITED STATES DISTRICT COURT
v.                         *    FOR
Doug Defendant,            *    THE DISTRICT OF MARYLAND
  Defendant                *    CASE NO.: 030163-X
*   *   *   *   *   *   *   *   *   *   *   *   *
```

COMPLAINT

The plaintiff, through his undersigned attorney, states the following:
1. The plaintiff is a resident of the State of Maryland, the defendant is a resident of the State of Delaware, and there is diversity of citizenship between the parties.
2. The amount in controversy exceeds $75,000, exclusive of interest and costs.
3. On July 1, 2004, plaintiff was exercising due care in taxiing his aircraft at Martin State Airport, when defendant negligently taxied his aircraft into defendant's aircraft.
4. As a result of defendant's negligence, plaintiff has incurred medical expenses of $30,000 and property damage to his aircraft in the sum of $50,000.

WHEREFORE, plaintiff claims judgment in the amount of $80,000, interest at the maximum legal rate, and costs of the action.

Respectfully submitted,

_____
Joe Counsel, Esq.
Attorney for Plaintiff
1001 W. Lombard St.
Baltimore, Maryland

Students should note that the complaint in this example is extremely simple. Most complaints are very detailed, and some run more than one hundred pages.

Once the complaint is filed with the court, the court will issue a summons. The summons is designed to notify the defendant that a complaint has been filed. The summons also serves to place the defendant on notice that a written response to the complaint must be filed within a specified period of time (usually around 30 days for in-state defendants and 60 days for out-of-state defendants). A summons that might be issued by the court in our example is depicted in Fig. 1-6. Typically the summons can be served on the defendant by certified mail, delivery by a sheriff or sheriff's deputy, or delivery by private process server.

Answer

Assuming the summons and complaint reach the defendant in a timely manner, the defendant will have to decide whether to respond. In some cases, a defendant will not answer a complaint if he does not wish to defend his position. If this happens, the defendant may have a default judgment entered against him for failure to respond to the plaintiff's claim. In our example, Doug decides to respond and his answer to Pete's complaint might look something like Fig. 1-7. Notice that it is important for each claim contained in Pete's complaint to be answered, paragraph by paragraph. Whenever Doug denies a claim by Pete, Pete must prove his assertions at a trial. If Doug admits a

> UNITED STATES DISTRICT COURT
> FOR THE DISTRICT OF MARYLAND
>
> WRIT OF SUMMONS
>
> CASE NO. 030163-X
>
> To: Doug Defendant
> 100 West End Avenue
> Dorchester, Delaware
>
> You are hereby summoned to file a written response by pleading or motion in this court to the attached complaint filed by:
>
> Pete Plaintiff
> 3617 North Main St.
> Woodcrest, Maryland
>
> within 30 days after the service of this summons against you.
> WITNESS, the Honorable Chief Judge of the United States District Court for the District of Maryland.
> To the person Summoned:
>
> 1. Failure to respond within the time allowed may result in a default judgment or the granting of the relief sought against you.
>
> 2. If you have been served by a scheduling order, your appearance is required pursuant to the scheduling order, regardless of the date your response is due.
> Date issued: November 23, 2004
>
> _____
>
> Clerk, United States District Court
> District of Maryland

Figure 1-6
Summons.

claim by Pete, Pete does not have to prove the matter. There may even be occasions on which an answer may contain affirmative defenses. An affirmative defense essentially alleges that the plaintiff's injuries or damages were caused by the plaintiff's own wrongdoing. For instance, in our example, Doug responded with an affirmative defense, alleging that Pete's injuries and damages were caused by his own negligence in taxiing his aircraft.

If appropriate, a defendant may also take the filing of an answer as an opportunity to file a claim against the plaintiff in a case. This type of claim by a defendant is known as a counterclaim. With a counterclaim in place, parties to litigation may be both plaintiffs and defendants in the same action. If a counterclaim is filed by a defendant, the plaintiff must respond with a timely reply. In our example, Doug might have responded with a counterclaim by suing Pete if Doug's aircraft had been damaged in the accident. However, after consultation with counsel, Pete determined that a counterclaim would not be appropriate in this action since he did not suffer any damages.

Before the trial

Once a complaint, an answer, and any counterclaims have been filed, the parties begin preparation for trial. All throughout this pretrial process, the door is always open for the parties to settle the case.

Figure 1-7
Answer.

```
Pete Plaintiff,              *    IN THE
Plaintiff                    *    UNITED STATES DISTRICT COURT
v.                           *    FOR
Doug Defendant,              *    THE DISTRICT OF MARYLAND
Defendant                    *    CASE NO.: 030163-X
  *     *     *     *     *     *     *     *     *     *     *     *

                            ANSWER

    The defendant, through his undersigned attorney, responds to the numbered para-
graphs in plaintiff's complaint as follows:
1. Admitted
2. Defendant has insufficient information to either admit or deny this charge
3. Denied
4. Denied

                      AFFIRMATIVE DEFENSE

    Defendant avers that any damages suffered by plaintiff in this case are the direct
cause of plaintiff's contributory negligence.
    WHEREFORE, defendant requests dismissal of the plaintiff's complaint with all court
costs of this action to be born by plaintiff.

                                            Respectfully submitted,

                                            _____
                                            Ann Lawyer, Esq.
                                            Attorney for Defendant
                                            2001 Charles Street
                                            Baltimore, Maryland
```

The most common chain of events prior to a trial is often the following:

- A motion for a judgment on the pleadings
- Discovery
- Settlement and scheduling conferences
- Motions for summary judgment

We outline each of these possible pretrial events below.

Motion for judgment on the pleadings Once the complaint and answer are filed, one or both of the parties may file a motion (a motion is a request to the court) for a judgment on the pleadings. Essentially a judgment on the pleadings requests that the court analyze the various complaints made to determine whether they are sufficient to be acted on in accordance with the law. The party filing this type of motion is arguing that even if the facts asserted by the complainant are true, they do not rise to the level of a legal wrong. In our example, this type of motion might be filed by Doug if he and his counsel thought that they could convince the judge that there is really no legally enforceable claim against Doug for the airplane taxi incident. If this type of motion is granted by the court, some of or all a plaintiff's claims may be dismissed. For the purposes of our example, let's assume that a judgment on the pleadings was not filed and the case continued to progress.

Discovery Another very important step in pretrial proceedings is discovery. Many students who are used to seeing television or movie trials in which surprise witnesses turn up dramatically in the final moments of a trial are often surprised to find that each party has the legal right to learn a great deal from the other party (including potential witnesses) long before a trial is scheduled to begin. This process of extracting relevant information regarding the case that your opponent plans to present is known as discovery.

Discovery may include any of the following procedures:

- Written questions (interrogatories) that the opposing party or other witnesses must reply to under oath
- Written requests for documents or other tangible evidence
- Court-ordered physical or mental examinations of the parties
- In-person questions of witnesses (usually referred to as depositions) in which answers are given under oath
- Written requests for specific admissions by the opposing parties

In the end, the purpose of discovery is to allow the parties to learn as much as possible about each other's cases in order to avoid surprise and confusion at trial. Discovery can be expensive because it forces the parties to incur significant amounts of attorney time. However, it can save money in the long run by forcing the parties to face the evidence and witnesses to be produced by the opposition. On many occasions things learned in discovery encourage one or both parties to seek settlement before trial.

In Pete and Doug's case, there would likely be discovery related to the chain of events leading to the accident, the extent of damage and injury, and the discovery of potential witnesses. If Pete and Doug still cannot reach a settlement after discovery, the next step may be a pretrial or scheduling conference with the court.

Pretrial conferences Some courts routinely require the parties to attend a pretrial conference. On other occasions, issues raised in a particular case may prompt a court to request such a conference. Typically, the conference will include a discussion of trial dates, completion of discovery (if it was not complete already), and jury selection. If discovery has already been completed, the court will typically attempt to persuade the parties to make admissions, stipulations of fact (agreements as to the facts of the case), and any required amendments to pleadings, to make time at trial more efficient and to focus on contested issues. For instance, prior to discovery, there may have been a question as to the extent of damage done to Pete's aircraft. If Pete has produced all his repair bills in a complete and timely manner, the court may encourage Doug to stipulate that the damages were the total amount of the bills produced by Pete. This would still leave open the matter of Doug's fault, but it would save the court valuable time in trying to determine the extent of Pete's damages. Depending on the results of discovery, the court may also encourage a settlement between the parties. If the parties are reluctant to settle, the court might require them to participate in mandatory mediation (see discussion of mediation below). If the parties settle at this point, the court will ask them to prepare an order documenting the settlement. If the court approves, the order will become the judgment in the case. If the parties cannot settle, the case moves on through the system.

If the facts of the case are so well framed that they are not in dispute, one or both of the parties may make a motion for summary judgment.

Motion for summary judgment If the facts of the case are not in dispute, the court may entertain a motion for summary judgment. A motion for summary judgment essentially argues that there is no need for a trial because the parties agree on all material facts. Therefore, the court can make its decision based on the law. If one party files a motion for summary judgment, the opposing party will be granted an opportunity to file a motion in opposition to the motion for summary judgment, outlining why the party thinks the motion is inappropriate.

When the court reviews a motion for summary judgment, it must review the facts in a light most favorable to the party opposing the motion. Therefore, if there is any credible hint that material facts remain in dispute, the motion must be denied. However, if it does appear that all material facts are in agreement, the court may grant a motion for summary judgment, and that ruling is binding on all parties just as if the case had gone to trial.

In the case between Pete and Doug we'll assume that there are two very different versions of the taxi collision emerging out of discovery—there is definitely no agreement on the material facts of the case. Therefore, neither party files a motion for summary judgment. Our case is now headed for trial.

The trial

In most cases, persons involved in litigation in the U.S. legal system are entitled to a jury trial. However, that does not mean that a jury trial is mandatory. In a jury trial, the jury serves as a fact finder, and the judge decides on issues of law. If the parties decide to waive a jury trial, the trial will be conducted by a judge who will decide on issues of both fact and law.

When a jury trial is selected, the first step in the trial process is the selection of a jury. Depending on the court selected for trial, prospective jurors will be questioned by counsel for the parties or by a judge. This examination of the jurors is called a *voir dire* (preliminary examination). The purpose of this examination is to allow the parties to exclude jurors that they believe will not be fair or impartial. Each party typically has an unlimited number of challenges for cause, which means that the party can exclude as many jurors as the party wishes if cause for the exclusion can be established. Additionally, each party may be permitted a set number of peremptory challenges. These peremptory challenges generally allow the parties to exclude a juror without cause. Jury selection in the United States has now become an art in and of itself. Sometimes lawyers hire jury consultants (typically sociologists, psychologists, and other social scientists) to assist in selecting a jury that will be most favorable to their case. These jury selection experts often create a jury profile for the lawyers to use during the selection process.

Once the trial is underway, each party will be entitled to make an opening statement. This will be an opportunity for each party to lay out the case she or he intends to present to the court, a sort of introduction to lay the framework for the argument. In our case, Pete is the plaintiff, and he will present his case first. His attorney will call whatever witnesses are available and question each of the witnesses while they are under oath. After each of

Pete's witnesses testifies, Doug's attorney will have an opportunity to cross-examine Pete's witnesses to challenge their veracity and credibility. In our hypothetical example, we can assume that Pete's witnesses will support his claim that Doug negligently taxied his aircraft into Pete's aircraft, causing the damages suffered by Pete. After Doug's attorney cross-examines each of Pete's witnesses, Pete's attorney may have an opportunity to conduct a redirect examination of the witnesses, to try to clear up any questions raised during their cross-examinations. This will usually open up the door to additional or recross- examination of Pete's witnesses by Doug's counsel. Throughout the proceedings, the judge may be called upon to rule on the appropriateness of evidence being offered by the parties including such things as testimony, documents, and physical evidence.

Once Pete's attorney has completed his presentation of the case, he will rest his case. At this point, Doug's attorney may decide to present a motion for a directed verdict. A directed verdict motion requests that the court view all the evidence presented by the plaintiff and determine whether there is substantial enough evidence to support a finding in favor of the plaintiff. Let's assume in our example that the judge rules that Pete has presented sufficient evidence to support a verdict in his favor and the judge wants to proceed with the trial to hear Doug's side of the story.

Doug's attorney will present his witnesses to support Doug's contention that he was not negligent in taxiing his aircraft and that it was Pete whose negligence caused the collision between the two aircraft. Just as before, each of Doug's witnesses will be subject to cross-examination, questions on redirect, and recross-examination.

At the close of all questioning, each party will have the opportunity to present closing arguments. Closing arguments usually include a summary of the evidence presented at trial and legal arguments supporting a verdict in favor of the party presenting the arguments. Typically, Pete's attorney will be allowed to make the first closing argument, followed by Doug's attorney, with an opportunity for a brief rebuttal by Pete's attorney. Now our case is ready for presentation to a jury or judge for a verdict.

In a jury trial, the lawyers and judge will hammer out instructions for the jury that will allow the jury to decide the case in accordance with the law. In a case decided by a judge, there will be no need for instructions, and the judge will issue a verdict in writing or orally. In Pete and Doug's case, the jury enters a ruling in favor of Pete for all damages sought.

Once a verdict is rendered, the losing party may make an immediate challenge to overturn the verdict. If this challenge fails, the losing party (in our example Doug) may decide to bring the case to the appeals court level. In this example, the appeal would be filed with the U.S. Circuit Court of Appeals.

Appeals
If Doug seeks an appeal, he will be referred to as the appellant. The prevailing party, Pete, who is defending on appeal, is referred to as an appellee.

Once Doug's attorney files a timely notice of appeal, Doug and his attorney will be provided with a certified record of the trial proceedings. The record will usually consist of transcripts of testimony, all the pleadings and motions proffered in the case, documentary and physical exhibits, and any other relevant evidence presented and accepted in the case at trial.

With these materials in hand, Doug's attorney will prepare a legal brief. A legal brief is a formal document in which the appellant attempts to persuade the appeals court that the decision reached by the trial court was in error. Appellate rules require that briefs contain a summary of the procedural history of the case, a brief summary of the facts, a statement outlining the issues on appeal, and argument. Doug can argue that the trial court was wrong on the facts or the law or both. After Doug's attorney files his brief, Pete's attorney will have an opportunity to file an answer to Doug's brief. Many appellate courts will also give Doug's attorney an opportunity to reply to Pete's answering brief.

After all the appellate briefs have been submitted, the parties will have an opportunity to present their cases to the appeals judges during oral argument. Usually each party is allotted only a short time (15 to 30 minutes) to make an oral presentation, and typically the oral argument sessions are loaded with questions from the judges to the lawyers. Once the oral argument is concluded, the appellate judges will decide the case and issue a written opinion.

In many cases, it will be difficult for an appellant to prevail on an argument that the trial court was wrong on the facts. Remember that the appellate court does not have a chance to actually see witnesses and evaluate their credibility. The appellate court has only transcripts and other evidence previously presented at the trial to evaluate. It is unlikely that an appellate court will overturn a judge's or jury's decision on the facts, because the trial court judge or jury had a first-hand opportunity to view the witnesses and evaluate their veracity. However, an appellate court will typically not hesitate to find that a trial court made an error with respect to the law (e.g., permitting improper evidence to be introduced at trial). These errors could result in the case being remanded (returned to the trial court to be determined in accordance with the appellate court ruling) or the trial court's ruling being reversed. If the appeals court does not believe that the trial court made any substantial errors, the trial court's verdict will be affirmed. In Doug and Pete's case, the appeals court affirms the trial court's verdict. Doug decides at this point that he does not wish to appeal to the U.S. Supreme Court (where the chances of having this type of case heard would be negligible), and so the judgment against Doug becomes final.

Enforcing judgments

After Pete wins his case, he will want to collect on his judgment. Doug may cooperate and pay immediately. However, if Doug does not cooperate, Pete will usually be able to enforce his judgment by several methods. The usual options available include

- Garnishing Doug's wages or salary
- Garnishing Doug's bank account(s)
- Seizing Doug's personal property or real estate

To pursue any of these options, Pete will need to find the assets necessary to satisfy his judgment against Doug. In many cases, Pete can get the information necessary through written interrogatories or questions that Doug will have to answer under oath. If this approach is unsuccessful, Pete may seek to have Doug ordered to court for an oral examination under oath. When and if Pete is finally successful in obtaining his monetary damages from Doug, he will be required to file a document with the court notifying the court that the judgment has been satisfied.

Alternative methods of dispute resolution

Due to the expense and time-consuming nature of litigation in the courts, many businesses and individuals are turning to alternative means to resolve disputes. As these alternative means become increasingly popular, it is important that aviation professionals become familiar with the various options available.

Arbitration

One widely used alternative to the litigation in the courts is arbitration. In arbitration, a neutral third party is selected to hear a case brought by the persons involved in a dispute. Usually, the arbitrator selected is an attorney who specializes in the area or areas of law touching on the dispute. The rules of procedure in arbitration are relatively relaxed when compared to the formal rules of civil procedure and evidence applied by the courts. Therefore, arbitration usually takes less time and money than court litigation.

There are different routes to arbitrations. In some cases the parties agree ahead of time (often in a contract) to submit any disputes to an arbiter. In other cases, the parties involved in a dispute mutually agree to submit their dispute to an arbiter. Both situations are examples of consensual arbitration (where the parties agree to arbitration as a means of dispute resolution).

In some instances, the law requires that the parties submit their dispute to arbitration. In these cases, federal or state law requires that the parties submit any disputes to arbitration. Usually labor disputes involving public employees, transportation workers (including pilots and air traffic controllers), police, teachers, or firefighters are subject to compulsory arbitration.

Agreements to arbitrate are enforceable by both federal and state law. The federal government has enacted the Federal Arbitration Act, and most states have adopted a uniform arbitration act.

As a general rule, the decision of an arbiter will be binding on the parties. However, the decision may be subject to review on very narrow grounds including the following:

- The decision was tainted by corruption, fraud, or other illegal means.
- There was bias on the part of the arbiter.
- The arbiter committed errors that prejudiced a party.
- The arbiter exceeded the scope of his or her authority.

With these very narrow grounds for appeal, most decisions by arbiters are upheld and are indeed binding.

Conciliation and mediation

Conciliation and mediation are processes that employ third parties to act as facilitators in an effort to bring the disputing parties to a settlement agreement. Many court systems now mandate that parties participate in a conciliation or mediation process before they get to the courtroom. Conciliators will typically explain issues, coordinate meetings, and act as intermediaries between the parties when the parties cannot or will not communicate directly. Mediators do the same work as conciliators. However, in addition, mediators will offer suggestions for resolving disputes between the parties. Because neither a conciliator nor a mediator renders a decision or an opinion in a matter, there is nothing binding in this process unless the parties agree to be bound by a settlement agreement.

Minitrials and summary jury trials

Minitrials and summary jury trials are attempts to present evidence to third parties in a less formal manner than would be seen in the courts. The third party (often a retired judge, lawyer, or group of privately selected jurors) then renders an opinion on the case. The opinion is nonbinding, and either party has the right to move the case to a court for a full trial.

BASICS OF LEGAL RESEARCH

Although your course work in aviation law is not designed to transform you into a lawyer, you may be exposed to research projects or cases that will require legal research. The purpose of this discussion is to acquaint you with some basic research tools that you might find helpful in your legal studies.

Case law

If you are looking for federal or state cases, there are a number of sources you can use to locate a case. When you are fortunate enough to have a law library available, there will be volumes of books (called *reporters*) for federal cases. Supreme Court, Federal Court of Appeals, and district court cases can be found in these reporters. The reporters for Supreme Court cases are typically found in the United States Reports. These volumes use the citation "U.S." There are other reporters for Supreme Court decisions including the Supreme Court Reporter (citation "S.Ct.") and Lawyer's Edition (citations "L.Ed." and "L.Ed.2d"). For Federal Court of Appeals (circuit court) decisions, the citation will usually be "F.2d." or "F.3d." For U.S. district court cases the citation will read "F.Supp." for the Federal Supplement reporter. When you analyze a case citation, you can use the representative citation found in Fig. 1-8 for guidance.

If you are looking at a citation for a case decided by a federal court of appeals, the citation will also include the deciding circuit court (e.g., Second Circuit). The same is true

Figure 1-8
Case citations.

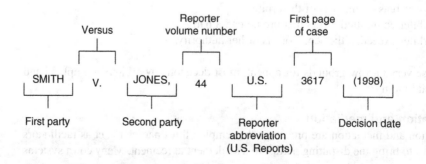

for U.S. district courts where the citation will include the location of the court (e.g., S.D.N.Y. for the U.S. District Court for the Southern District of New York).

If you are looking for state case law, you will have to find the specific state reporter you need. Each state has its own sets of official reporters. For instance, if you were looking for an Arizona case decided by the Arizona Supreme Court, you would look to Arizona Reports "Ariz." You could also find this case in the Pacific Reporter ("P.2d"). A Rhode Island Supreme Court decision can be found in the Rhode Island Reports ("R.I.") or the Atlantic Reporter ("A.2d").

If a law library is not readily available, there are still plenty of ways to get to federal cases via electronic means. Your college or university may subscribe to Lexis/Nexis or Westlaw or some other proprietary legal research engines. If so, your instructor can lead you through the various electronic tools available. Most of these electronic search engines are based on keyword searches. Even if these proprietary search engines are not available, you can still use the free legal research tools available at www.findlaw.com.

When you get to Chap. 5 and start looking at National Transportation Safety Board cases involving airman enforcement actions, you may find the NTSB web site (www.ntsb.gov) handy for locating cases. The NTSB web site catalogues cases dating back to January 1, 1992.

At times your research may require you to analyze a statute or regulation. Just as with cases, you can find federal and state statutes through proprietary research engines such as Westlaw or Lexis/Nexis. The web site www.findlaw.com also has a full complement of federal and state statutes available online. For regulations, you can typically go straight to the web site of the applicable regulatory agency. The FAA's web site at www.faa.gov contains all the FAA regulations and Federal Register pronouncements related to aviation matters. **Statutes and regulations**

Another research tool that you mind find helpful is the law review article. Law reviews are typically published by law schools. They contain articles on topical subjects by lawyers, law professors, and judges, and case summaries by law school students. Law review articles are heavily footnoted, and often the footnotes will lead you to primary sources such as statutes or cases. The footnotes may also lead you to other relevant articles. One of the most widely referred to law reviews in the aviation industry is the Southern Methodist University (SMU) *Journal of Air Law and Commerce*. You can log onto the SMU web site at www.smu.edu to review a listing of articles available. These articles can usually be accessed in full text by using a law library or a proprietary legal research engine such as Westlaw or Lexis/Nexis. **Law review articles**

CASES AND COMMENTARY

Author's Note: Toward the end of each chapter, you will be presented with edited versions of actual cases with aviation-related applications of the subject matter tackled in the chapter. It is important for you to gain some exposure to the actual language and analysis used by judges when they decide a case. The judge's opinions will also give you a sense of the interplay between the attorneys arguing the case and the judges who must make a decision on the cases. In the end, it is the attorneys who lay out the arguments that must be weighed by the judges, and ultimately those arguments find their way into court opinions.

Most of the cases selected will be cases before appellate courts. However, from time to time you will review cases that have been decided by trial court judges. On occasion your instructor may request that you "brief" a case. Guidance on how to prepare a case briefing is provided in App. A of this textbook.

For the two cases presented in this chapter, the facts have been laid out in summary form. In later chapters, you can gain some experience "unpacking" some complex fact patterns directly from the cases.

Case 1-1 deals with a question of federal jurisdiction. See if you agree with the court's determination on the issue of whether this case can be properly submitted to federal court jurisdiction.

CASE 1-1
LUCIA V. TELEDYNE
173 F. Supp. 2d 1253 (2001)

JUDGE: Richard W. Vollmer, Jr.

FACTS: *The Plaintiff in this case owns and operates two piston engine aircraft that use crankshafts manufactured by Defendant Teledyne Continental Motors Division ("Teledyne"). On April 19, 9999, Teledyne issued a "Critical Service Bulletin" ("CSB") noted cracks found in two of eight engine crankshafts it manufactured or reworked in 1998.*

Teledyne's CSB detailed an inspection process in which eligible crankshafts would be examined in field inspections by Teledyne representatives. The inspections call for ultrasound testing of the crankshafts. If problems were revealed, the engine would be removed and sent to Teledyne's facilities in Mobile, Alabama.

On April 22, 1999, the Federal Aviation Administration ("FAA") issued a "Priority Letter Airworthiness Directive" ("AD") pursuant to its authority under 49 U.S.C. Section 44701. The AD mandated ultrasonic inspections of the covered crankshafts and approved the technical contents of Teledyne's CSB.

One day after the issuance of the AD, Plaintiff Lucia filed a claim against Teledyne in the Circuit Court of Mobile County, Alabama requesting injunctive relief in the form of "magnaflux testing" of the crankshafts in question and compensatory damages "not in excess of $70,000 per class member, attorneys' fees from the common fund, interests and costs." The four causes of action asserted in the Plaintiff's complaint are (1) Misrepresentation, (2) Negligence, (3) Strict Liability, and (4) Breach of Express Warranty.

On May 16, 1999, Teledyne filed its Notice of Removal, claiming that this case should be removed to federal court based on grounds of subject matter jurisdiction over a federal question, or in the alternative, on the existence of complete diversity between the parties. In turn, the Plaintiff filed a Motion to Remand this case back to the state court where it was filed. The federal court now examines whether this case can be properly removed to a federal court.

OPINION

A) Diversity Jurisdiction: The foundation for federal court diversity jurisdiction — the power to decide cases between citizens of different states — is Article III of the United States Constitution. U.S. Const. art. III, § 2 [Citation]. However, when Congress created the lower federal courts, it limited their diversity jurisdiction to cases in which there was a minimum monetary amount in controversy between the parties. [Citation.] Today, the threshold amount in controversy for diversity jurisdiction, excluding interests and costs, is $75,000. [Citation.]

In [Citation] ("*Cohen II*"), the Eleventh Circuit held that a prior panel decision by the Former Fifth Circuit, [Citation], precluded aggregating punitive damages to establish diversity jurisdiction over a class action. *See Cohen II*, [Citation]. Though not addressing precisely the set of facts at issue, the holding is in accord with the rule that, generally speaking, when a set of plaintiffs join in one lawsuit, the value of their claims may not be added together, or "aggregated," to satisfy the amount in controversy requirement for diversity jurisdiction. [Citations.]

In this matter the compensatory, or actual damages of Plaintiff (or any other hypothetical class member) appear to be relatively small. According to Plaintiff, the alleged damages suffered include those costs incident to the proper inspection and repair of the allegedly faulty crankshafts, as well as those costs attaching to the loss of use of Plaintiff's aircraft. These costs Plaintiff values to be not in excess of $70,000. *See* Complaint, *supra*. In any event, assuming *arguendo* that a class could eventually be certified in this matter, the compensatory (or punitive if pled) damages claims could not be aggregated to form the basis for federal diversity jurisdiction, since any of the claims of this hypothetical class would arise from those class members' separate and individual agreements with Teledyne. [Citation.] Though the point is somewhat moot as only Plaintiff Lucia's claims are before this Court, such claims as would arise from various class members could not be aggregated for amount in controversy purposes. [Citation.]

The allegations by Teledyne that the aviation components are part of a single and distinct product line does not change the separate and distinct nature of the rights that could be asserted in this matter, assuming *arguendo* that a class could eventually be certified in this matter. [Citation.] ("Federal Rule of Civil Procedure, Rule 20 allows the joinder of parties plaintiff when there is a common question of law or fact and the claims of all plaintiffs arose out of the same transaction or occurrence. However, this joinder for convenience of the court affects in no way the entirely separate question of aggregation of claims to satisfy the jurisdictional amount."). Accordingly ... the Court finds that Teledyne has failed to demonstrate the amount in controversy for purposes of establishing diversity jurisdiction in this matter.

B) Federal Question Jurisdiction: Defendant also asserts that the Court has original subject matter jurisdiction over this case, pursuant to [Citation], because of the presence of a federal question. Teledyne argues that, since this lawsuit seeks compensatory and actual damages in an area of law which, according to Defendant, has been super-preempted by virtue of Congress' delegation to the Federal Aviation Administration, the Court therefore has proper jurisdiction. Defendant's field preemption argument operates on the assumption that any and all judicial intervention into actions even tenuously affecting issues involving the proper maintenance, repair and/or general safety regulation of aviation hardware must necessarily be of the federal type. This assumption fails to recognize, however, that the statutory grant to the FAA of exclusive federal regulatory power in the field extends only to those matters having the force and effect of law related to a price, route, or service of an air carrier that may provide air transportation under this subpart." [Citation.] The Court will address this argument more fully below. Alternatively, Defendant argues that by seeking injunctive relief to be issued by a state court in order to compel Teledyne to enact safety and inspection measures beyond the scope of those specifically approved in the FAA's AD of April 22, 1999, Plaintiff has bestowed federal question jurisdiction on the Court by operation of ordinary preemption.

CONCLUSION

Plaintiff's Complaint relies exclusively on state law causes of action. It does not on its face allege any basis for federal jurisdiction. Accordingly, under the well-pleaded complaint rule, the substance of the Complaint does not permit removal to federal court. As the Court has determined, the amount in controversy sufficient to establish diversity jurisdiction has not been met in this case. Therefore, Teledyne could have properly removed this case only if Plaintiff's state law claims are rendered federal in nature under either the doctrines of complete or ordinary preemption. As demonstrated above, neither doctrine is applicable to the matter at hand.

In addition to the state law claims for negligence, misrepresentation and breach of warranty, Plaintiff's Complaint in this case seeks injunctive relief against Defendant Teledyne in the form of a mandatory request for the detailed inspection of select aircraft components manufactured by Defendant. Plaintiff does not request that the state court either review or alter the regulatory decision of the FAA as manifested in its Advisory Directive, contrary to Defendant's attempts to characterize Plaintiff's pleadings as such. Rather, any question concerning the FAA's regulatory authority in this area of law arose only from Teledyne's defensive assertion of the purported preemptive effect of the FAA's regulatory activity on Plaintiff's request for injunctive relief. As the mere existence of a federal defense, even one involving federal (ordinary) preemption, cannot serve as the basis for federal subject matter jurisdiction, it follows that Defendant has improperly removed this matter to federal court. [Citation.]

Likewise, there is simply no foundation to support Teledyne's claims that the preemption provision of the Aviation Act, [Citation], indicates that Congress has chosen to occupy the field "so thoroughly ... as to make reasonable the inference that Congress left no room for the states to supplement federal law." [Citation.] As noted above, the preemption provision in this context merely creates a potential federal defense to Plaintiff's claims under state law, which the state court is more than capable of resolving under the doctrine of ordinary preemption. Further, the fact that the Aviation Act contains a provision preempting claims under state law which would affect airlines' rates, routes or services, [Citation], does not under the principles of complete preemption described above "create" removal jurisdiction. To the contrary, the Aviation Act does not contain the extraordinary preemptive force necessary to render Plaintiff's state claims federal in nature and removal proper.

Plaintiff, as the master of his claim, has successfully pleaded state law claims and is entitled to litigate his claims in state court, the forum of his choice. While his claim for injunctive relief may ultimately prove to be preempted by section 41713 of Title 49, the ordinary preemption issue is most appropriately decided by the state court and is not addressed here. [Citations.] For the foregoing reasons, the Court concludes that it lacks subject matter jurisdiction over this action. The court therefore REMANDS this matter to the Circuit Court of Mobile County, Alabama. The clerk is DIRECTED to take all steps necessary to effectuate this remand, each party to bear its own costs.

Fundamentals of the U.S. Legal System | 31

Notice that in this case the defendant, Teledyne, was seeking to have the case moved from a state trial court to a federal trial court (the U.S. district court). Why, do you think, Teledyne would spend so much time and money to attempt to have the case moved to a federal court?

Teledyne raises an interesting argument that this case should be heard before a federal court because it has a potential defense based on preemption (we will discuss the issue of preemption in greater detail in Chap. 2). However, the court rejects this argument. Why does the court reject this argument? Do you think the court made the right decision?

You are encouraged to find and read the full text of this case. The full text contains a rich and full discussion of the law relating to federal jurisdiction.

The next case, Case 1-2, involves a question of personal jurisdiction. This case was deemed to be so significant that the U.S. Supreme Court elected to review the case. Just how much contact must a foreign helicopter transportation company have with the state of Texas to be subject to the jurisdiction of the Texas court system? See what the Supreme Court thinks in Case 1-2.

CASE 1-2
HELICOPTEROS NACIONALES DE COLOMBIA, S.A. v. HALL ET AL.
466 U.S. 408 (1984)

FACTS: *"Helicol", the Petitioner, a Colombian corporation, entered into a contract to provide helicopter transportation for a Consorcio/WSH, a Peruvian company closely related to a joint venture that had its headquarters in Houston, Tex., during the consortium's construction of a pipeline in Peru for a Peruvian state-owned oil company. Petitioner has no place of business in Texas and never has been licensed to do business there. Its contacts with Texas consisted of sending its chief executive officer to Houston to negotiate the contract with the consortium, accepting into its New York bank account checks drawn by the consortium on a Texas bank, purchasing helicopters, equipment, and training services from a Texas manufacturer, and sending personnel to that manufacturer's facilities for training. After a helicopter owned by petitioner crashed in Peru, resulting in the death of respondents' decedents—United States citizens who were employed by the consortium—respondents instituted wrongful-death actions in a Texas state court against the consortium, the Texas manufacturer, and petitioner. Denying petitioner's motion to dismiss the actions for lack of in personam jurisdiction over it, the trial court entered judgment against petitioner on a jury verdict in favor of respondents. The Texas Court of Civil Appeals reversed, holding that in personam jurisdiction over petitioner was lacking, but in turn was reversed by the Texas Supreme Court.*

OPINION: JUSTICE BLACKMUN delivered the opinion of the Court.

In ruling that the Texas courts had *in personam* jurisdiction, the Texas Supreme Court first held that the State's long-arm statute reaches as far as the Due Process Clause of the Fourteenth Amendment permits. Thus, the only question remaining for the court to decide was whether it was consistent with the Due Process Clause for Texas courts to assert *in personam* jurisdiction over Helicol.

The Due Process Clause of the Fourteenth Amendment operates to limit the power of a State to assert *in personam* jurisdiction over a nonresident defendant. Due process requirements are satisfied when *in personam* jurisdiction is asserted over a nonresident corporate defendant that has "certain minimum contacts with [the forum] such that the maintenance of the suit does not offend 'traditional notions of fair play and substantial justice.'" [Citation.]

When a controversy is related to or "arises out of" a defendant's contacts with the forum, the Court has said that a "relationship among the defendant, the forum, and the litigation" is the essential foundation of in *in personam* jurisdiction. [Citation.]

Even when the cause of action does not arise out of or relate to the foreign corporation's activities in the forum State, due process is not offended by a State's subjecting the corporation to its *in personam* jurisdiction when there are sufficient contacts between the State and the foreign corporation. [Citations.] In *Perkins*, the Court addressed a situation in which state courts had asserted general jurisdiction over a defendant foreign corporation. During the Japanese occupation of the Philippine Islands, the president and general manager of a Philippine mining corporation maintained an office in Ohio from which he conducted activities on behalf of the company. He kept company files and held directors' meetings in the office, carried on correspondence relating to the business, distributed salary checks drawn on two active Ohio bank accounts, engaged an Ohio bank to act as transfer agent, and supervised policies dealing with the rehabilitation of the corporation's properties in the Philippines. In short, the foreign corporation, through its president, "[had] been carrying on in Ohio a continuous and systematic, but limited, part of its general business," and the exercise of general jurisdiction over the Philippine corporation by an Ohio court was "reasonable and just." [Citation.]

All parties to the present case concede that respondents' claims against Helicol did not "arise out of," and are not related to, Helicol's activities within Texas. We thus must explore the nature of Helicol's contacts with the State of Texas to determine whether they constitute the kind of continuous and systematic general business contacts the Court found to exist in *Perkins*. We hold that they do not.

It is undisputed that Helicol does not have a place of business in Texas and never has been licensed to do business in the State. Basically, Helicol's contacts with Texas consisted of sending its chief executive officer to Houston for a contract-negotiation session; accepting into its New York bank account checks drawn on a Houston bank; purchasing helicopters, equipment, and training services from Bell Helicopter for substantial sums; and sending personnel to Bell's facilities in Fort Worth for training.

The one trip to Houston by Helicol's chief executive officer for the purpose of negotiating the transportation-services contract with Consorcio/WSH cannot be described or regarded as a contact of a "continuous and systematic" nature, as *Perkins* described it, [Citation], and thus cannot support an assertion of *in personam* jurisdiction over Helicol by a Texas court. Similarly, Helicol's acceptance from Consorcio/WSH of checks drawn on a Texas bank is of negligible significance for purposes of determining whether Helicol had sufficient contacts in Texas. There is no indication that Helicol ever requested that the checks be drawn on a Texas bank or that there was any negotiation between Helicol and Consorcio/WSH with respect to the location or identity of the bank on which checks would be drawn. Common sense and everyday experience suggest that, absent unusual circumstances, the bank on which a check is drawn is generally of little consequence to the payee and is a matter left to the discretion of the drawer. Such unilateral activity of another party or a third person is not an appropriate consideration when determining whether a defendant has sufficient contacts with a forum State to justify an assertion of jurisdiction. [Citations.]

The Texas Supreme Court focused on the purchases and the related training trips in finding contacts sufficient to support an assertion of jurisdiction. We do not agree with that assessment, for the Court's opinion in [Citation] makes clear that purchases and related trips, standing alone, are not a sufficient basis for a State's assertion of jurisdiction.

The defendant in *Rosenberg* was a small retailer in Tulsa, Okla., who dealt in men's clothing and furnishings. It never had applied for a license to do business in New York, nor had it at any time authorized suit to be brought against it there. It never had an established place of business in New York and never regularly carried on business in that State. Its only connection with New York was that it purchased from New York wholesalers a large portion of the merchandise sold in its Tulsa store. The purchases sometimes were made by correspondence and sometimes through visits to New York by an officer of the defendant. The Court concluded: "Visits on such business, even if occurring at regular intervals, would not warrant the inference that the corporation was present within the jurisdiction of [New York]." [Citation.]

This Court in *International Shoe* acknowledged and did not repudiate its holding in *Rosenberg*. [Citation.] In accordance with *Rosenberg*, we hold that mere purchases, even if occurring at regular intervals, are not enough to warrant a State's assertion of *in personam* jurisdiction over a nonresident corporation in a cause of action not related to those purchase transactions. Nor can we conclude that the fact that Helicol sent personnel into Texas for training in connection with the purchase of helicopters and equipment in that State in any way enhanced the nature of Helicol's contacts with Texas. The training was a part of the package of goods and services purchased by Helicol from Bell Helicopter. The brief presence of Helicol employees in Texas for the purpose of attending the training sessions is no more a significant contact than were the trips to New York made by the buyer for the retail store in [Citation].

We hold that Helicol's contacts with the State of Texas were insufficient to satisfy the requirements of the Due Process Clause of the Fourteenth Amendment. Accordingly, we reverse the judgment of the Supreme Court of Texas. It is so ordered.

Students should note that in this case the U.S. Supreme Court determined that the Texas Supreme Court incorrectly ruled that the contacts of Helicol were sufficient to assert jurisdiction over the corporation. This case was not easily decided (as evidenced by the disagreement between the Texas Supreme Court and the U.S. Supreme Court). In fact, there was a dissenting (minority) opinion published by Justice Brennan in this case. In his dissenting opinion, Justice Brennan states: "… I believe that the undisputed contacts in this case between … Helicol … and the State of Texas are sufficiently important, and sufficiently related to the underlying cause of action, to make it fair and reasonable for the State to assert personal jurisdiction over Helicol.…" Justice Brennan goes on to state that "… maintenance of this suit in Texas courts 'does not offend traditional notions of fair play and substantial justice,'" citing *International Shoe Co v. Washington*[4] as the basis for his dissent.

The *International Shoe* case is a landmark case on the issue of personal jurisdiction. Find the case and read the opinion. After reading the opinion, do you still agree with the opinion of the majority in this case or do you side with Justice Brennan and the Texas Supreme Court?

DISCUSSION CASES

1. Captain Smith was a pilot for Windswept Airlines ("Windswept"). In 2003, Smith was forced to resign by the airline. Smith sued Windswept, alleging that his employment was terminated due to his pro-union activities, and not because of substandard job performance as claimed by the airline. During discovery, a memo written by a Windswept manager was produced that stated: "More than a few crew members claimed that Smith professed to being a leftist-activist. His over-activity in the local pilots' union, coupled with inquiries regarding company files to our secretary, led to the conclusion that potential trouble could be avoided by the acceptance of Smith's resignation." Smith claimed that the report is evidence of the true reason for his discharge. Windswept files a motion for summary judgment with the trial court. Should the motion for summary judgment be granted? Why or why not?

2. The plaintiff corporation, incorporated in Michigan, hired a defendant from Florida to operate a helicopter to spray agricultural chemicals on fields in Ohio. In a contract written and signed by the plaintiff in Michigan and later signed by the defendant in Florida, the defendant agreed that if he left the employment of the plaintiff, he would not create or become employed by a competing business. After two years, the defendant left the plaintiff's employ and immediately began a competing business in Ohio. The plaintiff sued in Ohio to enforce the covenant not to compete. Such a covenant was void under Michigan law but valid and enforceable if reasonable under the laws of Florida and Ohio. Which state's law should be applied in the Ohio court? Discuss.

3. While taxiing his aircraft in Virginia, Dodd, a resident of North Carolina, struck Paul, a resident of Alaska. As a result of the accident, Paul incurred more than $80,000 in medical expenses. Paul would like to know, if he personally serves the proper papers to Dodd, whether he can obtain jurisdiction against Dodd for damages in the following courts:
    (a) Alaska state trial court
    (b) Federal Circuit Court of Appeals for the Ninth Circuit (includes Alaska)
    (c) Virginia state trial court
    (d) Virginia federal district court
    (e) Federal Circuit Court of Appeals for the Fourth Circuit (includes Virginia and North Carolina)
    (f) Virginia equity court
    (g) North Carolina state trial court

4. Gus Gullible, a resident of Kansas, and Ned Naïve, a resident of Missouri, each purchased an $85,000 used aircraft in their home states from Slick Sam, Inc.("Slick Sam"), an aircraft broker incorporated in Delaware with a principal place of business in Kansas. Both Gullible and Naïve believe they were cheated by Slick Sam on the aircraft sales, and they would like to sue Slick Sam for fraud. Assuming that there is no federal question at issue, assess the accuracy of the following statements.
    (a) Gullible can sue Slick Sam in a Kansas state trial court.
    (b) Gullible can sue Slick Sam in a federal district court in Kansas.
    (c) Naïve can sue Slick Sam in a Missouri state trial court.
    (d) Naïve can sue Slick Sam in a federal district court in Missouri.

5. Peter Plaintiff brought an action in an Illinois trial court to recover damages for breach of warranty against Digital Defendant, Inc. ("Digital Defendant"). (A warranty is a duty assumed by a seller of goods that relates to the quality of the goods.) Digital Defendant manufactures weather radar devices for light aircraft. Digital Defendant is an Indiana corporation with a primary place of business in Decatur, Indiana. Digital Defendant has no office in Illinois and no agent authorized to do business on its behalf in Illinois. Peter Plaintiff saw Digital Defendant's weather radar equipment on display at the Chicago Aviation Exposition. In addition, promotional literature related to Digital Defendant's equipment was circulated at the exposition. Several aviation trade magazines, delivered to Peter Plaintiff in Illinois, contained advertisements for Digital Defendant's equipment. Eventually, Peter Plaintiff purchased a Digital Defendant weather radar device at Fly-By-Night Aviation, Inc., a fixed-base operator located at an airport in Illinois. A written warranty was issued by Digital Defendant and delivered to Peter Plaintiff in Illinois. Digital Defendant seeks to have Peter Plaintiff's suit dismissed on the grounds that Digital Defendant should not be subject to the jurisdiction of the Illinois courts. Will Digital Defendant's argument succeed? Explain.

6. Joe Pilot brought a lawsuit in a California trial court against his former employee Pacific Skyways in a dispute over whether certain overtime pay was due to Pilot. When he applied for employment at Pacific Skyways, Pilot signed a form that contained an arbitration agreement clause. When Pacific Skyways sought to open arbitration proceedings, Pilot refused to arbitrate, citing Section 229 of California law that states that actions commenced for the collection of wages may be maintained "without regard to the existence of any private agreement to arbitrate." Pacific Skyways files a petition with the California trial court to compel arbitration under the Federal Arbitration Act. Who will win on the issue of arbitration? Discuss.

7. The defendant, Design Maintenance, Ltd., rebuilt an aircraft engine for the plaintiff, Perfect Landings, Inc., sometime in July 2004. On January 15, 2005, the defendant's president was served with a summons and complaint in a lawsuit filed by Perfect Landings. Perfect Landings claimed that Design Maintenance's defective work on the engine forced Perfect Landings to spend in excess of $25,000 in repairs. Under court rules, an answer was due from the defendant on February 16, 2005. Design Maintenance attempted to answer on February 18, 2005, but the court refused to accept the filing because it was untimely. The court entered a default judgment against the defendant and set a hearing to establish damages. Design Maintenance filed a motion to set aside the default judgment on the grounds of "excusable neglect," pointing out that its president had received two summonses in the same week and believed that Perfect Landings had been served at the same time as the other summons on January 18, 2005. Should the court set aside the default judgment? Why or why not?

8. Plaintiff Parsons & Company filed suit against defendant Destinations Unlimited, an air freight company, claiming $100,000 in damages from the alleged mishandling of fragile cargo by Destinations. The lawsuit was filed in a New York state court. As part of the discovery process in the lawsuit, Parsons sought to take the deposition of Destinations' president. An attorney for Destinations told the court that Destinations' president lived in California, was 75 years old, and was in bad health (as supported by affidavits by physicians). Must Destinations' president travel to New York for the deposition? Explain.

ENDNOTES

1. *Blacks Law Dictionary,* 5th ed. (St. Paul: West Publishing Co., 1979).
2. Huntington, Cairns, *Legal Philosophy from Plato to Hegel* (Baltimore, Md.: Johns Hopkins University Press, 1949).
3. For an interesting discussion of how the law deals with the question of whether an affirmative duty to assist another in peril exists, see *Soldano v. O'Daniels*, 141 Cal. App. 3d 443 (1983).
4. 326 U.S. 310 (1945).

2 The U.S. Constitution and Aviation

OVERVIEW OF THE U.S. CONSTITUTION   38
Structure and organization of the federal government   39
Distribution of authority between the federal and state governments   41
Basic rights of individuals and businesses   49

CASES AND COMMENTARY   52

DISCUSSION CASES   62

ENDNOTES   64

When the United States first gained its independence, it was faced with a difficult chore of establishing a federal government. The Continental Congress took on this chore in 1778 by adopting the Articles of Confederation. With the passage of time, there was a sense that the Articles of Confederation were not adequate in creating an effective federal government—the United States was really never "united" by the articles. The Articles of Confederation tended to perpetuate the sense that the United States was a collection of individual states without a strong central government. A clear example of this is found in Article III of the Articles of Confederation where it states that the individual states would form a "firm league of friendship ... for their common defense." It is now hard to imagine the U.S. military as a collection of individual state militias.

In response to the weaknesses inherent in the Articles of Confederation, a Constitutional Convention was convened in Philadelphia, Pennsylvania, in 1787. At the convention there was great debate over just how strong the federal government of the United States should be—and how the power of the individual states would interplay with a stronger central government. The delegates at the Constitutional Convention eventually agreed that a new constitution was necessary for the United States. After months of debate a new Constitution was presented to the U.S. Congress in 1787. All the individual states in the United States ratified the new U.S. Constitution by the end of 1788.

It is a testament to the wisdom of those who drafted the U.S. Constitution that the document, as the supreme law of the land, has been able to evolve and grow to meet the ever-changing demands of our legal system and society. Through amendments and court interpretations, the U.S. Constitution remains as relevant today as it was the day it was written.

Note that every state in the United States has its own state constitution. In many cases these documents contain provisions that are very similar in nature to the U.S. Constitution. An individual state's constitution is the supreme law of that state. However, if an individual state's constitutional provisions are in conflict with the U.S. Constitution, those provisions will fail.

It is quite remarkable that a document written long before the airplane was ever invented could have such a significant influence on aviation in the United States. But that is indeed the case—the U.S. Constitution has a powerful impact on how aviation has developed in this country.

In the first part of this chapter we will explore the basics of the U.S. Constitution. Some of the topics addressed will have aviation-related applications that are discussed later in this chapter. Other topics discussed will surface again in aviation contexts throughout this book.

OVERVIEW OF THE U.S. CONSTITUTION

Before going any further in this chapter, you should take some time to review the U.S. Constitution. A copy of the Constitution is found in App. B of this textbook. A careful reading of the U.S. Constitution reveals that the document addresses three very significant issues in the law:

- The organization of the federal government into legislative, executive, and judicial branches
- How authority will be distributed between the federal and state governments
- The protection of basic individual and business rights through limitations on federal and state government authority

It is also very important to note that the very first line of the U.S. Constitution unambiguously states that the authority for the Constitution comes from people of the United States. The fact that the Constitution comes from the people gives it the unquestioned authority to serve as the basis for our rule of law.

Structure and organization of the federal government

In the Constitution, Article I deals with the legislative authority granted to the U.S. Congress; Article II outlines the executive authority granted to the President of the United States; and Article III addresses the workings of the Supreme Court and lower federal courts in the United States. To a large extent, the separation of powers between the various branches of the federal government reflects the desire of the constitutional framers to strengthen the federal government—but not to allow any one group or person to dominate the others. In fact, this system of creating legislative, executive, and judicial branches of the federal government was designed in large part to create a system of checks and balances among the various branches.

Some prominent examples of how our system of checks and balances works include the following:

- The legislative branch creates laws, and the executive branch has veto power over those laws.
- The executive branch appoints federal judges (including Supreme Court judges); however, the judicial branch has the power to interpret the law and review the constitutionality of any actions taken by the Congress or executive branch.
- The legislative branch confirms appointments to both the judicial branch and the executive branch of the federal government.

While most of these checks and balances are easily discernible within the text of the U.S. Constitution, one that is not so readily apparent is the power of the courts to review the constitutionality of laws passed by the legislative branch and the enforcement of those laws by the executive branch. There is no express language in the Constitution that grants the power to determine what is constitutional and what is not to the judicial branch. In the early days of the United States, the question of which branch of the federal government had the power to determine constitutionality was unsettled. However, for all practical purposes, that question was answered when U.S. Supreme Court assumed (some would say seized) the power of judicial review in the case of *Marbury v. Madison*.[1] Interestingly, the Supreme Court's assumption of this very important function has never been challenged by the other branches of our federal government. To some extent, the lack of challenge may be because there is, arguably, a certain logic to leaving questions of constitutionality within the courts. Federal judges are typically tenured for life, and therefore they are less likely to be influenced or pressured by the politics of the day. This allows for a more stable platform upon which questions of constitutionality can be addressed. Therefore, it is fair to say that the federal courts, and especially the U.S. Supreme Court, as final authority, serve as stewards of the U.S. Constitution. Also note that federal courts have authority to review the action of state and local courts, legislatures, and executives to determine whether they have created and enforced laws in a manner consistent with the U.S. Constitution.

A graphic overview of some of the checks and balances established in the U.S. Constitution is illustrated in Fig. 2-1.

Although the Constitution seeks to maintain a system of checks and balances that separate power, it would be a mistake to infer that this separation is absolute. On many occasions it is necessary and practical for the various branches of government to engage in reasonably overlapping functions.

An example of these overlapping functions includes the occasions on which Congress confers quasi-judicial powers on government agencies. For instance, Congress has granted power to the National Transportation Safety Board (NTSB), an independent government agency, to adjudicate cases involving alleged airman violations and questions of qualification. Another example would be the delegation of judicial like functions to the Social Security Administration (SSA), an executive branch agency, in determining the eligibility of claimants for disability benefits. On other occasions, Congress takes on certain executive functions when it conducts investigations. Many also argue that the courts engage in lawmaking—a constitutionally delegated legislative power—when they interpret laws and the judicial interpretations become ingrained in the law as precedent.

Sometimes this overlap of government functions has been referred to as *borrowing*. Engaging in such borrowing has generally been tolerated as long as (1) it is deemed to be reasonably necessary and incidental to the primary function of the branch of government doing the borrowing and (2) the borrowing does not inappropriately enlarge one branch of government over any other(s).

Figure 2-1
Constitutional checks and balances.

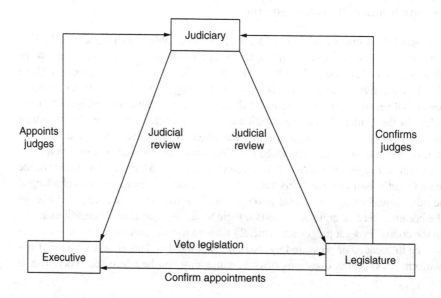

However, they are limits to how far any borrowing or overlap of functions may go. For example, in *Clinton v. City of New York*,[2] the U.S. Supreme Court tackled the question of the Line-Item Veto Act. This act gave the President of the United States the authority to eliminate or cancel certain tax and spending legislation after the President had signed such measures into law. After reviewing the case, the Supreme Court decided that granting such authority to the executive branch violated Article I, Section 7, clause 2 of the U.S. Constitution (often referred to as the *presentment clause*) and the doctrine of separation of powers. Specifically, the Supreme Court held that if the Line-Item Veto Act were held to be valid, it would allow the executive branch to effectively create a statute that was not voted on by either house of Congress.

Another example of where the lines may blur between the various branches of government is illustrated in cases where Congress expressly delegates legislative authority to the executive and/or judicial branches. For instance, Congress expressly delegates rule-making power to the Federal Aviation Administration (FAA), an executive branch agency. A further discussion of this type of delegation to an administrative agency will be found in Chap. 5. However, suffice it to say that such express delegations of legislative authority will be deemed constitutionally valid as long as Congress clearly indicates its basic policy objectives and provides reasonable guidance as to how the agency is to exercise the delegated powers.

Distribution of authority between the federal and state governments

As indicated earlier in this chapter, state governments held a significant share of the authority when the United States first came into being. This tilting of authority to state governments was manifest in the Articles of the Confederacy. Even when it was decided that a stronger federal government was necessary, there was still a significant question of just how much power the federal government should have.

In drafting the U.S. Constitution, this difficult chore was taken on. When the original 13 states (formerly colonies) agreed to the framework of the U.S. Constitution, they also agreed to cede significant sovereign powers to the U.S. federal government. It is easy nowadays to take this for granted. However, this transfer of substantial powers by the states was a significant step and an extraordinary risk, as seen in those times. Every state that has joined the union since the creation of the U.S. Constitution has been required to transfer those same powers to the U.S. federal government.

Under our system of federalism, the federal government holds only those powers that have been delegated to it by the states through the U.S. Constitution. These powers are often referred to as delegated powers or enumerated powers. The powers that have not been granted to the federal government through the U.S. Constitution remain reserved to the states. These powers of the states are often referred to as reserved powers.

Perhaps the most significant delegated powers are found in Article 1, Section 8 of the U.S. Constitution. In this section, the U.S. Congress is granted, among others, the following authority:

- To impose and collect taxes
- To borrow money
- To regulate interstate commerce

- To declare war
- To establish a post office
- To coin money

As you review these delegated powers, it should become clear that what the framers of the Constitution were seeking to do is to identify those powers that are best left at the federal government level. Printing money, declaring war, and establishing a system for the delivery of mail required a strong federal government—that's why these types of powers were delegated to the federal government.

While the powers delegated to the federal government were far-reaching, the powers reserved to the states were equally broad. Keep in mind that if the federal government was not specifically delegated a power, that power was meant to remain with state governments.

The specific delegation of authority to the federal government still left the states with significant authority to regulate the health, safety, morality, and general welfare of their citizens. Laws related to property rights, crime and punishment, operation of motor vehicles, education, and a plethora of other activities are all regulated primarily by the states. This broad general power of the states is often referred to as the state police power.

While it is often clear as to whether a power is delegated to the federal government or reserved to a state, sometimes the lines are blurred and conflicts arise. The remainder of this section will focus on three significant constitutional concepts that are often called in to play when there is a question as to whether federal or state law should apply.

In reviewing the interplay between federal and state laws, first we will take a look at the concept of federal preemption. Next, we will review the history and impact of the commerce clause. Last, we will look at the Constitution's full faith and credit clause. These three constitutional provisions, particularly federal preemption and the commerce clause, have a significant impact on aviation activities in the United States.

Federal preemption

Review Article VI of the Constitution. Notice that the second paragraph and clause clearly indicate that the Constitution and the laws of the federal government, including any U.S. treaties, are the supreme law of the land, binding on all judges in every state. This clause of the U.S. Constitution is often referred to as the *supremacy clause*.

Over time, it has been recognized that the supremacy clause can come into play in three situations:

- Where federal law expressly prohibits any state legislation with regard to a particular subject. This is often referred to as express preemption.
- Where federal law fails to expressly prohibit state law involvement but the nature of the subject area regulated clearly lends itself solely to federal authority. This is often referred to as implied preemption.
- Where federal and state law come into direct conflict with each other.

Express preemption   Express preemption is rather uncommon. It would require that Congress specifically state that no state or local government can adopt a law similar to the federal law being passed. Even though Congress has the power of express preemption, it rarely makes use of the power. Indeed, Congress often expressly opens the door for similar or related state or local legislation as long as such lawmaking does not come into direct conflict with the federal law.

Implied preemption   In most cases federal lawmakers will regulate activity but won't indicate whether similar or related state or local laws are permissible. In these cases, anyone who wishes to challenge state or local law on the grounds that such law is preempted must typically provide convincing evidence that

- Federal regulation of the activity or activities in question is comprehensive. This means a challenger will have to establish the federal regulation of the activity is so pervasive in nature that it forecloses any state or local regulation.
- A strong need exists for uniform federal regulation of the activity in question. Most likely this means demonstrating that state or local laws regulating the activity are likely to interfere in an adverse manner.

In most instances, both of these tests will have to be met to convince a court that state or local regulations are preempted by federal legislation.

Questions of implied preemption frequently arise in the context of aviation activities. One landmark case exploring the ability of local government to regulate aircraft noise is found in Case 2-2 below (*City of Burbank v. Lockheed Air Terminal*[3]).

Direct conflict   While preemption by direct conflict is often viewed on as a separate category of preemption, it can also be looked upon as just a narrower form of implied preemption. This type of preemption will apply to a specific state or local law, not to an entire field of activity. It is generally recognized that on two occasions preemption by direct conflict may be recognized:

- It is impossible to comply with both federal and state laws.
- A state law materially interferes with the purpose of a federal law.

In the realm of aviation, an example of a state law that might directly conflict with a federal law or regulation is a state statute that requires all employers to refrain from mandatory retirement of any employee younger than age 65. This sort of state requirement would come into direct conflict with federal regulations requiring that air carriers retire any pilot in command on reaching age 60. In such a case, the state law would be preempted by federal regulations.

An example of material interference might be a state law that permits a governor to step in and put a halt to a labor strike when state economic interests are threatened. This state law might be looked upon as interfering with a federal law that allows the President to step in and halt a strike only in very narrowly defined situations where national defense or national interests are at stake.

In the end note that with the exception of express preemption, there is generally a presumption against federal preemption in the U.S. legal system. Therefore, if there is no express preemption (and there rarely is), it is up to a challenger to prove that there is implied preemption or direct conflict.

Commerce clause

As indicated earlier, Article I, Section 8 of the Constitution grants Congress broad authority to "regulate commerce with foreign nations, and among the several states." This provision in the Constitution is often referred to as the *commerce clause*.

As you might guess, this provision was drafted—at least in part—as an attempt to address the problems created by the Articles of Confederation. Under the Articles of Confederation, a collection of essentially sovereign states made it very difficult for the United States to benefit from the advantages of free trade among and between the various states. This Balkanization (fragmentation) of the U.S. economy also made it very difficult for the United States to operate in the international community with one voice when it came to issues of trade.

Despite the seeming simplicity of the commerce clause, it has generated more than its fair share of legal controversy since the drafting of the Constitution. Typically, legal questions involving the commerce clause fall into one (or more) of the following categories:

- The scope of federal power over commerce
- Discrimination against interstate commerce
- Undue burdening of interstate commerce

We'll approach our study of the commerce clause by reviewing each of these three subject areas as they developed within the context of the commerce clause.

Scope of federal commerce power   As is often the case, the Constitution leaves room for interpretation. What exactly was meant when the framers gave Congress the right to regulate commerce among the several states? This question has been debated since the inception of the Constitution, and the debate continues in modern-day cases.

In reviewing the history of the commerce clause, it becomes apparent that until the late 1930s the Supreme Court applied a very strict interpretation of what constituted commerce among the several states. In case after case the Supreme Court struck down federal laws that were applied to businesses or industries that were engaged in strictly intrastate commerce. The strict Supreme Court interpretations made it difficult, if not impossible, for Congress to regulate working conditions (including hours and wages) for manufacturing employees—at a time when many employers were exploiting children and others in the labor force by forcing them to work long hours in dangerous conditions for low pay.

The pendulum began to sway in the opposite direction in 1942 with the Supreme Court case of *Wickard v. Filburn*.[4] The *Wickard* case involved a farmer in Ohio named Filburn who planted 23 acres of wheat, disregarding a federal limit of approximately 11 acres of wheat. The federal government assessed a penalty against Filburn for failure to comply with the federal limits. Filburn fought the penalty by arguing that since he used almost all the wheat for his own farm and sold the rest at a local grain elevator within Ohio, he was in no way engaged in interstate commerce. Filburn's argument noted that since he was not engaged in interstate commerce he should not be subject to any federal restriction on the amount of wheat that he could plant.

Prior to the *Wickard* case, there would be little doubt that Filburn's argument would prevail before the Supreme Court. However, the Court took a 180-degree turn and ruled against Filburn. The Court first noted that the federal limitations were put in place to regulate pricing—the intent being to support the price of wheat by limiting the supply of wheat. With this being the purpose of the federal regulation, the Court was able to take the next step. The Court could now argue that if all small farmers such as Filburn were to disregard the federal limitations, their actions would have a sort of ripple effect that would bring wheat prices down throughout the United States.

Armed with this rationale, the Supreme Court continued to promulgate an expansive view of the commerce clause. This allowed the federal government to enact workplace safety rules, minimum wage standards, antitrust laws, securities laws, environmental protection rules, and laws prohibiting employment discrimination.

Despite the sense that the commerce clause (after *Wickard*) had virtually no limits, the pendulum may have started to turn back a bit to a pre-*Wickard* understanding of the commerce clause. Two relatively recent cases that have lawyers, judges, and academics discussing the commerce clause again are *United States v. Lopez*[5] and *United States v. Morrison.*[6]

In the *Lopez* case, the federal government attempted to prosecute a 12th-grade student, charging the student was knowingly possessing a firearm in a "school zone." The United States charged that this violated the Gun Free School Zone Act of 1990. The accused argued that the act exceeded the power of the federal government to legislate such activities. The federal government relied upon the commerce clause to validate the law and pursue a conviction. In a close 5-to-4 vote the Supreme Court held that the Gun Free School Zone Act of 1990 was an unconstitutional overreach by Congress to regulate an area in which it had no constitutional authority.

In the *Morrison* case, a woman university student filed a civil suit in federal court, alleging that she had been raped by two fellow students—and that the attack violated federal law which provided a civil remedy for victims of gender-motivated violence. The United States intervened in this case to defend the federal statute's validity under the Constitution. In another close 5-to-4 vote the Supreme Court rule held that the federal government had overreached its authority under the Constitution with this law.

The *Lopez* and *Morrison* cases taken together leave us with some questions as to where the commerce clause is headed. Is the power to regulate "commerce … among the several states" as it appears in Article I, Section 8 to be interpreted narrowly so as to restrict the ability of Congress to pass sweeping national legislation? Or is it wiser to allow for a broader interpretation of this clause to allow Congress to pass regulations that create national uniformity? It appears that the current Supreme Court will be less likely to give the federal government an easy pass on legislation simply on a claim of commerce clause power. Only time and additional cases will help determine which direction the Supreme Court will take when it comes to determining the scope of federal authority under the commerce clause.

Discrimination against interstate commerce   The commerce clause gives primary authority to the federal government when it comes to regulating interstate commerce. Therefore, any state action that might expressly or effectively discriminate against interstate commerce might be deemed unconstitutional. This means that a state may be prohibited from passing laws that will give industry or citizenry within its borders an unfair advantage over out-of-state competitors.

Sometimes states pass laws that expressly discriminate against commerce from out of state. Such is the case in the 1992 Supreme Court decision *Fort Gratiot Sanitary Landfill, Inc. v. Michigan Department of Natural Resources*,[7] a case involving a Michigan law that prohibited privately owned landfills within the state of Michigan from accepting solid waste from any source outside the county in which the landfill was located (with the exception of those cases in which a county government granted express permission to do so). Fort Gratiot, a landfill operator, was denied an application to allow it to accept out-of-state solid waste. The landfill operator sued the county involved and the state of Michigan. Fort Gatriot argued that the Michigan law in question violated the commerce clause of the Constitution. The Supreme Court ultimately agreed with Fort Gatriot that the Michigan law discriminated against interstate commerce and was therefore unconstitutional. In making its ruling, the Court rejected the state and county government's argument that solid waste is not an article of interstate commerce. The Court stated in its ruling that "the restrictions enacted by Michigan authorized each of its 83 counties to isolate itself from the national economy. The Court has consistently found parochial legislation of this kind to be constitutionally invalid."

In other cases a law may not specifically or expressly discriminate against interstate commerce. However, the circumstances and effect of the law may very clearly indicate discrimination against interstate commerce. An example of this type of case is *Hunt v. Washington State Apple Advertising Commission*.[8] This case involved a North Carolina law requiring all apples sold in North Carolina to have the U.S. Department of Agriculture grade stamped on the apple crates. The law also specifically prohibited any state grades from being stamped on the crates. This law was unique to North Carolina. For years before this law was enacted, the state of Washington had been stamping its apple crates with its special state grade. The state of Washington had spent many years developing a reputation for its special state grade. During the course of the legal proceedings, it became apparent that the North Carolina law was specifically designed to protect North Carolina growers from competition with growers from the state of Washington. The Court concluded that this type of intentional discrimination against interstate commerce was unconstitutional. Therefore the North Carolina law was invalid.

Undue burdening of interstate commerce   Sometimes state regulators may not intend to discriminate against interstate commerce. However, the results or impact of their legislation or regulations may create an undue burdening of interstate commerce. A review of Supreme Court cases clearly demonstrates that state laws may be invalidated if they have the effect of impeding the free flow of commerce between the states.

In many cases where the state laws are challenged based on an undue burdening of interstate commerce, the courts will need to apply a balancing test or analysis. There is no easily quantifiable formula indicating how this balancing test and supporting analysis are executed. Nonetheless, there are some common threads that run through the court cases.

As a general rule, the courts will weigh the state or local interest supported by the law in question against the amount or degree of burden the law would place on the free flow of commerce among the states. If the state's interest is very high, there is a possibility that the court will allow for some measure of burden on interstate commerce to permit the law to remain effective.

In cases where a state's economic interests are the driving force behind the law in question, the courts may scrutinize the law more carefully. If the law reflects a state interest in the protection of its citizens' health, safety, or welfare, the courts may be more likely to defer on the side of the state law. However, this does not mean that the courts will roll over and accept a state law simply because the state asserts that the law is designed to protect the health, safety, or welfare of its citizens.

In the landmark case of *Kassell v. Consolidated Freightways, Corp.*[9] the state of Iowa passed a law that prohibited the use of 65-foot double trailers on its highways. At the same time 55-foot single trailers and 60-foot double trailers were allowed. The states surrounding Iowa all allowed for 65-foot double trailers on their highways. The state of Iowa asserted that the law was a safety measure. Contrary to this assertion, evidence showed that there was no relationship between truck length and accident rates. In fact, statistics indicated that the law could create a situation in which increased numbers of smaller trucks would be required to deliver the same goods and cargo. In turn, this would increase the accident rate—a result quite contrary to Iowa's stated intent. In the end, the Supreme Court found that this law was unconstitutional because it would unduly burden interstate commerce by requiring the 65-foot double trailers to either go around the state of Iowa or unload into smaller trucks prior to entering the state.

More than a fair share of the "undue burden" cases involve taxation. Most of these cases involve state or local property and/or sales and use taxes. In the typical case, the state asserts that it has the right to tax the value of property located within its borders or property that is purchased or used within its borders. The underlying assumption is that it is fair for the owner of the property to pay a tax to the state to contribute to the cost of infrastructure needed to protect and preserve that property (i.e., police and fire protection, roads, etc.).

One of the most widely cited cases on the issue of undue burden in taxation is the case of *Complete Auto Transit, Inc. v. Brady.*[10] In the *Complete Auto Transit* case, the Supreme Court reviewed whether a Mississippi tax on the privilege of doing business within the state created an undue burden on interstate commerce. The case involved Complete Auto

Transit, a company that transported motor vehicles from out-of-state manufacturers between points in the state of Mississippi. More important than the case itself was the test created by the Supreme Court in deciding the *Complete Auto Transit* case. In the end, the court held that the state tax could withstand a commerce clause challenge of undue burden if it

1. Is applied to an activity with a substantial nexus to the taxing state
2. Is fairly apportioned
3. Does not discriminate against interstate commerce
4. Is fairly related to the services provided by the state

The "nexus" test is concerned with whether or not property that is being taxed has enough of a connection with the state that is taxing the property. For instance, under this test, a state could not pursue property tax on an aircraft that had only stopped for a brief refueling within the state. However, if the aircraft was based and used primarily within that state, the aircraft could be subject to taxation.

The apportionment test is often the subject of litigation. Normally it requires a formula that allows the state to tax only the degree of connection that property may have within that state. This may mean the property owners will be required to carefully record and report the number of days an item of property was in a particular state. Apportionment questions often arise in the context of transportation equipment such as aircraft and railroad cars.

Whether a tax discriminates against interstate commerce is essentially the same question that was addressed above. A tax cannot be greater for items used in interstate commerce than for items that are used only within the taxing state.

The test of whether a tax is fairly related to the services provided by the state is seldom invoked by the courts. However, this test would likely arise in the context of a question as to whether the tax applies a fair rate or amount to the property being taxed.

Full faith and credit clause

Article IV, Section 1 of the Constitution is often referred to as the full faith and credit clause. It states (in pertinent part) that "… full faith and credit shall be given in each State to the public acts, records, and judicial proceedings of every other State." This clause of the Constitution has the impact of requiring every state court to recognize and honor the court rulings and public laws of every other state.

This clause can have a significant impact in the world of aviation because quite often the subject of a claim may be an aircraft—personal property that can be moved at a moment's notice. This section will allow a creditor who obtains a valid judgment against a debtor in New York and who is able to locate the debtor's property in Maryland (including an aircraft) to enforce the creditor's judgment in Maryland.

While the full faith and credit clause functions within the various states of the United States, it does not give validity to U.S. judgments in foreign courts or foreign court judgments in the United States. However, there is a concept in international law known as the *doctrine of comity* that calls for nations to respect and enforce the court judgments of other nations.

Basic rights of individuals and businesses

The third prong of our review of constitutional law includes a look at the protection of fundamental individual and business rights built into the Constitution. The Constitution contains numerous provisions designed to limit the authority of federal and state governments from interfering with the affairs of individuals and businesses. Some of these provisions are found in the body of the Constitution. Most are found in the amendments to the Constitution. The first 10 amendments to the Constitution are most commonly known as the Bill of Rights. At this point you should take another look at the Constitution to familiarize yourself with the Bill of Rights.

In reviewing the basic rights provided by the Constitution, it is important to note two points from the beginning. First, you need to recognize that the Bill of Rights as it was initially drafted and understood applied to actions of only the federal government. It did not apply to state or local government actions. Interestingly, in 1791 when the Bill of Rights was drafted, there was no intention of holding states accountable for rights such as freedom of religion or freedom of the press. However, all this changed after 1868 (shortly after the Civil War) and the adoption of the Fourteenth Amendment to the U.S. Constitution. In Section 1 of the Fourteenth Amendment, the well-known language was drafted that says that no state shall deprive any person of life, liberty, or property without due process of law or deny to any person within its jurisdiction the equal protection of the laws. The Supreme Court has consistently held that the Fourteenth Amendment requires that all states respect the protections built into the Bill of Rights.

Second, all the basic rights protected under the Constitution involve protections against government actions. These basic rights do not apply to protection against individuals or private businesses. Therefore, while the Bill of Rights will protect individuals against government intrusions on their right to free speech, these same protections did not prohibit a private employer from imposing restrictions on the free-speech rights of his or her employees while they were on business premises.

It is beyond the scope of this book to review all the basic rights protected in the U.S. Constitution. Therefore, we will focus on some of the provisions that come into play for the aviation industry and aviation activities.

Fifth Amendment takings clause

The Fifth Amendment clearly indicates that the government may not take private property for public use without just compensation being provided to the property owner. As indicated earlier in this chapter, this provision originally applied to only the federal government. However, after ratification of the Fourteenth Amendment, the Fifth Amendment became applicable to states as well as to the federal government.

Most readings of the cases involving the Fifth Amendment seem to indicate that the courts rely on a very broad definition of what constitutes property for purposes of the takings clause. Property clearly includes land and buildings. However it has also been interpreted to mean mineral rights below the surface, air rights above the surface of the land, and intangible personal property such as equipment, aircraft, and automobiles.

As clearly indicated in the Fifth Amendment, if the government wants to take private property, the government must do so for a public purpose. However, over the years, this public purpose has not been rigorously questioned by the courts. Just about any purpose will do for the purposes of the Fifth Amendment. To a large extent this deference to the government's taking goes back to the long recognized principle that the government may take private property for public use. This power has often been referred to as the power of eminent domain. When the government takes property, especially real property, it is often said that the government has condemned the property.

Often it is obvious if a government taking has occurred. For instance, a state or local government may decide it needs to condemn certain farmland to make use of the land for purposes of waste disposal. Sometimes it is not so obvious that a taking has occurred. Suppose, for instance, that a state or local government extends an airport runway so that takeoffs and landings now occur at a very low altitude over a private residence. If the owner of the private residence feels that the value of his or her property has been diminished and rendered unusable by the government's taking, the owner may decide to sue the government for taking property and violating her or his Fifth Amendment rights by failing to compensate. This sort of action is often referred to as *inverse condemnation*. Case 2-1 addresses this very issue.

Very recently, the Supreme Court tackled a controversial case involving the Fifth Amendment takings clause. The case, *Kelo v. City of New London, Connecticut*,[11] involved the question of whether a local government could take private property earmarked for development by private developers. The owners of the property to be taken argued that the City of New London could not take their homes because the property was not being taken by the government for a public use. The City of New London argued that its plan of economic rejuvenation for the community in question justified the taking. In a close 5-to-4 vote, the Supreme Court held that although the city could not take a private home simply to confer a benefit on a private developer, the city could take property under the Fifth Amendment taking clause if the taking were part of a carefully considered development plan designed to meet public needs.

Due process
The Fifth and Fourteenth Amendments of the Constitution protect individuals and businesses by prohibiting the federal government and state government from depriving anyone of life, liberty, or property without the due process of law. Due process embodies both substantive due process and procedural due process.

Procedural due process can best be viewed as the method(s) or process employed by the courts and the government to apply substantive laws. The requirements of procedural due process will be met if the government can establish that it provided for a fair process that allowed a person to state his or her case before the government could deprive the person of life, liberty, or property. Therefore, before the government can exercise eminent domain over property, it must provide the property owner with a fair hearing to determine the validity of the government's taking and, if necessary, the valuation of the property. Likewise, a fair hearing with appropriate protections must be provided to a defendant in a criminal case where his or her life or liberty is in jeopardy. The intricacies of due process as it applies to criminal cases will be dealt with in greater detail in Chap. 3. Keep in mind throughout all this, however, that procedural due process is not a requirement unless the person involved faces deprivation of life, liberty, or property. If government action might adversely affect the person, but doesn't deprive her or him of life, liberty, or property, then due process is not required.

Substantive due process does not involve the methods or process of the law. Instead, substantive due process is concerned with the question of whether a law, in and of itself, passes constitutional muster. Is the law arbitrary or capricious? Is the law so vague or difficult to enforce as to render it unconstitutional? When determining whether a law meets the requirements of substantive due process, the courts have devised two tests. The first test asks whether the law in question has a rational relationship to legitimate government interests. If so, the law will be deemed constitutional. This rational relationship test is typically applied in cases that involve economic and/or social legislation.

A tougher, second test will be applied when the government is attempting to legislate restrictions on individuals' fundamental rights under the Constitution. Under the so-called strict scrutiny test, the government needs to demonstrate that there is some overriding or compelling state interest that it wishes to promote if its legislation will have an impact on basic constitutional rights (e.g., freedom of speech, religion, the press, etc.).

Equal protection
The equal protection clause is found in the Fourteenth Amendment to the Constitution. It states: "nor shall any State ... deny to any person within its jurisdiction the equal protection of the laws." Even though the wording of this amendment applies to only the individual states, it has been interpreted to apply to the federal government as well. The essence of the equal protection clause is a guarantee that the government will provide equal treatment for similarly situated persons.

Under government laws either directly or indirectly having the result of classifying persons, there is the potential for the equal protection clause to come into play. When the courts review an equal protection challenge to the law, three possible tests might be employed.

The first possible test is the rational relationship test. This test typically applies to economic regulation or legislation. It is the least rigorous of all three tests. For this test to be successfully overcome, the government merely needs to show that the classification of persons bears some rational relationship to a legitimate government interest. Because this test is so difficult to overcome for challengers of government action, it is very rare to see a court overturn a government action based on this test.

The second equal protection test is often referred to as the intermediate scrutiny test. This test is typically applied when government actions or loss classified individuals based on gender. This test requires that the government show that the classification has some substantial relationship to an important government interest. This test has been used to strike down laws that prohibited courts from awarding alimony to men but not to women (*Orr v. Orr*[12]). In a landmark 1996 case, the Supreme Court invalidated a state university's admission policies that prohibited women (*United States v. Virginia*[13]).

The third test applied in equal protection cases is known as the strict scrutiny test. When this test is applied, the courts do not defer to the government. This test is typically applied when government action or laws affect individuals' basic rights or appear to involve suspect classifications (e.g., classifications based on race or ethnicity). Probably the most famous case applying the strict scrutiny test was *Brown v. Board of Education of Topeka*[14] in which the Supreme Court ruled that the equal protection clause was violated by segregated school systems.

Freedom of speech and the press

Take a look at the First Amendment to the Constitution in App. B. The First Amendment clearly states that the government cannot abridge the freedom of speech or of the press.

Despite the fact that the protection of speech and press granted by the First Amendment appears to be absolute, it is not. There are a few forms of speech such as obscenity that receive no protection under the Constitution.

As a general matter, application of the First Amendment is relatively straightforward—the protections benefit the speaker. However, this is not always the case.

Defamation involves speech intended to discredit another. This sort of speech is typically unprotected by the First Amendment. However, there is one important exception to the usual rule when it comes to defamation. In the case of *New York Times Co. v. Sullivan*,[15] the Supreme Court held that if a public figure is defamed, he or she could recover only if it were proved that the defamatory statements were made with malice.

Another type of speech that is not wholly protected by the First Amendment is commercial speech. Commercial speech is speech related to the economic interests of the person speaking to her or his audience. Usually, commercial speech comes in the form of advertisements or promotions. This sort of speech will be protected under the First Amendment as long as it does not relate to illegal activity and is not misleading. Because of this interpretation, the government may regulate and/or prohibit false or misleading advertisements.

CASES AND COMMENTARY

The first case for review (Case 2-1) is a Supreme Court case from 1946. This case is a landmark case in the aviation industry because it was the first to tackle the question of whether airport operations can result in a "taking" of property within the meaning of the Fifth Amendment.

CASE 2-1
UNITED STATES V. CAUSBY ET UX.
328 U.S. 256 (1946)

OPINION: MR. JUSTICE DOUGLAS delivered the opinion of the Court.

This is a case of first impression. The problem presented is whether respondents' property was taken, within the meaning of the Fifth Amendment, by frequent and regular flights of army and navy aircraft over respondents' land at low altitudes.

Respondents own 2.8 acres near an airport outside of Greensboro, North Carolina. It has on it a dwelling house, and also various outbuildings which were mainly used for raising chickens. The end of the airport's northwest-southeast runway is 2,220 feet from respondents' barn and 2,275 feet from their house. The path of glide to this runway passes directly over the property—which is 100 feet wide and 1,200 feet long. The 30 to 1 safe glide angle approved by the Civil Aeronautics Authority passes over this property at 83 feet, which is 67 feet above the house, 63 feet above the barn and 18 feet above the highest tree. The use by the United States of this airport is pursuant to a lease executed in May, 1942, for a term commencing June 1, 1942 and ending June 30, 1942, with a provision for renewals until June 30, 1967, or six months after the end of the national emergency, whichever is the earlier.

Various aircraft of the United States use this airport—bombers, transports and fighters. The direction of the prevailing wind determines when a particular runway is used. The northwest-southeast runway in question is used about four per cent of the time in taking off and about seven per cent of the time in landing. Since the United States began operations in May, 1942, its four-motored heavy bombers, other planes of the heavier type, and its fighter planes have frequently passed over respondents' land and buildings in considerable numbers and rather close together. They come close enough at times to appear barely to miss the tops of the trees and at times so close to the tops of the trees as to blow the old leaves off. The noise is startling. And at night the glare from the planes brightly lights up the place. As a result of the noise, respondents had to give up their chicken business. As many as six to ten of their chickens were killed in one day by flying into the walls from fright. The total chickens lost in that manner was about 150. Production also fell off. The result was the destruction of the use of the property as a commercial chicken farm. Respondents are frequently deprived of their sleep and the family has become nervous and frightened. Although there have been no airplane accidents on respondents' property, there have been several accidents near the airport and close to respondents' place. These are the essential facts found by the Court of Claims. On the basis of these facts, it found that respondents' property had depreciated in value. It held that the United States had taken an easement over the property on June 1, 1942, and that the value of the property destroyed and the easement taken was $2,000.

I. The United States relies on the Air Commerce Act. Under [the Air Commerce Act] the United States has "complete and exclusive national sovereignty in the air space" over this country. [The Air Commerce Act] grant[s] any citizen of the United States "a public right of freedom of transit in air commerce through the navigable air space of the United States." And "navigable air space" is defined as "airspace above the minimum safe altitudes of flight prescribed by the Civil Aeronautics Authority." And it is provided that "such navigable airspace shall be subject to a public right of freedom of interstate and foreign air navigation." It is, therefore, argued that since these flights were within the minimum safe altitudes of flight which had been prescribed, they were an exercise of the declared right of travel through the airspace. The United States concludes that when flights are made within the navigable airspace without any physical invasion of the property of the landowners, there has been no taking of property. It says that at most there was merely incidental damage occurring as a consequence of authorized air navigation. It also argues that the landowner does not own superadjacent airspace which he has not subjected to possession by the erection of structures or other occupancy. Moreover, it is argued that even if the United States took airspace owned by respondents, no compensable damage was shown. Any damages are said to be merely consequential for which no compensation may be obtained under the Fifth Amendment.

It is ancient doctrine that at common law ownership of the land extended to the periphery of the universe—*Cujus est solum ejus est usque ad coelum*. But that doctrine has no place in the modern world. The air is a public highway, as Congress has declared. Were that not true, every transcontinental flight would subject the operator to countless trespass suits. Common sense revolts at the idea. To recognize such private claims to the airspace would clog these highways, seriously interfere with their control and development in the public interest, and transfer into private ownership that to which only the public has a just claim.

But that general principle does not control the present case. For the United States conceded on oral argument that if the flights over respondents' property rendered it uninhabitable, there would be a taking compensable under the Fifth Amendment. It is the owner's loss, not the taker's gain, which is the measure of the value of the property taken. Market value fairly determined is the normal measure of the recovery. And that value may reflect the use to which the land could readily be converted, as well as the existing use. If, by reason of the frequency and altitude of the flights, respondents could not use this land for any purpose, their loss would be complete. It would be as complete as if the United States had entered upon the surface of the land and taken exclusive possession of it.

We agree that in those circumstances there would be a taking. Though it would be only an easement of flight which was taken, that easement, if permanent and not merely temporary, normally would be the equivalent of a fee interest. It would be a definite exercise of complete dominion and control over the surface of the land. The fact that the planes never touched the surface would be as irrelevant as the absence in this day of the feudal livery of seisin on the transfer of real estate. The owner's right to possess and exploit the land—at is to say, his beneficial ownership of it—would be destroyed.

There is no material difference between the supposed case and the present one, except that here enjoyment and use of the land are not completely destroyed. But that does not seem to us to be controlling. The path of glide for airplanes might reduce a valuable factory site to grazing land, an orchard to a vegetable patch, a residential section to a wheat field. Some value would remain. But the use of the airspace immediately above the land would limit the utility of the land and cause a diminution in its value.

The fact that the path of glide taken by the planes was that approved by the Civil Aeronautics Authority does not change the result. The navigable airspace which Congress has placed in the public domain is "airspace above the minimum safe altitudes of flight prescribed by the Civil Aeronautics Authority." If that agency prescribed 83 feet as the minimum safe altitude, then we would have presented the question of the validity of the regulation. But nothing of the sort has been done. The path of glide governs the method of operating—of landing or taking off. The altitude required for that operation is not the minimum safe altitude of flight which is the downward reach of the navigable airspace. The minimum prescribed by the

Authority is 500 feet during the day and 1,000 feet at night for air carriers, and from 300 feet to 1,000 feet for other aircraft, depending on the type of plane and the character of the terrain. Hence, the flights in question were not within the navigable airspace which Congress placed within the public domain. If any airspace needed for landing or taking off were included, flights which were so close to the land as to render it uninhabitable would be immune. But the United States concedes, as we have said, that in that event there would be a taking. Thus, it is apparent that the path of glide is not the minimum safe altitude of flight within the meaning of the statute. The Civil Aeronautics Authority has, of course, the power to prescribe air traffic rules. But Congress has defined navigable airspace only in terms of one of them—the minimum safe altitudes of flight.

We have said that the airspace is a public highway. Yet it is obvious that if the landowner is to have full enjoyment of the land, he must have exclusive control of the immediate reaches of the enveloping atmosphere. Otherwise buildings could not be erected, trees could not be planted, and even fences could not be run. The principle is recognized when the law gives a remedy in case overhanging structures are erected on adjoining land. The landowner owns at least as much of the space above the ground as he can occupy or use in connection with the land. The fact that he does not occupy it in a physical sense—by the erection of buildings and the like—is not material. As we have said, the flight of airplanes, which skim the surface but do not touch it, is as much an appropriation of the use of the land as a more conventional entry upon it. We would not doubt that, if the United States erected an elevated railway over respondents' land at the precise altitude where its planes now fly, there would be a partial taking, even though none of the supports of the structure rested on the land. The reason is that there would be an intrusion so immediate and direct as to subtract from the owner's full enjoyment of the property and to limit his exploitation of it. While the owner does not in any physical manner occupy that stratum of airspace or make use of it in the conventional sense, he does use it in somewhat the same sense that space left between buildings for the purpose of light and air is used. The superadjacent airspace at this low altitude is so close to the land that continuous invasions of it affect the use of the surface of the land itself. We think that the landowner, as an incident to his ownership, has a claim to it and that invasions of it are in the same category as invasions of the surface.

In this case, the damages were not merely consequential. They were the product of a direct invasion of respondents' domain. As stated in [a previous Supreme Court case] "… it is the character of the invasion, not the amount of damage resulting from it, so long as the damage is substantial, that determines the question whether it is a taking."

The airplane is part of the modern environment of life, and the inconveniences which it causes are normally not compensable under the Fifth Amendment. The airspace, apart from the immediate reaches above the land, is part of the public domain. We need not determine at this time what those precise limits are. Flights over private land are not a taking, unless they are so low and so frequent as to be a direct and immediate interference with the enjoyment and use of the land. We need not speculate on that phase of the present case. For the findings of the Court of Claims plainly establish that there was a diminution in value of the property and that the frequent, low-level flights were the direct and immediate cause. We agree with the Court of Claims that a servitude has been imposed upon the land.

Notice that the Supreme Court's holding in the *Causby* case made the United States, the owner and operator of the aircraft involved, liable for the taking under the Fifth Amendment. The question of who is responsible for the taking arrives at the Supreme Court once again in 1962 in the case of *Griggs v. Allegheny County*.[16]

In the *Griggs* case, Allegheny County owned and operated the Greater Pittsburgh Airport. The county planned and constructed the airport in accordance with the standards approved by the Civil Aeronautics Board (CAA) [precursor to the Federal Aviation Administration (FAA)]. As in the *Causby* case, one of the approach/departure zones took aircraft very low over an adjacent residential property, forcing the owner and his family to move. The Supreme Court held, consistent with *Causby*, that a taking within the meaning of the Fifth Amendment had occurred. However, the issue that came to the forefront in this case was the determination of who was liable for the taking.

In the end, the Supreme Court ruled that Allegheny County, the airport proprietor, was liable for the taking. In making this ruling, the Court rejected arguments that the taking was done by someone other than Allegheny County. In rejecting the county's arguments that the airlines or the CAA was responsible for the taking, the Court reasoned that it was Allegheny County that decided where the airport would be located, what runways would be needed, and their direction and length. Two of the Supreme Court's Justices (Black and Frankfurter) dissented from the majority opinion in the *Griggs* case. Both agreed that a Fifth Amendment taking had occurred. However, both Black and Frankfurter argued that the U.S. government was responsible for the taking because the airport had to be designed, constructed, and approved by the federal government.

What's your reaction to the *Causby* and *Griggs* cases? Should the courts find a taking when property has not been physically taken by an airport? If a Fifth Amendment taking has occurred, who should be responsible?

Another interesting constitutional question that arises in the aviation environment relates to aircraft noise regulation. How much authority does a state or local government possess when it comes to regulating aircraft noise? In the following landmark case (Case 2-2), the Supreme Court tackled this issue in 1972.

CASE 2-2
CITY OF BURBANK ET AL. V. LOCKHEED AIR TERMINAL, INC. ET AL.
411 U.S. 624 (1973)

OPINION: MR. JUSTICE DOUGLAS delivered the opinion of the Court.

This suit asked for an injunction against the enforcement of an ordinance adopted by the City Council of Burbank, California, which made it unlawful for a so-called pure jet aircraft to take off from the Hollywood-Burbank Airport between 11 p.m. of one day and 7 a.m. the next day, and making it unlawful for the operator of that airport to allow any such aircraft to take off from that airport during such periods. The only regularly scheduled flight affected by the ordinance was an intrastate flight of Pacific Southwest Airlines originating in Oakland, California, and departing from Hollywood-Burbank Airport for San Diego every Sunday night at 11:30.

The District Court found the ordinance to be unconstitutional on both Supremacy Clause and Commerce Clause grounds. The Court of Appeals affirmed on the grounds of the Supremacy Clause both as respects pre-emption and as respects conflict. The case is here on appeal. We noted probable jurisdiction. We affirm the Court of Appeals.

The Federal Aviation Act of 1958, and the regulations under it, are central to the question of pre-emption.

Section 1108 [Citation to the Federal Aviation Act], provides in part, "The United States of America is declared to possess and exercise complete exclusive national sovereignty in the airspace of the United States...." By Sections [Citations to the Federal Aviation Act] the Administrator of the Federal Aviation Administration (FAA) has been given broad authority to regulate the use of the navigable airspace, "in order to insure the safety of aircraft and the efficient utilization of such airspace ..." and "for the protection of persons and property on the ground...."

The Solicitor General, though arguing against pre-emption, concedes that as respects "airspace management" there is pre-emption. That, however, is a fatal concession, for as the District Court found: "The imposition of curfew ordinances on a nationwide basis would result in a bunching of flights in those hours immediately preceding the curfew. This bunching of flights during these hours would have the twofold effect of increasing an already serious congestion problem and actually increasing, rather than relieving, the noise problem by increasing flights in the period of greatest annoyance to surrounding communities. Such a result is totally inconsistent with the objectives of the federal statutory and regulatory scheme." It also found "the imposition of curfew ordinances on a nationwide basis would cause a serious loss of efficiency in the use of the navigable airspace."

Curfews such as Burbank has imposed would, according to the testimony at the trial and the District Court's findings, increase congestion, cause a loss of efficiency, and aggravate the noise problem. FAA has occasionally enforced curfews. But the record shows that FAA has consistently opposed curfews, unless managed by it, in the interests of its management of the "navigable airspace."

As stated by Judge Dooling in [Citation]: "The aircraft and its noise are indivisible; the noise of the aircraft extends outward from it with the same inseparability as its wings and tail assembly; to exclude the aircraft noise from the Town is to exclude the aircraft; to set a ground level decibel limit for the aircraft is directly to exclude it from the lower air that it cannot use without exceeding the decibel limit."

There is, to be sure, no express provision of pre-emption in the 1972 Act. That, however, is not decisive.

It is the pervasive nature of the scheme of federal regulation of aircraft noise that leads us to conclude that there is pre-emption. As Mr. Justice Jackson stated, concurring in [Citation]: "Federal control is intensive and exclusive. Planes do not wander about in the sky like vagrant clouds. They move only by federal permission, subject to federal inspection, in the hands of federally certified personnel and under an intricate system of federal commands. The moment a ship taxis onto a runway it is caught up in an elaborate and detailed system of controls."

If we were to uphold the Burbank ordinance and a significant number of municipalities followed suit, it is obvious that fractionalized control of the timing of takeoffs and landings would severely limit the flexibility of FAA in controlling air traffic flow. The difficulties of scheduling flights to avoid congestion and the concomitant decrease in safety would be compounded. In 1960 FAA rejected a proposed restriction on jet operations at the Los Angeles airport between

10 p.m. and 7 a.m. because such restrictions could "create critically serious problems to all air transportation patterns." The complete FAA statement said: "The proposed restriction on the use of the airport by jet aircraft between the hours of 10 p.m. and 7 a.m. under certain surface wind conditions has also been reevaluated and this provision has been omitted from the rule. The practice of prohibiting the use of various airports during certain specific hours could create critically serious problems to all air transportation patterns. The network of airports throughout the United States and the constant availability of these airports are essential to the maintenance of a sound air transportation system. The continuing growth of public acceptance of aviation as a major force in passenger transportation and the increasingly significant role of commercial aviation in the nation's economy are accomplishments which cannot be inhibited if the best interest of the public is to be served. It was concluded therefore that the extent of relief from the noise problem which this provision might have achieved would not have compensated the degree of restriction it would have imposed on domestic and foreign Air Commerce."

This decision, announced in 1960, remains peculiarly within the competence of FAA, supplemented now by the input of EPA. We are not at liberty to diffuse the powers given by Congress to FAA and EPA by letting the States or municipalities in on the planning. If that change is to be made, Congress alone must do it.

The *Burbank* case requires careful analysis. It is easy to walk away from the case with the impression that the Supreme Court intends to preempt any state or local action that might tend to regulate aircraft noise. However, a careful review of the *Burbank* decision indicates that while the Court clearly indicates that the federal government preempts state and local regulations over aircraft noise, the decision does not prohibit all types of noise regulation that might be imposed by state or local government. Indeed, the Court more narrowly held that the federal government, most notably through the Federal Aviation Administration and the Environmental Protection Agency (EPA), has full control over aircraft noise by preempting state and local control under a state's police power. The Supreme Court expressly left open the question as to how much authority state or local government would have over noise regulation if the state or local government were an airport proprietor (see footnote 14 in the full text of the case).

Subsequent to *Burbank*, other courts have recognized what is often referred to as the *airport proprietor exception* to the *Burbank* case. This exception allows an airport proprietor to impose reasonable, nonarbitrary, and nondiscriminatory regulations relating to noise at an airport and its immediate surroundings (see *British Airways Board v. Port Authority of New York*[17]). The logic in permitting this airport proprietor exception relates to the liability that an airport owner or proprietor might have for aircraft noise. Essentially, the exception allows an owner of an airport to mitigate or decrease his or her exposure to damages over aircraft noise by permitting the airport proprietor to issue and enforce reasonable noise regulations. In some measure, this exception may have been rendered necessary by the Supreme Court's earlier decisions in *Causby* and *Griggs,* where it was held that airport owners were liable for Fifth Amendment takings when airport noise impeded the use of neighboring properties. In effect, airport owners and proprietors needed some tool to limit their exposure to liability that had been created by *Griggs*.

Of course, the requirement of reasonableness will often raise questions. There have been several cases in which airport proprietor regulation of aircraft noise has been reviewed by the courts. Consider the following situations. How do you think a court would hold if it were evaluating the following noise regulations by an airport proprietor?

- A curfew on all-night flight operations regardless of the noise level emitted (see *United States v. State of New York*[18])
- A ban on all-night touch-and-go aircraft traffic (see *Santa Monica Airport Association, et al. v. City of Santa Monica*[19])
- A ban on all helicopter flight training (see *Santa Monica*)
- A ban on low approaches during weekends (see *Santa Monica*)
- A ban on all jet aircraft (see *Santa Monica*)
- A ban on supersonic air transport aircraft (see *British Airways Board v. New York Port Authority*[20])

The next case for review (Case 2-3) is a 1954 Supreme Court decision addressing a thorny issue of taxation. Is it a violation of the interstate commerce clause when a state taxes flight equipment owned by an airline that does not have a permanent situs in that state? The case also questions whether the federal government's preemption of air commerce prohibits a state from regulating an interstate air carrier through taxation of its flight equipment.

CASE 2-3
BRANIFF AIRWAYS, INC. V. NEBRASKA STATE BOARD OF EQUALIZATION AND ASSESSMENTS ET AL.
347 U.S. 590 (1954)

OPINION: MR. JUSTICE REED delivered the opinion of the Court.

The question presented by this appeal from the Supreme Court of Nebraska is whether the Constitution bars the State of Nebraska from levying an apportioned ad valorem tax [a tax imposed on the value of property] on the flight equipment of appellant, an interstate air carrier. Appellant is not incorporated in Nebraska and does not have its principal place of business or home port registered in that state. Such flight equipment is employed as a part of a system of interstate air commerce operating over fixed routes and landing on and departing from airports within Nebraska on regular schedules.

Appellant does not challenge the reasonableness of the apportionment prescribed by the taxing statute or the application of such apportionment to its property. It contends only that its flight equipment used in interstate commerce is immune from taxation by Nebraska because without situs in that state and because regulation of air navigation by the Federal Government precludes such state taxation.

The home port registered with the Civil Aeronautics Authority and the overhaul base for the aircraft in question is the Minneapolis-St. Paul Airport, Minnesota. All of the aircraft not undergoing overhaul fly regular schedules upon a circuit ranging from Minot, North Dakota, to New Orleans, Louisiana, with stops in fourteen states including Minnesota, Nebraska and Oklahoma.

No stops were made in Delaware [the state where Braniff was incorporated]. The Nebraska stops are of short duration since utilized only for the discharge and loading of passengers, mail, express, and freight, and sometimes for refueling. Appellant neither owns nor maintains facilities for repairing, reconditioning, or storing its flight equipment in Nebraska, but rents depot space and hires other services as required.

It is stipulated that the tax in question is assessed only against regularly scheduled air carriers and is not applied to carriers who operate only intermittently in the state. The statute defines "flight equipment" as "aircraft fully equipped for flight," and provides that "any tax upon or measured by the value of flight equipment of air carriers incorporated or doing business in this state shall be assessed and collected by the Tax Commissioner." A formula is prescribed for arriving at the proportion of a carrier's flight equipment to be allocated to the state.

The statute uses the allocation formula of the "proposed uniform statute to provide for an equitable method of state taxation of air carriers" adopted by the Council of State Governments upon the recommendation of the National Association of Tax Administrators in 1947. Use of a uniform allocation formula to apportion air-carrier taxes among the states follows the recommendation of the Civil Aeronautics Board in its report to Congress. The Nebraska statute provides for reports, levy, and rate of tax by state average.

Required reports filed [by Braniff] for 1950 show that about 9% of its revenue and 11 1/2% of the total system tonnage originated in Nebraska and about 9% of its total stops were made in that state. From these figures, using the statutory formula, the Tax Commissioner arrived at a valuation of $118,901 allocable to Nebraska, resulting in a tax of $4,280.44. Since Mid-Continent filed no return for 1951 the same valuation was used and an increased rate resulted in assessment of $4,518.29. The Supreme Court of Nebraska held the statute not violative of the Commerce Clause and dismissed appellant's petition.

[Braniff] argues that federal statutes governing air commerce enacted under the commerce power pre-empt the field of regulation of such air commerce and preclude this tax. Congress, by the Civil Aeronautics Act of 1938, enacted: "The United States of America is declared to possess and exercise complete and exclusive national sovereignty in the air space above the United States, including the air space above all inland waters and the air space above those portions of the adjacent marginal high seas, bays, and lakes, over which by international law or treaty or convention the United States exercises national jurisdiction."

This provision originated in the Air Commerce Act of 1926. The 1938 Act also declares "a public right of freedom of transit" for air commerce in the navigable air space to exist for any citizen of the United States.

The provision pertinent to sovereignty over the navigable air space in the Air Commerce Act of 1926 was an assertion of exclusive national sovereignty. The convention between the United States and other nations respecting international civil aviation accords. The Act, however, did not expressly exclude the sovereign powers of the states. The Civil Aeronautics Act of 1938 gives no support to a different view. After the enactment of the Air Commerce Act, more than twenty states adopted the Uniform Aeronautics Act. It had three provisions indicating that the states did not consider their sovereignty affected by the National Act except to the extent that the states had ceded that sovereignty by constitutional grant. The recommendation of the National Conference of Commissioners on Uniform State Laws to the states to enact this Act was withdrawn in 1943. Where adopted, however, it continues in effect. Recognizing this "exclusive national sovereignty" and right of freedom in air transit, this Court in *United States* v. *Causby*, [Citation], nevertheless held that the owner of land might recover for a taking by national use of navigable air space, resulting in destruction in whole or in part of the usefulness of the land property.

These Federal Acts regulating air commerce are bottomed on the commerce power of Congress, not on national ownership of the navigable air space, as distinguished from sovereignty. In reporting the bill which became the Air Commerce Act, it was said:

"The declaration of what constitutes navigable air space is an exercise of the same source of power, the interstate commerce clause, as that under which Congress has long declared in many acts what constitutes navigable or nonnavigable waters. The public right of flight in the navigable air space owes its source to the same constitutional basis which, under decisions of the Supreme Court, has given rise to a public easement of navigation in the navigable waters of the United States, regardless of the ownership of the adjacent or subjacent soil."

The commerce power has comprehended navigation of streams. Its breadth covers all commercial intercourse. But the federal commerce power over navigable streams does not prevent state action consistent with that power. Since, over streams, Congress acts by virtue of the commerce power, the sovereignty of the state is not impaired. The title to the beds and the banks are in the states and the riparian owners, subject to the federal power over navigation. Federal regulation of interstate land and water carriers under the commerce power has not been deemed to deny all state power to tax the property of such carriers. We conclude that existent federal air-carrier regulation does not preclude the Nebraska tax challenged here.

Nor has appellant demonstrated that the Commerce Clause otherwise bars this tax as a burden on interstate commerce. We have frequently reiterated that the Commerce Clause does not immunize interstate instrumentalities from all state taxation, but that such commerce may be required to pay a nondiscriminatory share of the tax burden. And appellant does not allege that this Nebraska statute discriminates against it nor, as noted above, does it challenge the reasonableness of the apportionment prescribed by the statute.

The argument upon which appellant depends ultimately, however, is that its aircraft never "attained a taxable situs within Nebraska" from which it argues that the Nebraska tax imposes a burden on interstate commerce. In relying upon the Commerce Clause on this issue and in not specifically claiming protection under the Due Process Clause of the Fourteenth Amendment, appellant names the wrong constitutional clause to support its position. While the question of whether a commodity en route to market is sufficiently settled in a state for purpose of subjection to a property tax has been determined by this Court as a Commerce Clause question, the bare question whether an instrumentality of commerce has tax situs in a state for the purpose of subjection to a property tax is one of due process. However, appellant timely raised and preserved its contention that its property was not taxable because such property had attained no taxable situs in Nebraska. Though inexplicit, we consider the due process issue within the clear intendment of such contention and hold such issue sufficiently presented.

[The] situs issue devolves into the question of whether eighteen stops per day by appellant's aircraft is sufficient contact with Nebraska to sustain that state's power to levy an apportioned ad valorem tax on such aircraft. We think such regular contact is sufficient to establish Nebraska's power to tax even though the same aircraft do not land every day and even though none of the aircraft is continuously within the state. "The basis of the jurisdiction is the habitual employment of the property within the State." Appellant rents its ground

> facilities and pays for fuel it purchases in Nebraska. This leaves it in the position of other carriers such as rails, boats and motors that pay for the use of local facilities so as to have the opportunity to exploit the commerce, traffic, and trade that originates in or reaches Nebraska. Approximately one-tenth of appellant's revenue is produced by the pickup and discharge of Nebraska freight and passengers. Nebraska certainly affords protection during such stops and these regular landings are clearly a benefit to appellant.
>
> *Affirmed.*

The *Braniff Airways* case serves as a reminder that unless there is express preemption of a specific area of the law by the federal government, there is generally a presumption against preemption.

Notice also that the *Braniff Airways* case came before the Supreme Court's articulation of a more detailed test for state and local taxation in the *Complete Auto Transit* (see discussion above) case in 1977. Would the Nebraska state law examined in *Braniff Airways* have passed muster under the *Complete Auto Transit* test?

For aircraft owners and operators, questions of state and local taxation typically arise in the context of personal property taxes and sales and use taxes. Many states impose personal property taxes on tangible personal property. Aircraft are typically considered to be tangible personal property due to their obvious physical presence and ability to be moved. Tangible personal property is also subject to state sales and use taxes. Sales taxes are taxes on the sale of goods (such as an aircraft) within a state. Use taxes are taxes on the use of goods (such as an aircraft) within a state.

While it is plainly recognized that states have the authority to impose personal property taxes and sales and use taxes on aircraft, it must be kept in mind that the states imposing the tax are held accountable to the standards articulated in *Complete Auto Transit*.

DISCUSSION CASES

1. Forgang County owns and maintains the Greater Smith Airport. The airport is located on land that the county purchased to provide air transportation services and air transport facilities. The airport was designed for public use and conformed to all federal statutes and regulations related to airport development. Ned Neighbor owns a home not far from the approach end of a runway at the Greater Forgang Airport Regular and continuous daily flights, often several minutes apart, were made by aircraft directly over and near Neighbor's home. During these overflights, the noise level in the home reached such high levels that it was often impossible to converse or use the telephone. Windows would rattle, and plaster often fell from walls and ceilings. Neighbor brings a case to his state supreme court against Forgang County, and the state supreme court rules that if there

was a taking of Neighbor's property, it was not a taking by Forgang County. In making its ruling, the state supreme court clearly implied that although a taking of Neighbor's property may have occurred, it was the United States that had taken the property by approving the design and location of the airport and its runways, not Forgang County. The case has now been appealed to the U.S. Supreme Court. Who wins? Explain.

2. Stanton Airport, located in Davidson County, New York, is a general aviation airport designated as a "reliever airport." In this capacity, Stanton Airport relieves New York's major commercial airports, JFK International Airport, LaGuardia, and Newark, from excessive use by noncommercial aircraft. The airport is one of only three New York City area reliever airports capable of all-weather operations. New York State owns the airport and enacts a curfew that extends to all aircraft operating in or out of Stanton Airport. The curfew prohibits aircraft operations at Stanton Airport between the hours of 11:00 p.m. and 7:00 a.m. regardless of the amount of noise a particular aircraft may emit during takeoff or landing. The federal government brings a lawsuit against New York State, claiming that New York's curfew is unlawful. The State of New York argues that as airport proprietor, it has the right to enact the curfew to directly control the sources of aircraft noise. Who will win this lawsuit? Why?

3. The City of Ryan, California, owns and operates Ryan Municipal Airport. Ryan Municipal is a general aviation airport with significant training and corporate operations. Residents of Ryan have begun complaining about aircraft noise. The residents seem particularly concerned about noise coming from helicopter training operations. Ryan City aldermen are considering an ordinance banning all helicopter flight training. As part of their deliberations, they turn to the Ryan City Attorney's Office for legal guidance. One of the aldermen has read about the *Burbank* and *Causby* cases, and he wants to determine whether a ban on helicopter training will be legally defensible. He turns to you as the new Ryan City Attorney for guidance. Can you develop any arguments that will support a ban on all helicopter training operations at Ryan City Airport? Please draft a brief memo, outlining your arguments and citing relevant legal precedent.

4. Robinson Manufacturing, Inc., produces automobile parts at its plant in Zeek Falls, Indiana. To meet the demands of customers scattered throughout the United States, Robinson purchases a business jet that can be used to transport its executives to customers and vendors for meetings and sales conferences. The business jet is purchased in Florida. Robinson paid the 6 percent Florida sales tax at the time the aircraft was purchased in Florida. After some minor repairs and discrepancies were ironed out in Florida, the aircraft was flown to its home base in Indiana. A few months later, Robinson was contacted by the Indiana sales and use tax authorities, who demanded that Robinson pay an Indiana use tax of 5 percent. Robinson protested the assessment of the 5 percent Indiana sales and use tax and argued that it should be credited with having paid a 6 percent sales tax to Florida. Robinson points to an Indiana statute requiring the state to provide a credit against use taxes for sales taxes paid to another state. Indiana responds to this argument by pointing out that the same statute also provides that the credit for sales taxes paid to other states "does not apply to the use tax on boats, automobiles, and *aircraft*" (emphasis added). This case is now at a hearing before your court. You are the judge. How will you decide? Explain your reasoning.

5. Fox City, Utah, opened a new terminal in 2005 at its growing municipal airport. The upper level has a secure concourse area for passengers boarding and leaving airplanes as well as a "preconcourse" area that includes a gift shop, restaurant, bar, rest rooms, and public telephones. The lower level of the terminal houses rental car vendors, baggage claim areas, and taxi services. Horace Horst publishes a magazine for singles. His typical method of distributing his magazines is through news racks. After the new terminal opened, Horst placed a new rack with his magazines in the public area of the terminal without the approval of the airport or Fox City administrators. The Fox City airport administrator removed the news rack with Horst's magazines and sent him a letter suggesting that he sell his magazines through the gift shop. Horst refused this request and filed a lawsuit against Fox City and the airport administrator. In his lawsuit, Fox argues that the airport is a public forum. Therefore, his First Amendment rights have been violated by the removal of the news rack containing his magazines. A trial court grants a judgment in favor of Horst. The case has now been appealed, and you are the appeals judge deciding the case. How will you decide? Explain.

6. Podonk Airfield is a small, privately owned, general aviation airfield with a short grass strip located in the center of Nebraska. The airport occasionally has visitors from out of state, but virtually all its air traffic is from local pilots with aircraft based at the airfield or pilots from nearby who want to train on grass fields. Podonk Airport typically has two to three employees who deal with the administration and maintenance of the airfield. The owner of Podonk Airfield is being sued for discrimination based on race and color by a former employee who was fired from his position as an airfield maintenance technician. The lawsuit is based on the Federal Civil Rights Act of 1964. Podonk's owner claims that the federal statute cannot be applied because he is not engaged in interstate commerce. You are the U.S. district court judge assigned to this case. Can the U.S. government regulate Podonk Airfield under the interstate commerce clause? Why or why not?

7. Brocato Entertainment, Inc., wishes to lease space at a city-owned airport terminal so it can operate an amusement machine parlor. A city ordinance required that operators of coin-operated amusement machines be licensed and provided that the police chief was to determine whether an applicant had "any connections with criminal elements." The city manager, after receiving the report of the chief of police and reports from the building inspector and the city planner, would then decide whether to issue the license. If the application were denied, the applicant could then petition the city council for a license. What defenses could you think of for a rejected applicant? Do you think they would succeed? Why or why not?

ENDNOTES

1. 5 U.S. 137 (1803).
2. 524 U.S. 417 (1998).
3. 411 U.S. 624 (1973).
4. 317 U.S. 111 (1942).
5. 514 U.S. 549 (1995).

6. 529 U.S. 598 (2000).
7. 504 U.S. 353 (1992).
8. 432 U.S. 333 (1977).
9. 450 U.S. 662 (1981).
10. 430 U.S. 274 (1977).
11. 125 S.Ct. 2655 (2005).
12. 440 U.S. 268 (1979).
13. 518 U.S. 515 (1996).
14. 347 U.S. 483 (1954).
15. 376 U.S. 254 (1964).
16. 369 U.S. 84 (1962).
17. 558 F.2d 75 (2nd Cir. 1977).
18. 552 F. Supp. 255 (N.D.N.Y. 1982).
19. 659 F. 2d. 100 (9th Cir. 1981).
20. 437 F. Supp. 804 (S.D.N.Y. 1977).

3 Impact of Criminal Law on Airmen and Air Carriers

BASICS OF CRIMINAL LAW   68
Classifications of criminal law   68
Elements   69
Criminal procedure   69
Constitutional protections for defendants   71

CRIMINAL LAWS AFFECTING AVIATION ACTIVITIES   74
Federal provisions   74
State provisions   76

CASES AND COMMENTARY   76

DISCUSSION CASES   88

ENDNOTES   89

In the United States we live by the rule of law. One of the primary functions of the law is to ensure that citizens conform to fundamental social norms. Criminal law is often the first legal line of defense for a civilized society against behavior and actions that might threaten civilized society.

Crimes are wrongs committed against society in general. When someone is accused of a criminal offense, the "plaintiff" or complainant is the federal, state, or local government (representing society in general). As a general rule, the government is seeking to punish the wrongdoer.

In criminal cases, the burden on the government to prove its case is substantial. For a person or entity to be found guilty of a crime, the evidence produced by the government must establish *beyond a reasonable doubt* that the person accused committed the alleged crime. While it is not easy to explain what *beyond a reasonable doubt* might mean to different individuals, it might be fairly looked upon as a standard requiring a juror to be 99 percent certain that the defendant committed a crime before a conviction can be made. This being the case, there may be occasions on which a jury or judge might strongly believe that it is more likely than not that a defendant committed a crime. However, the government may not have proved its case to the point that guilt is established beyond a reasonable doubt. Therefore, the jury or judge would be duty-bound to acquit the accused. It might be fair to say that our legal and court system is designed in such a way that if an error occurs, the system prefers a guilty person to be set free than an innocent person to be deprived of life, liberty, or property.

In this chapter first we will review the basics of criminal law, including the different classes of crimes, general elements of crimes, basics of criminal procedure, and constitutional protections granted to individuals accused of crimes. Then we'll explore the various criminal law statutes and regulations related to airmen and aviation businesses and industry. Finally, we will expose you to some cases that shed light on how criminal law affects the aviation environment.

BASICS OF CRIMINAL LAW

All states and the federal government have criminal statutes or codes that detail the crimes and the penalties for crimes within their respective jurisdictions. A violator of the criminal law can be faced with punishment including imprisonment or death, fines, restitution payments, probation, community service, and a host of other remedies allowed by the law.

Classifications of criminal law

The most serious crimes are typically classified as felonies. Felonies are usually crimes that involve inherently bad acts—lawyers sometimes refer to this type of act as *malum in se,* which means inherently evil. Examples of felonies include murder, rape, armed robbery, embezzlement, and bribery. Within the classification of felony, there may be further subclassifications such as first, second, or third degree. Usually these subclassifications indicate different levels of severity and often correspond to the range of penalties for a guilty party.

Crimes that are less serious than felonies are often referred to as misdemeanors. Often crimes solely against property may be classified as misdemeanors. In some jurisdictions, misdemeanors carry penalties such as fines or imprisonment that do not exceed 1 year and/or $1000.

On the lowest end of the criminal law classifications are violations. Violations often involve traffic infractions, jaywalking, and other similar crimes. Usually, violations are punished solely by fines and/or a short jail term.

Elements

For a person to be convicted of a crime, two elements must be proved by the government. First, there must be a criminal act. Second, there must be criminal intent.

Lawyers and judges often refer to the criminal act as *actus reas* (the guilty act). It is important that a criminal act be proved because in our society we do not convict someone of a crime merely for thinking about the commission of the crime. Sometimes the act that must be established is an actual act—shooting a gun that harms or injures another. Other times the criminal act might be an act of omission—failure to file a required report to the government.

Criminal intent, or *mens reas*, is required to establish that a person accused of a crime had the necessary criminal evil intent at the time the criminal act occurred. Sometimes a criminal act must be accompanied by specific intent. Specific intent exists when the defendant intentionally and knowingly committed the act prohibited by criminal law. In other situations, the criminal act need only be accompanied by general intent. General intent is found when the defendant's reckless or indifferent conduct leads to the commission of a crime (e.g., driving a car while intoxicated and killing or injuring another person).

Although most crimes require both a criminal act and criminal intent, some crimes do not require criminal intent. These crimes are often referred to as strict or absolute liability crimes. In these cases, a crime can be committed where there is no intent, only an act. One example of this type of strict liability crime might be a violation of a state's environmental laws in which an underground fuel tank at an airport leaks jet fuel. The state may impose criminal sanctions on the persons responsible for the tank even though the persons involved did not have any intention of committing a crime.

In the Model Penal Code, drafted by the American Law Institute, several standards of fault are recognized and organized to match associated crimes. Figure 3-1 provides an outline of the standards of fault applied in many U.S. jurisdictions.

Criminal procedure

Criminal procedure is a very intricate set of rules and processes designed to ensure that a criminal defendant gets the due process required by the U.S. Constitution. Criminal procedure involves activities that occur before, during, and after a trial. A very brief outline of these procedures is provided below.

Figure 3-1
Levels of fault in criminal cases.

Classification of fault	Possible levels of fault	Applicable crimes
Subjective	Knowing, purposeful, reckless	• Assault • Theft • Falsification/fraud
Objective	Careless, negligent	• Careless operations of motor vehicles or aircraft • Writing bad checks
No fault (strict liability)	None	• Environmentally hazardous spills/activities • Alcohol sales to minors

Arrest

In most cases, prior to an arrest, police must obtain an arrest warrant from the courts. This requirement is meant to protect individuals from unjustifiable detentions by police. To obtain an arrest warrant, police officers must be able to persuade a judge or magistrate that they have probable cause to believe that the person they seek to arrest either committed a crime or intends to commit a crime. In some cases, an arrest warrant is unnecessary. For instance, the police may be in pursuit of someone who is fleeing a crime scene. Police officers may also be able to avoid the need for a warrant if they are legitimately concerned that significant evidence might be destroyed if they are forced to take the time to obtain an arrest warrant.

Once an arrest has been made, the person arrested is taken into the custody of the police. This process is often referred to as booking and involves paperwork recording the arrest, fingerprinting, photographing, etc.

Indictment

For serious crimes, a person must be indicted before he or she can be brought to a trial. This process involves a grand jury hearing in which the prosecution presents its side of the story to a group of citizens (usually from 6 up to 24). The grand jury decides whether there is enough evidence to take the person accused to a trial. If an indictment is issued, the defendant is brought to trial. If the grand jury does not believe enough evidence exists to bring a defendant to trial, the case will be dismissed.

In cases of lesser crimes, a grand jury indictment is usually not necessary. In these instances, a defendant may be brought to trial on a magistrate's information statement. Similar to the procedure for an indictment, if a magistrate does not believe the evidence is sufficient to go to trial, the case against the defendant will be dismissed.

It is important to note that the right to an indictment by a grand jury is not absolute. States have not been required to meet the requirements of a grand jury through the application of the Fourteenth Amendment.

Arraignment

After an indictment or information is issued, an arraignment takes place at the trial court. At the arraignment, first the person accused is informed of the charges being made against her or him. Next, the court requests that the defendant enter a plea. In most cases, the plea can be guilty, not guilty, or no contest (otherwise referred to as *nolo contendere*). With a *nolo contendere* plea, a defendant is not admitting to guilt; instead the defendant is saying that he or she will agree to a penalty or punishment but will not admit to any guilt (this plea can come in handy because it will prevent the criminal proceedings from being used in a civil case against the defendant later).

Plea bargains

Often prior to a trial, intense bargaining and negotiation between the prosecutors and defense can take place in an effort to settle a case. Sometimes, this negotiation takes the form of plea bargaining in which a prosecutor may be willing to accept a lesser charge and/or penalty in exchange for a guilty plea. This process may appear to be a bit unseemly. However, plea bargaining does address the practical problems of prison overcrowding, the limited resources of prosecutors and defendants, and the ever-present risk and uncertainty involved in litigating a case.

Trial

The trial of a criminal case is run in much the same way as a trial is described in Chap. 1. One of the unique features of a criminal trial is that for a criminal conviction to occur, all the jurors must agree. If any juror (or all the jurors) does not agree that the defendant is guilty (the juror has reasonable doubt), the defendant will be acquitted. In some instances, if some of the jurors do not find the defendant guilty, the verdict is characterized as "not guilty." If all the jurors agree that the defendant should not be convicted, the verdict is referred to as innocent.

The defendant who is found guilty will still have an opportunity to file an appeal with the appropriate appeals court. However, if the government does not prevail at a trial, it may not appeal the verdict. If the jury encounters differences in opinion that do not allow for reconciliation, this may result in a hung jury. With a hung jury, the government may retry the case.

Constitutional protections for defendants

As indicated earlier, the U.S. Constitution was drafted with great thought given to protecting the rights of the accused in criminal proceedings. Sir William Blackstone, in *Commentaries on the Laws of England* (1809), wrote: "It is better that ten guilty persons escape, than that one innocent suffer." Similarly, our legal system builds in many important safeguards that might allow guilty parties to go free. As Chief Justice Cardozo writes in *People v. DeFore*,[1] "The criminal is to go free because the constable has blundered."

Some of the provisions of the U.S. Constitution designed to protect citizens' rights in criminal matters are outlined below.

Fourth Amendment

The applicable portions of the Fourth Amendment indicate that citizens are to be protected from: "… unreasonable searches and seizures." The Fourth Amendment goes on to mandate that any search warrant issued be based upon "probable cause ... and particularly describing the place to be searched, and the persons or things to be seized."

Over centuries of court decisions, it is now established that some searches without warrants may be valid. Some of the more notable exceptions to the general rule requiring a warrant are as follows:

- The search is incident to an arrest.
- The evidence sought is in plain sight of the police.
- There is a high probability that the evidence may be destroyed or damaged if action is not taken immediately.
- The search involves items that are in "plain sight" or held out to the public.
- Searches performed by drug dogs.
- Inventory searches are made of an arrested person's personal belongings.

If evidence is obtained in a manner contrary to the Fourth Amendment's requirements, the evidence may be excluded at a trial or hearing to which the person searched is a party. This rule is often referred to as the *exclusionary rule*. The rule has very few exceptions, but one notable exception occurs when police were operating on a good faith belief that an appropriate search warrant had been issued. Importantly, highly regulated businesses such as airlines, charter operations, and repair shops may be subject to searches without warrants if applicable regulatory or statutory requirements are met.

Fifth Amendment

There are two significant protections in the Fifth Amendment that relate to criminal cases. The first protection noted is the privilege of every citizen against self-incrimination. The second is the protection against double jeopardy. Each of these protections is discussed below.

The language in the Fifth Amendment related to self-incrimination states that no person "shall be compelled in any criminal case to be a witness against himself." Judicial interpretations of this clause indicate that this means a person may not be required to provide testimonial evidence while in the custody of police or in the midst of a criminal judicial proceeding. However, this privilege does not extend to evidence that is not testimonial in nature, for instance, DNA samples, fingerprints, bloodstains, and other physical evidence. Fifth Amendment privileges may also be waived in exchange for a prosecutor's grant of immunity. Of course, the Fifth Amendment is the foundation of the famous *Miranda* rights that must be read to any criminal suspect whom the police take into custody.

The double-jeopardy language in the Fifth Amendment provides that no person shall "... be subject for the same offense to be twice put in jeopardy of life or limb." Essentially, this provision prohibits the government from putting someone on trial twice for the same crime. Therefore, if an airline is tried for the crime of mail fraud and acquitted, the government cannot try to bring the case again. Nonetheless, if the alleged mail fraud involved several different crimes, the accused can be tried for each separate crime without violating the spirit or intent of the double-jeopardy provisions. The Fifth Amendment prohibition against doublejeopardy would not be violated if a defendant were granted a new trial as a result of his or her successful appeal.

Sixth Amendment

The Sixth Amendment requires the following rights and protections for the accused in a criminal case:

- Speedy, impartial, and public trial
- Jury trial
- Information as to the nature and cause of the crime charged
- Right to confront and cross-examine witnesses testifying against the accused
- Ability to compel favorable witnesses to attend and testify at trial
- Assistance of counsel

Most of the Sixth Amendment rights relate to criminal procedure and the conduct of criminal trials. The clause requiring assistance of counsel in criminal proceedings was initially interpreted to require counsel for defendants only in cases involving capital offenses, in *Betts v. Brady*.[2] However, in the landmark case of *Gideon v. Wainwright*,[3] the U.S. Supreme Court overturned the *Betts* case and held that the Sixth Amendment right to assistance of counsel applied to all criminal prosecutions.

While there is no specific mention of the time required for a "speedy trial" in the U.S. Constitution, federal statutory law requires that a trial begin no more than 70 days after a defendant is indicted.[4]

Eighth Amendment

The Eighth Amendment to the U.S. Constitution provides that no "cruel or unusual punishments" may be inflicted on criminal defendants. The understanding of what constitutes cruel and unusual punishment has been modified from time to time. It has always served as a prohibition against torture. However, the death penalty has been a source of great debate by legal scholars and politicians for decades. Currently, the Supreme Court holds that the Eighth Amendment does not prohibit capital punishment.[5]

A summary of the primary constitutional protections for defendants is found in Fig. 3-2.

Constitutional amendment	Rights/protections granted
Fourth	• No unreasonable searches or seizures
Fifth	• Right to due process • No double-jeopardy • Freedom from self-incrimination • Federal right to indictment for capital crimes
Sixth	• Public trial • Speedy trial • Right to confront and present witnesses • Right to competent counsel
Eighth	• No excessive bail • No cruel or unusual punishment

Figure 3-2
Summary of constitutional protections in criminal cases.

CRIMINAL LAWS AFFECTING AVIATION ACTIVITIES

There are myriad criminal statutes to which aviation professionals and businesses are subject. Some of these laws are based on federal statutes, and others originate from state statutes. We will take some time now to explore how the more significant criminal provisions in federal and state law impact the aviation community.

Federal provisions

Federal criminal law is wide-reaching, and all aviation professionals should be aware of the following provisions. All these criminal statutes are found in the United States Code (abbreviated U.S.C.).

Document falsification

Compliance with the complex web of regulations that all aviation activities are subject to is often evidenced primarily by documents and statements and assertions in those documents. Questions related to the airworthiness of an aircraft typically begin with an inspection of aircraft maintenance logbooks. Determinations regarding a pilot's medical fitness to fly are dependent on truthful responses to an application for a medical certificate. To determine whether a pilot is in compliance with flight currency regulations prior to a flight requires reference to that pilot's entries in a logbook.

This reliance on documentary evidence for safety places a premium on honesty and accuracy in aviation-related documentation. Failure to properly comply with document requirements can invite problems with the Federal Aviation Administration (FAA). However, falsification of documentary requirements can also lead to criminal prosecution. 18 U.S.C. § 1001 prohibits knowingly and willfully making a materially false statement regarding facts within the purview of a federal regulatory agency. Penalties for violation of the statute call for fines and terms of imprisonment up to 5 years.

It is important to note that although the statute only prohibits "knowingly and willfully" making false statements, in some cases the courts have held that someone may have acted willfully or knowingly if she or he acted with a reckless disregard for the truth tied to a conscious effort to avoid the truth.[6]

Government authorities regularly prosecute individuals for violation of 18 U.S.C. § 1001. Cases typically involve falsification of certificates and logbooks and false statements or omissions in applications for medical or airman certificates. The importance of accurate and honest paperwork and statements in the aviation environment should never be underestimated by aviation professionals and businesses.

Mail and wire fraud

The United States has two statutes (18 U.S.C. §§ 1341 and 1343) that provide for criminal penalties for anyone involved in a fraudulent behavior using the United States mail or wire communications. Wire communications include contacts by telephone, facsimile, Internet, and television (among others). Sections 1341 and 1343 call for fines, restitution, and prison sentences for up to 5 years.

In one case, the government indicted Eastern Airlines in 1991 for wire fraud. The government charged that Eastern and its employees made several false maintenance entries in computers that were meant to establish that regularly scheduled maintenance had been performed when in fact it had not been. The wire fraud came into play because the government alleged that the airline (through marketing and other outlets) led potential customers and passengers to believe that it was faithful to a rigorous maintenance program for its aircraft. Eastern pled guilty to the charges and paid a $3.5 million fine.[7]

Hazardous Materials Transportation Act
The Hazardous Materials Transportation Act is found at 49 U.S.C. § 5101 *et seq*. This provision prohibits persons from willfully violating the act and its detailed requirements for packaging, storing, and labeling hazardous materials. The case of *United States v. SabreTech, Inc.*[8] illustrates a government prosecution of such a case. For an edited version of the case and commentary, please see Case 3-2 later in the chapter.

Resource Conservation and Recovery Act
United States Code § 6928 imposes fines and/or prison terms for anyone who knowingly violates the provisions of the Resource Conservation and Recovery Act (RCRA) related to the disposal, treatment, transportation, and/or storage of hazardous waste products. Fines can reach up to $50,000 per day with prison terms up to 5 years for violations of this law.

The RCRA has a great impact on the aviation industry. For better or for worse, aviation businesses deal regularly in hazardous waste such as petroleum products, hydraulic fluid, and paints. Anyone operating in the aviation industry should be aware of the necessity for careful compliance with the RCRA's provisions.

Terrorism
Federal law prohibits certain acts such as willfully damaging an aircraft or aviation facility (see 18 U.S.C. § 32). A violation of this code section can result in fines, restitution, and/or up to 20 years in prison. If the violation results in a death, the death penalty is possible.

This provision in the law was clearly designed to deter international or domestic terrorism targeting aircraft or airports. However, it has been used in circumstances that might not be so apparent from a plain reading of the statute. Notice that in the *SabreTech* case presented later in this chapter, the government charged SabreTech with violation of this section; however, the jury at the trial court level acquitted SabreTech of the charge.

Miscellaneous criminal provisions
There are several miscellaneous, but important, criminal provisions related to aviation activities in Title 49 of the United States Code. Some of the more significant infractions and related code sections are listed here:

- Certificate forgery, operating an unregistered aircraft, unauthorized fuel tank modifications[9]
- Failure to comply with an NTSB or Department of Transportation (DOT) subpoena to testify[10]

- Willful violations of security provisions of the Federal Aviation Program[11]
- Knowing and willful violations of certain air commerce safety and statutes and regulations[12]

State provisions

Although it does not happen often, states sometimes assert their police power by criminalizing certain aviation-related conduct. To date, approximately 35 states have criminal laws prohibiting the operation of aircraft in a careless or reckless manner or while under the influence of alcohol or drugs. Some state laws go beyond the general prohibitions of careless or reckless operations and become very specific. For instance, a law in Texas prohibits taking off or landing an aircraft on a road or highway except in an emergency.

CASES AND COMMENTARY

The question often raised by state and local regulation of aviation activities is, Just how far can a state or local government go without being preempted by federal regulation of aviation? This is the question that arises in Case 3-1.

CASE 3-1
ROBERT DAVID WARD v. STATE OF MARYLAND
374 A. 2d 1118 (1977)

OPINION BY: ORTH

OPINION: For over a quarter of a century it has been a crime under the laws of Maryland "to operate an aircraft in the air, or on the ground or water, while under the influence of intoxicating liquor, narcotics, or other habit-forming drug, or, ... in a careless or reckless manner so as to endanger the life or property of another." [Citations.] The statute has never been before the appellate courts of this State. Robert David Ward presents it to us in this appeal. In challenging the judgment entered against him upon being found guilty of operating an aircraft in a reckless manner, he does not dispute the accuracy of the facts and circumstances resulting in his arrest and prosecution by the State. Nor does he present any question as to the sufficiency of the evidence to prove beyond a reasonable doubt the corpus delicti of the offense and his criminal agency. And he offers no claim that the penalties imposed were in any way at variance with the sanctions authorized. The issue he raises is that Maryland was precluded from prosecuting him because the statute proscribing the conduct was preempted by federal law. He, therefore, claims error in the denial by the trial court of his motion to dismiss the charges.

The issue for decision is narrow in lineation but sweeping in implication. The great majority of the states have enacted legislation similar to the Maryland statute, some in identical language, making it a crime to operate an aircraft so as to endanger the person or property of another. If Ward's view that the state statutes are preempted prevails there could be no criminal prosecution, under existing laws, by any state, or, as we shall see, by the federal government, for operating an aircraft in a careless or reckless manner or while under the influence of intoxicating liquor, narcotics, or other habit-forming drug, no matter how flagrant the conduct proscribed and regardless of the seriousness of the danger to life and property occasioned by that conduct. We have not been referred to a decision of another jurisdiction, state or federal, which addresses the matter of preemption of a statute prohibiting such operation of an aircraft. There being no decision precisely on point to bind or persuade us, we shall follow the path laid out by the decisions of the Supreme Court of the United States concerning the doctrine of preemption to determine whether the Maryland statute is enforceable.

I. [Facts]: To put the issue for decision in perspective we recount what Ward did and what happened as a result. Ward was the holder of a valid student aircraft pilot license and medical certificate issued by the Federal Aviation Administration (FAA), pursuant to the applicable Federal Aviation Regulations (FAR). On the late afternoon of 18 May 1975 he made an instructional flight from Hyde's Airfield in Clinton, Prince George's County, Maryland. Thereafter, he and his instructor pilot enjoyed a quiet dinner complete with pre- and postprandial liquid refreshment in an amount sufficient, when Ward was examined some hours later, to show a blood alcohol level of 0.17 percent. Later that evening, after a visit with Ward's parents, Ward and his instructor returned to the airport. Ward slept in his car for some hours. On awakening in the early morning he decided to practice some "touch and gos"—taking off and landing without leaving the airport pattern. Inadvertently entering clouds, he climbed above the cloud layer and circled what he assumed to be the Chesapeake Bay area, until breaks were visible in the overcast. It was about 6:00 A.M. when he descended through a hole in the clouds. He established that his position was over Greenbelt, Maryland, and in an apparent burst of euphoria he "buzzed" nearby apartments. His activities were observed by a Maryland State Police officer in a helicopter who followed the plane back to the airport. Ward was arrested upon landing.

Ward was tried in the District Court of Maryland in Prince George's County on a charge of operating an aircraft in a reckless manner, convicted, and appealed from the judgment entered to the Circuit Court for Prince George's County. Tried de novo by the judge in the circuit court, he was again found guilty. He was fined $500 and costs and sentenced to imprisonment for a term of 90 days. The prison sentence was suspended "upon special condition that he not operate an aircraft within the State for a period of one year from September 17, 1975." Upon Ward's petition we issued a writ of certiorari to review the judgment of the Circuit Court for Prince George's County.

In a civil proceeding, the Federal Aviation Administration determined that Ward had violated seven Federal Aviation Regulations in his operation of the aircraft on 19 May 1975, including FAR 91.9, prohibiting the careless or reckless operation of an aircraft so as to endanger the life or property of another. It ordered that his airman certificate be revoked. Subsequently, however, "after a thorough investigation of Mr. Ward, his instructor, his level of aeronautical knowledge and the flight leading to his prosecution, FAA granted permission to reissue his license six (6) months after surrender to it."

* * *

We have no difficulty whatever in deciding that Congress has not occupied the entire field of aeronautics by the Federal Aviation Act of 1958. There is no such command explicitly stated in the language of the Act, and we do not find it implicitly contained in the statute's structure and purpose. Generally accepted, for example, is that in the economic area of aeronautics all state legislation is not excluded. [Citations.] The Supreme Court has held that the Federal Aviation Administrator in conjunction with the Environmental Protection Agency "has full control over aircraft noise, preempting state and local control." [Citing City of Burbank v. Lockheed]. But that conclusion was not because Congress had "unmistakably ordained" that its enactments alone were to regulate air commerce in its entirety, but because the 1968 amendment [Citation] and the 1972 amendment (The Noise Control Act, 86 Stat. 1234) to the 1958 Act established "the pervasive nature of the scheme of federal regulation of aircraft noise." We construe the holding as limited to that aspect of aeronautics. [Citation.]

Maryland has expressly retained its police powers with respect to crimes related to aeronautics. It declared in its Aviation Act, § 3-307 that "[a]ll crimes ... committed by or against an airman or passenger while in flight over this State shall be governed by the laws of this State...." Congress has recognized the right of a state to do so. In 1961 the Federal Aviation Act of 1958 was amended by designating criminal penalties for specific crimes. [Citation.] House Report 958 on the proposed legislation observed that "in the case of crimes committed in the airspace over States of the United States, most of the acts with which this legislation deals would be violations of the laws of one or more of such States." [Citation.] It declared: "The offenses punishable under this legislation would not replace any State jurisdiction but would, where both Federal and State law provided for punishment for the same act, be in addition to the State criminal law." [Citation.] In discussing the need for the legislation, the report stated, [Citation]:

> As is well known, the Federal Government does not provide a general criminal code for all crimes committed in the United States. That is the province of the various States. However, criminal codes of the States are at times supplemented by Federal Law in fields where the Federal Government has responsibilities.
>
> We wish to emphasize that it is not our intent to divest the States of any jurisdiction they now have. This legislation merely seeks to give the Federal Government concurrent jurisdiction with the States in certain areas where it is felt that concurrent jurisdiction will contribute to the administration of justice and protect air commerce.

The Report asserted: This, we want to make clear, does not preempt any State jurisdiction but would only supplement it. [Citation.]

We reject Ward's suggestion that "the entire [Maryland Aviation] Act should be struck down."

Having determined that Congress has not excluded all state action in the field of aeronautics or even in the more limited field of air safety, we center on § 10-1002 of the Maryland Aviation Act to ascertain whether it conflicts with the federal law in the field. Although the Maryland statute and the federal regulation proscribe identical conduct with respect to the careless or reckless operation of an aircraft, the two laws are not coterminous because of the sanctions prescribed. When the relationship between the Maryland and federal laws as they are written, as they are interpreted and as they are applied, is considered, it is manifest that the Maryland law does not stand as an obstacle to the accomplishment and execution of the full purposes and objectives of Congress, namely, "to provide for the regulation and promotion of civil aviation in such manner as to best foster its development and safety, and to provide for the safe and efficient use of the airspace by both civil and military aircraft." [Citation.] On the contrary, the Maryland law enhances the Congressional purposes and objectives by deterring through criminal sanctions the operation of an aircraft in such manner as to endanger the life and property of others. It would be incongruous indeed, in light of the federal purposes and objectives, if Maryland were to be constitutionally precluded from the criminal prosecution of a person for such conduct because a federal regulation authorized civil penalties.

We are not persuaded that there is a conflict of constitutional dimension simply because the State court may prohibit a violator of its statute from operating an aircraft within Maryland for up to one year even though the violator holds a federal airman certificate which would permit him to operate an aircraft. Section 10-1003 of the State Statute expressly provides:

> In no event shall this subsection be construed as warrant for the court or any other agency or person to take away, impound, hold or mark any federal airman or aircraft certificate, permit, rating or license....

In any event, we do not see how prohibiting, for a limited time, the operation of an aircraft within Maryland by a person who has been found guilty in a court of law of operating an aircraft in a careless or reckless manner so as to endanger the life or property of another could be said to stand as an obstacle to the Congressional purposes and objectives.

In short, it was the clear intent of Congress that criminal prosecution of the proscribed conduct be left with the states, and there is no conflict within the contemplation of the doctrine of preemption between the State and federal laws.

* * *

We hold, therefore, that the Circuit Court for Prince George's County did not err in denying the motion to dismiss the charges.

Note that the *Ward* case was appealed to the U.S. Supreme Court on a *writ of certiori*. However, the Supreme Court denied the writ and refused to hear the case. This left the *Ward* case as the prevailing law in Maryland and a source of persuasive authority in other states.

One of the more curious aspects of the *Ward* case is the court's restriction that prohibited the pilot from flying in the state of Maryland for 1 year—even though the pilot would have had his certificate back from the FAA enforcement proceeding before the 1-year period expired.

The Maryland Court of Appeals decision effectively barred the pilot from Maryland airspace for the 1-year period. Note that two of the seven judges on the Maryland Court of Appeals panel wrote dissenting opinions on this point. The dissenting judges argued that while Maryland's criminal penalties may not have been preempted by federal law, the provision in the Maryland law that prohibited a convicted person from operations in Maryland crossed the line and should be subject to preemption. What do you think?

Whether a regulatory violation crosses the line into the realm of criminal violations is a difficult question to answer at times. In Case 3-2, the court must wrestle with whether a maintenance supplier can be held criminally liable for acts and omissions that led to a terrible air tragedy in the early 1990s.

CASE 3-2
UNITED STATES OF AMERICA v. SABRETECH
271 F. 3d 1018 (2001)

OPINION: DUBINA, Circuit Judge.

On May 11, 1996, a ValuJet commercial airliner crashed in the Florida Everglades and all persons on board perished. It was a tragic accident that could have been avoided. Following an investigation, the government, for the first time, indicted an aviation repair station, SabreTech ("SabreTech"), and several of its employees for various violations related to the transportation of hazardous materials. The record reflects that these aviation repair station personnel committed mistakes, but they did not commit crimes. The jury found SabreTech not guilty of willful violations of the Hazardous Materials Transportation Act ("HMTA"). [Citation.] The jury did, however, find SabreTech guilty of recklessly causing the transportation of hazardous material in air commerce. [Citation.] We hold that the government and the district court improperly relied upon hazardous materials regulations that had not been authorized by the Federal Aviation Act ("FAA"), as required by 49 U.S.C. § 46312, to support the reckless counts. As will be discussed *infra*, these counts are a legal nullity. Accordingly, we affirm in part, vacate and remand in part.

I. BACKGROUND

From July 1995 until June 1999, SabreTech engaged in the repair, modification, and maintenance of commercial aircraft. In January 1996, ValuJet Airlines delivered three used McDonnell Douglas MD-80 aircraft to SabreTech's facility in Miami, Florida, for major modification and maintenance prior to their introduction into the ValuJet fleet. In the course of overhauling these aircraft, SabreTech's mechanics determined that many of the oxygen generators had exceeded their 12 year service life. ValuJet personnel issued written work orders instructing the mechanics to remove the old oxygen generators and install new ones. The work orders listed steps for SabreTech's mechanics to follow when replacing the old oxygen generators with new ones. The work orders contained a warning that unexpended oxygen generators could generate extremely high temperatures.

In the process of removing the old oxygen generators, mechanic John Taber ("Taber") noticed the absence of shipping caps. Taber asked his supervisor David Wiles ("Wiles") about the shipping caps, and Wiles told him to set the old generators aside and continue working. Taber, along with fellow mechanics Robert Rodriguez ("Rodriguez") and Eugene Florence ("Florence"), then proceeded to wrap the lanyards tightly around the firing pins of the generators and tape the ends of the lanyards to the body of the generators to prevent the release of the trigger mechanism. In March 1996, new generators, bearing yellow, diamond-shaped stickers, arrived at the SabreTech facility. The mechanics installed the new generators and tagged the old generators with green "unserviceable" tags on which they wrote "out-of-date" as the reason for removal. The mechanics boxed the generators and stored them in the hangar where they remained for six weeks.

On May 10, 1996, SabreTech shipping clerk Andy Salas ("Salas") re-packed the generators with bubble wrap and placed them in boxes that he sealed with tape. He placed ValuJet "COMAT" labels on the boxes, indicating that the boxes contained ValuJet "company materials." The shipping ticket described the contents as "5 boxes" of "Oxy Canisters Empty." The next day, a SabreTech driver took the boxes to the ValuJet ramp area where Flight 592 was scheduled to depart for Atlanta. ValuJet personnel placed the boxes in the forward section of the aircraft's cargo compartment. Shortly after take-off, a fire erupted on the plane.

Flight 592 crashed into the Everglades and killed all 110 persons on board.

On July 13, 1991, a grand jury returned a 24-count indictment against SabreTech and employees Daniel Gonzalez ("Gonzalez"), Florence, and Mario Valenzuela ("Valenzuela"). * * *

II. ISSUES

1. Whether the district court erred in denying SabreTech's motion to dismiss the 49 U.S.C. § 46312 counts.
2. Whether the evidence was sufficient to convict SabreTech on the reckless counts and the failure to train count.

* * *

IV. DISCUSSION
A. Validity of Reckless Counts
SabreTech argues that the statutory and regulatory history creates a dichotomy between the HMTA and the FAA that cannot support as a crime the reckless violations of the hazardous materials regulations. It contends that the reckless counts are a legal nullity. In these counts, the government alleged that SabreTech should be punished pursuant to 49 U.S.C. § 46312—the criminal penalty provision of the former FAA—for violation of certain hazardous materials regulations. Section 46312(a)(2) makes it a crime to recklessly cause to be transported hazardous material in violation of any regulation or requirement prescribed under the FAA. SabreTech posits that the regulations it was convicted of recklessly violating were not enacted under the FAA. These regulations were promulgated under a different statutory authority—the HMTA. That statute penalizes only willful violations of its regulations. 49 U.S.C. § 5124. Therefore, SabreTech contends that the government improperly charged it with a crime. We agree.

1. Statutory Overview
* * *

Shortly after the passage of the HMTA, the Secretary of Transportation created the Materials Transportation Bureau ("MTB") to implement the new law. The MTB amended and reissued the authority citations for the hazardous materials regulations. The MTB deleted each of the previous authority citations for the hazardous materials regulations and replaced them with references to the regulatory authority for the new HMTA, 49 U.S.C. §§ 1803, 1804, 1808. Thus, by January 1977, Congress had eliminated any authority for the regulations which may have existed under the FAA, and placed that authority under the HMTA.

* * * [I]n 1994 ... Congress modified this section [Section 46312] to read as follows:

A person shall be fined under title 18, imprisoned for not more than 5 years, or both, if the person, in violation of a regulation or requirement related to the transportation of hazardous material prescribed by the Secretary of Transportation under this part —

1. willfully delivers, or causes to be delivered, property containing hazardous material to an air carrier or to an operator of a civil aircraft for transportation in air commerce; or
2. recklessly causes the transportation in air commerce of the property.

* * *

Thus, at the time of the ValuJet crash, criminal liability existed for willful violations of the hazardous materials regulations promulgated pursuant to the HMTA. Criminal liability also existed for reckless violation of regulations authorized by Part A—Air Commerce and Safety, Subtitle VII, an entirely separate subtitle. The criminal penalties were distinct. Therefore, only regulations adopted under the authority of the FAA in Part A could provide the predicate for prosecuting a reckless violation. *See* 49 U.S.C. § 46312. However, the government charged SabreTech with recklessly violating regulations promulgated pursuant to the HMTA. That Act contains no criminal liability for recklessly violating the hazardous materials regulations.

2. Regulatory Overview
The Department of Transportation's Research and Special Programs Administration ("RSPA"), the successor agency to the MTB, administered a comprehensive scheme for the regulation of hazardous materials transportation. These hazardous materials regulations are contained in 49 C.F.R. §§ 171–180. At the beginning of each of these parts, the RSPA cites 49 U.S.C. §§ 5101–5127 as the only authority for promulgating each part. The RSPA does not cite any statutory authority found in Part A—Air Commerce and Safety. Thus, during the pertinent timeframe involved here, none of the regulations cited in the reckless counts were issued pursuant to the statutory authority contained in Part A—Air Commerce and Safety.

The statutory and regulatory history demonstrate that at the time of the crash, the HMTA was the only authority for the hazardous materials regulations under which the government indicted SabreTech. That Act punishes willful violations, not reckless ones. In sum, none of the hazardous materials regulations in existence in May 1996, and relied upon in the indictment, were based upon statutory authority contained within Part A—Air Commerce and Safety. Therefore, the reckless counts are invalid, and the district court erred in denying SabreTech's motion to dismiss them.

B. Sufficiency of the Evidence
Based on our review of the record, we conclude there was sufficient evidence presented at trial to support SabreTech's conviction for willfully failing to train its employees in accordance with the hazardous materials regulations. The evidence demonstrates that SabreTech had a manual instructing personnel with regard to the handling and packaging of hazardous material. Thus, the jury could infer that SabreTech knew about the regulations dealing with the handling and packaging of hazardous material and knew that it should inform its employees of these regulations. Moreover, several SabreTech employees testified that they received no hazardous materials training while employed with SabreTech. In light of this evidence, we affirm SabreTech's conviction on Count XXIII. [Citation.] (All reasonable inferences and credibility choices must be made in favor of the verdict.)

V. CONCLUSION
As stated previously, this was a tragic accident that needlessly claimed the lives of over 100 people. That loss is irreplaceable. However, the record is clear that SabreTech and its employees did not intend to kill these people when it packed the old oxygen canisters and transported them to the ValuJet aircraft. Furthermore, the statutory and regulatory history demonstrates that the regulations cited as predicates for the alleged criminal activity were invalid. Because the district court improperly relied on the hazardous materials regulations as predicates for the alleged criminal activity, we vacate SabreTech's convictions on the reckless counts. Because there was sufficient evidence presented to the jury on the willful failure to train count, we affirm that conviction. Finally, we remand this case to the district court for re-sentencing.

As discussed in this chapter, the elements of a crime include an act and intent. The element of intent can often be established by evidence that points to a reckless indifference. One of the difficult problems addressed in the *SabreTech* case is the distinction between negligence and recklessness. How would you define a test to help with this distinction? Can such a test be developed?

Case 3-3 addresses the following questions: (1) Is a medical certificate an airman certificate, and (2) for criminal law purposes is a private pilot certificate with a single-engine rating the same as a private pilot certificate with a multiengine rating? The statute in the spotlight in this case (49 U.S.C. § 1472) is the predecessor to current-day 49 U.S.C. § 46306. The statute in question makes it an offense to "knowingly and willfully serve, or attempt to serve, in any capacity as an airman without a valid airman's certificate authorizing such person to serve in such capacity...."

CASE 3-3
UNITED STATES V. EVINGER
919 F. 2d 381 (1990)

OPINION BY: PER CURIAM

OPINION: Defendant-Appellee, John Lee Evinger, holder of an airman certificate authorizing him to act as a private pilot of a single-engine aircraft, was charged in a three-count indictment with one count of serving as a pilot of a single-engine aircraft without a valid airman's certificate, 49 U.S.C. App. § 1472(b)(1)(E), because his medical certificate had expired, and two counts of serving as a pilot of a twin-engine aircraft without a valid airman's certificate in connection with the transportation of marijuana, 49 U.S.C. App. § 1472(b)(1)(E), (b)(2)(B), in connection with incidents alleged to have occurred in July and November of 1989, after the medical certificate had expired. Following a hearing, the district court granted Evinger's motion to dismiss and the government timely filed its notice of appeal. Finding no reversible error, we affirm.

I. [Facts]

An airman's certificate was issued to Evinger in 1982, after he complied with all requirements imposed by statute and regulation. That certificate, authorizing Evinger to exercise the privileges of a private pilot, has been neither revoked nor suspended since its issuance. The rating on the certificate authorizes operation of "airplane single-engine land." Evinger's most recent medical certificate was obtained in September of 1986, but expired in September of 1988. The alleged offenses occurred in 1989.

II. [Procedural History]

Defendant was charged by grand jury indictment filed November 21, 1989, three separate violations. The specific acts complained of in the indictment were that Evinger operated aircraft "without a valid airman's certificate authorizing him to operate such aircraft." On March 15, 1990, Evinger submitted a motion to dismiss the charges against him, claiming that, as a matter of law, he held a valid airman's certificate at the time of the alleged offenses and that the district court had no authority to enforce a term, condition or limitation of an airman's certificate unless requested to do so by the Federal Aviation Authority in accordance with 49 U.S.C. App. § 1487. Evinger's motion was heard by the district court on April 9, 1990. Following argument of counsel and submission of exhibits, the court took the matter under advisement and, on May 22, 1990, granted Evinger's motion. The district court held that § 1472 provides criminal penalties for forged or fraudulent airmen's certificates, and that operating a twin-engine aircraft while holding a single-engine rating, and flying an aircraft without a medical certificate, were safety violations subject to civil penalties under 49 U.S.C. App. §§ 1430 and 1471(a). The government appealed only the dismissal of counts II and III, piloting a twin-engine aircraft without a valid airman's certificate in connection with the transportation of marijuana.

III. [Analysis]

The government argues that fraud and forgery are not elements of a violation under 49 U.S.C. App. § 1472(b)(1)(E). It also argues that the medical certificate is an "airman certificate" and therefore flying without one establishes all of the elements of the offense.

Resolution of the issue depends on the proper interpretation of §§ 1471 and 1472. "In ascertaining the plain meaning of the statute, the court must look to the particular statutory language at issue, as well as the language and design of the statute as a whole." [Citations.] If the statute is unambiguous, the court does not look beyond its express terms. [Citation.] "'We must take the intent of Congress ... to be that which its language clearly sets forth.'" [Citation.]

Section 1471(a)(1) provides civil penalties for violation of the safety regulations in Subchapter VI; section 1472(a) specifically excludes violations of Subchapter VI from the criminal penalties in that section. The violations the government alleges, flying a twin-engine aircraft when rated for a single-engine aircraft and flying without a valid medical certificate, are violations of the Subchapter VI safety regulations. [Citations.] Congress has provided civil penalties for the safety violations alleged, and therefore the district court correctly determined that § 1472 was inapplicable.

The government argues that the medical certificate is an airman certificate, relying on the definition of airman certificate in 49 C.F.R. § 821.1. This definition, however, is specifically limited to that part of the regulations (Practice and Procedure before the National Transportation Safety Board) and is not relevant to this discussion.

The government's reliance on *Bullwinkel v. U.S. Dep't of Transp., FAA*, 787 F. 2d 254 (7th Cir. 1986) and *King v. National Transp. Safety Bd.*, 766 F. 2d 200 (5th Cir. 1985) is also misplaced. *Bullwinkel* holds only that a medical certificate is a license within the meaning of the Equal Access to Justice Act. *Bullwinkel*, at 256–57.

King likewise does not support the government's position. In *King* the Federal Aviation Administration had suspended proceedings to revoke King's pilot license for knowingly carrying marijuana on an aircraft in violation of 49 C.F.R. § 91.12(a) (1985) pending the outcome of his appeal of a state court conviction for possession of marijuana. This court merely held that the delay in bringing the proceedings was not unreasonable under the circumstances.

Under the statutory scheme set forth by Congress, Evinger is subject to the civil penalties under § 1471 for his violations of the safety regulations. He had a valid airman certificate that permitted him to act in the capacity of private pilot and therefore he did not violate § 1472(b)(1)(E). The district court's order dismissing the indictment for lack of jurisdiction is AFFIRMED.

Do you agree with the court's analysis and conclusion in this case? Did the court handle the single-engine/multiengine certificate the same as it handled the medical certificate issue? If not, what differences in argument can you identify? Are the differences legitimate, in your opinion?

The final case for this chapter, Case 3-4, addresses an intriguing Fourth Amendment question: Should evidence obtained by law enforcement officers flying over private property be admissible as evidence in a criminal case? The Supreme Court tackles that question in Case 3-4.

CASE 3-4
CALIFORNIA V. CIRAOLO
476 U.S. 207 (1986)

OPINION BY: BURGER

OPINION: CHIEF JUSTICE BURGER delivered the opinion of the Court.

We granted certiorari to determine whether the Fourth Amendment is violated by aerial observation without a warrant from an altitude of 1,000 feet of a fenced-in backyard within the curtilage of a home.

I

On September 2, 1982, Santa Clara Police received an anonymous telephone tip that marijuana was growing in respondent's backyard. Police were unable to observe the contents of respondent's yard from ground level because of a 6-foot outer fence and a 10-foot inner fence completely enclosing the yard. Later that day, Officer Shutz, who was assigned to investigate, secured a private plane and flew over respondent's house at an altitude of 1,000 feet, within navigable airspace; he was accompanied by Officer Rodriguez. Both officers were trained in marijuana identification. From the overflight, the officers readily identified marijuana plants 8 feet to 10 feet in height growing in a 15- by 25-foot plot in respondent's yard; they photographed the area with a standard 35mm camera.

On September 8, 1982, Officer Shutz obtained a search warrant on the basis of an affidavit describing the anonymous tip and their observations; a photograph depicting respondent's house, the backyard, and neighboring homes was attached to the affidavit as an exhibit. The warrant was executed the next day and 73 plants were seized; it is not disputed that these were marijuana.

After the trial court denied respondent's motion to suppress the evidence of the search, respondent pleaded guilty to a charge of cultivation of marijuana. The California Court of Appeal reversed, however, on the ground that the warrantless aerial *observation* of respondent's yard which led to the issuance of the warrant violated the Fourth Amendment. [Citation.] That court held first that respondent's backyard marijuana garden was within the "curtilage" of his home, under *Oliver* v. *United States*, 466 U.S. 170 (1984). The court emphasized that the height and existence of the two fences constituted "objective criteria from which we may conclude he manifested a reasonable expectation of privacy by any standard." [Citation.]

Examining the particular method of surveillance undertaken, the court then found it "significant" that the flyover "was not the result of a routine patrol conducted for any other legitimate law enforcement or public safety objective, but was undertaken for the specific purpose of observing this particular enclosure within [respondent's] curtilage." [Citation.] It held this focused observation was "a direct and unauthorized intrusion into the sanctity of the home" which violated respondent's reasonable expectation of privacy. [Citation.] The California Supreme Court denied the State's petition for review.

We granted the State's petition for certiorari, [Citation]. We reverse.

The State argues that respondent has "knowingly exposed" his backyard to aerial observation, because all that was seen was visible to the naked eye from any aircraft flying overhead. The State analogizes its mode of observation to a knothole or opening in a fence: if there is an opening, the police may look.

The California Court of Appeal, as we noted earlier, accepted the analysis that unlike the casual observation of a private person flying overhead, this flight was focused specifically on a small suburban yard, and was not the result of any routine patrol overflight. Respondent contends he has done all that can reasonably be expected to tell the world he wishes to maintain the privacy of his garden within the curtilage without covering his yard. Such covering, he argues, would defeat its purpose as an outside living area; he asserts he has not "knowingly" exposed himself to aerial views.

II

The touchstone of Fourth Amendment analysis is whether a person has a "constitutionally protected reasonable expectation of privacy." *Katz* v. *United States*, 389 U.S. 347, 360 (1967).... *Katz* posits a two-part inquiry: first, has the individual manifested a subjective expectation of privacy in the object of the challenged search? Second, is society willing to recognize that expectation as reasonable? [Citation.]

Clearly—and understandably—respondent has met the test of manifesting his own subjective intent and desire to maintain privacy as to his unlawful agricultural pursuits. However, we need not address that issue, for the State has not challenged the finding of the California Court of Appeal that respondent had such an expectation. It can reasonably be assumed that the 10-foot fence was placed to conceal the marijuana crop from at least street-level views. So far as the normal sidewalk traffic was concerned, this fence served that purpose, because respondent "took normal precautions to maintain his privacy." [Citation.]

Yet a 10-foot fence might not shield these plants from the eyes of a citizen or a policeman perched on the top of a truck or a two-level bus. Whether respondent therefore manifested a subjective expectation of privacy from *all* observations of his backyard, or whether instead he manifested merely a hope that no one would observe his unlawful gardening pursuits, is not entirely clear in these circumstances. Respondent appears to challenge the authority of government to observe his activity from any vantage point or place if the viewing is motivated by a law enforcement purpose, and not the result of a casual, accidental observation.

We turn, therefore, to the second inquiry under *Katz*, *i.e.*, whether that expectation is reasonable. In pursuing this inquiry, we must keep in mind that "[the] test of legitimacy is not whether the individual chooses to conceal assertedly 'private' activity," but instead "whether the government's intrusion infringes upon the personal and societal values protected by the Fourth Amendment." [Citation.]

Respondent argues that because his yard was in the curtilage of his home, no governmental aerial observation is permissible under the Fourth Amendment without a warrant. The history and genesis of the curtilage doctrine are instructive. "At common law, the curtilage is the area to which extends the intimate activity associated with the 'sanctity of a man's home and the privacies of life.'" [Citations.] The protection afforded the curtilage is essentially a protection of families and personal privacy in an area intimately linked to the home, both physically and psychologically, where privacy expectations are most heightened. The claimed area here was immediately adjacent to a suburban home, surrounded by high double fences. This close nexus to the home would appear to encompass this small area within the curtilage. Accepting, as the State does, that this yard and its crop fall within the curtilage, the question remains whether naked-eye observation of the curtilage by police from an aircraft lawfully operating at an altitude of 1,000 feet violates an expectation of privacy that is reasonable.

That the area is within the curtilage does not itself bar all police observation. The Fourth Amendment protection of the home has never been extended to require law enforcement officers to shield their eyes when passing by a home on public thoroughfares. Nor does the mere fact that an individual has taken measures to restrict some views of his activities preclude an officer's observations from a public vantage point where he has a right to be and which renders the activities clearly visible. [Citations.] "What a person knowingly exposes to the public, even in his own home or office, is not a subject of Fourth Amendment protection." [Citation.]

The observations by Officers Shutz and Rodriguez in this case took place within public navigable airspace, [Citation], in a physically nonintrusive manner; from this point they were able to observe plants readily discernible to the naked eye as marijuana. That the observation from aircraft was directed at identifying the plants and the officers were trained to recognize marijuana is irrelevant. Such observation is precisely what a judicial officer needs to provide a basis for a warrant. Any member of the public flying in this airspace who glanced down could have seen everything that these officers observed. On this record, we readily conclude that respondent's expectation that his garden was protected from such observation is unreasonable and is not an expectation that society is prepared to honor.

The dissent contends that the Court ignores Justice Harlan's warning in his concurrence in *Katz v. United States*, 389 U.S., at 361–362, that the Fourth Amendment should not be limited to proscribing only physical intrusions onto private property. [Citation.] But Justice Harlan's observations about future electronic developments and the potential for electronic interference with private communications, see *Katz, supra*, at 362, were plainly not aimed at simple visual observations from a public place. Indeed, since *Katz* the Court has required warrants for electronic surveillance aimed at intercepting private conversations. [Citation.]

Justice Harlan made it crystal clear that he was resting on the reality that one who enters a telephone booth is entitled to assume that his conversation is not being intercepted. This does not translate readily into a rule of constitutional dimensions that one who grows illicit drugs in his backyard is "entitled to assume" his unlawful conduct will not be observed by a passing aircraft—or by a power company repair mechanic on a pole overlooking the yard. As Justice Harlan emphasized, "a man's home is, for most purposes, a place where he expects privacy, but objects, activities, or statements that he exposes to the 'plain view' of outsiders are not 'protected' because no intention to keep them to himself has been exhibited. On the other hand, conversations in the open would not be protected against being overheard, for the expectation of privacy under the circumstances would be unreasonable." [Citation.]

One can reasonably doubt that in 1967 Justice Harlan considered an aircraft within the category of future "electronic" developments that could stealthily intrude upon an individual's privacy. In an age where private and commercial flight in the public airways is routine, it is unreasonable for respondent to expect that his marijuana plants were constitutionally protected from being observed with the naked eye from an altitude of 1,000 feet. The Fourth Amendment simply does not require the police traveling in the public airways at this altitude to obtain a warrant in order to observe what is visible to the naked eye.

Reversed.

Do you agree or disagree with the Supreme Court's determination in this case? If you disagree, you are in good company. The Court was split in a 5-to-4 decision. Strong dissents were published in this case, and you may wish to read the dissenting opinions to get a comprehensive feeling for all the arguments presented in this very difficult case.

DISCUSSION CASES

1. Could increasing the threat of criminal penalties targeting aviation professionals and businesses have a negative impact on aviation safety? Why or why not? Consider the following issues as you consider the question.
    (a) Will criminal penalties for certain types of violations create a stronger deterrent?
    (b) How will potential criminal penalties impact the cooperation of airmen and aviation businesses in safety investigations?
    (c) Would it be appropriate for the government to emphasize the use of criminal action only in highly publicized cases?

2. A captain and first officer for a National Airways flight were pushing back from the gate at LaGuardia International when the flight was recalled to the gate. It turns out that officials at a security checkpoint smelled alcohol on the pilots' breath and immediately filed a report with New York transit police. The police contacted the Transportation Security Administration, and air traffic controllers recalled the flight. Upon disembarking, the pilots were given a breathalyzer test. Both pilots failed the test with blood alcohol levels exceeding the FAR § 91.17 limit of 0.04 and the New York State criminal law limit of 0.08. However, the pilots did not exceed the federal criminal standard of 0.10 that applies to transportation workers in the rail industry, airlines, water, and bus transportation. The pilots were fired, and the FAA revoked their airman certificates. Later the State of New York filed criminal charges for violation of New York's law related to the operation of an aircraft while under the influence of alcohol. The pilots filed a motion to dismiss the New York criminal charges on the grounds that the New York criminal statute is preempted by federal law in the area of pilot qualifications and ability to operate aircraft in interstate commerce. How would you decide on the motion to dismiss? Explain.

3. Brandy is the pilot in command of a Maule when an aircraft crash causes the death of his passenger. The State of Florida charges Brandy with murder (the charge is later reduced to manslaughter before trial). Florida supports its efforts to convict Brandy of manslaughter by citing witnesses' statements that prior to the crash Brandy (1) was flying so close to the ground that he had to pull up to avoid palm trees and antennas and (2) buzzed a crowded motel at an altitude of 40 or 50 feet. Brandy claimed that the accident was caused by a mechanical defect in the aircraft. Brandy's claim regarding the mechanical difficulty was later thrown out by the courts as being false. The Florida statute in question defines manslaughter as an act demonstrating a "careless disregard of safety and welfare of the public" involving "reckless indifference to the rights of others which is equivalent to an intention of violation of [those rights]." Should Brandy be convicted of manslaughter for the death of his passenger? Explain.

4. Karl, a German citizen, applies for a first-class medical certificate from the FAA in preparation for his U.S. Airline Transport Pilot (ATP) certificate and a job interview with a major airline. On the FAA medical application a question asks, "Have you ever … been convicted of any offense involving drugs or alcohol?" More than 12 years ago, Karl was convicted in Germany for driving while intoxicated. As a first-time offender, he was fined and committed to a rehabilitation center for one month. Under German law, Karl's conviction for the charge of driving while intoxicated was expunged from his criminal record, and he was no longer required to ever reference the conviction. Karl decides that he does not need to report the German conviction on his medical application. The FAA and U.S. District Attorney later learn about Karl's prior conviction and his failure to report it in his application for a medical certificate. Karl's medical certificate is revoked by the FAA, and the U.S. District Attorney's office charges Karl with a violation of 18 U.S.C. § 1001, which makes it a crime to "knowingly and willfully … make any materially false … representation." Should Karl be convicted of making a false statement under 18 U.S.C. § 1001? Why or why not?

5. Freddie is a used-aircraft distributor. To boost profits, he purchased used aircraft, rolled back the Hobbs meter, and subsequently sold the aircraft to dealers and fixed based operators (FBOs) for resale. The dealers and FBOs were not aware of Freddie's tampering with the aircraft engine Hobbs meters. The dealers, in turn, sold the aircraft at inflated prices based on lower indicated hours for engine time. To complete the resale of each aircraft, the dealer or FBO would submit, by mail, a bill of sale to the FAA in Oklahoma City on behalf of the ultimate purchaser of the aircraft. The receipt of the bill of sale was required for transferring title and obtaining aircraft registration from the FAA. Law enforcement authorities became aware of Freddie's fraud, and he was convicted on 15 counts of mail fraud. Freddie appealed and argued that the mailings that were relied on for the indictment—the mailings to the FAA—were not in furtherance of his fraudulent scheme. You are the appeals court judge drafting the decision for this case. Is Freddie guilty of mail fraud? Why or why not?

ENDNOTES

1. 242 N.Y. 13, 15 N.E. 585 (1926).
2. See *Betts v. Brady*, 316 U.S. 455 (1942).
3. 37 U.S. 335 (1963).
4. 18 U.S.C. § 316(d)(1).
5. See *Baldwin v. Alabama*, 472 U.S. 372 (1985).
6. See *United States v. Tamargo*, 637 F. 2d 346 (5th Cir. 1981).
7. See NTSB Bar Association, Select Comm. On Aviation Public Policy, "Aviation Professionals and the Threat of Criminal Liability—How Do We Maximize Aviation Safety?" 67 J. Air L. & Com 3 (Summer 2002).
8. 271 F. 3d 1018 (11th Cir. 2001).
9. 49 U.S.C. § 46306.
10. 49 U.S.C. § 46313.
11. 46 U.S.C. § 46307.
12. 46 U.S.C. § 46316.

4
Tort Liability and Air Commerce

INTENTIONAL TORTS 93
Intentional torts against persons 93
Intentional torts against property 95

NEGLIGENCE 97
Elements of negligence 97
Defenses to negligence 102

WRONGFUL DEATH 104

STRICT PRODUCT LIABILITY 104
Elements 104
Defenses to strict product liability 105

CASES AND COMMENTARY 108

DISCUSSION CASES 114

ENDNOTES 116

As you learned in Chap. 3, criminal law involves wrongs committed against society that must be redressed by the wrongdoers. Tort law takes a different perspective and allows individuals redress for wrongs committed that caused them damage. Unlike criminal cases in which the plaintiff is the government, in tort cases the plaintiff (or complainant) is typically an individual or a business. Another important distinction between criminal and tort law is the type of redress sought. In criminal cases, the typical redress may be incarceration or fines paid to the government. In tort cases, the remedy sought is most often monetary damages. Sometimes the monetary damages are intended to compensate the injured party for any harm done (e.g., medical expenses, lost wages, property damage). In certain situations the monetary damages are also designed to punish the wrongdoer and set an example to prevent future bad acts. This type of monetary award is often referred to as punitive damages.

From an evidentiary standpoint, perhaps the most critical difference between criminal and tort cases is the burden of proof placed on the plaintiff or complainant. In a criminal case, the government must prove its case beyond a reasonable doubt. In tort cases, the plaintiff has a lesser burden of proof. The plaintiff in a tort case must meet the burden of proving her case by a preponderance of the evidence. This standard of proof for tort cases requires that a plaintiff prove that it is more likely than not that the wrong was committed.

Many times criminal law and tort law overlap, and the same act can be subject to both criminal and tort law actions. One very public case in recent history was the O. J. Simpson murder case. Note that in that case, Simpson was acquitted of criminal charges where the burden of proof required proof beyond a reasonable doubt. However, in a subsequent tort case brought by the victims' families, Simpson was found to be liable for the deaths of the victims under the lesser standard of proof requiring a preponderance of evidence. He was consequently ordered by the courts to pay civil damages to the victims' families.

As a general matter, tort law has the following objectives:

- To compensate persons who are injured by another's misconduct
- To require those who caused harm to pay for the injuries
- To prevent future injury

It is fair to generalize that the elements of a tort include the following: (1) there is a breach of a duty that one person owes to another, (2) the breach is the proximate cause of injury, and (3) there is injury or damage to a person with an interest that is legally protected.

Tort law has a pervasive effect on the aviation industry. Aviation has evolved into a very safe activity. However, when accidents occur, the potential for significant damages is high.

In this chapter, we will focus attention on tort law issues that directly affect aviation commerce and the aviation industry as a whole. Specifically, we will address the following topics:

- Intentional torts
- Negligence
- Wrongful death
- Strict product liability

Toward the end of the chapter, you will review selected cases that address tort law in the aviation environment. Commentary following some of the cases addresses efforts at tort reform in the aviation industry. Note that throughout this chapter, there will be various references to the *Restatement of Torts*. Because the body of tort law is primarily driven by case or common law, the sources of tort law are as scattered and diverse as courts throughout the United States. The American Law Institute (an organization consisting of legal scholars, judges, and lawyers) has developed a summary of sorts that discusses and dissects the prevailing current law of torts. For the purposes of this discussion, we will refer to the most current *Restatement of Torts* as the *Restatement*.

INTENTIONAL TORTS

The first category of torts we will review is intentional torts. Intentional torts are torts in which the wrongdoer intends to do harm. Within this category of torts, first we will study intentional torts against persons. Next we will review intentional torts against property.

The law attempts to protect all persons from uninvited and unauthorized touching or contact (including restraint). Within the context of aviation, we will take a look at five possible intentional torts against persons:

Intentional torts against persons

- Assault
- Battery
- False imprisonment
- Intentional infliction of emotional distress
- Defamation

Each of these torts must be established by proving that certain "elements" were present at the time the alleged wrong was committed. A review and a brief explanation of the elements for intentional torts against persons are found below:

Assault
The elements of assault are relatively straightforward. A person has been assaulted if he has been subject to (1) the threat of immediate harm or offensive contact or (2) any threat or action that causes him to be reasonably fearful or apprehensive of imminent harm.

Notice that there is no requirement for physical contact with assault. Also, the threat of future harm does not rise to the level of assault.

For example, if a passenger checking in for a flight has an altercation with an airline ticketing agent over the weight or size of baggage, there is no assault if the passenger threatens the agent by stating that "you should watch over your shoulder in the future—because I may be there...." However, if the agent is a 5-foot, 105-pound woman and the passenger is a 6-foot 5-inch, 250-pound man menacing her with a closed fist, there is likely an assault. Conversely, if the ticketing agent is a 6-foot 5-inch, 250-pound man and a frail 85-year-old woman passenger is threatening him with a closed fist, there may not be an assault because it could be found that he would not be reasonably apprehensive of imminent harm.

Battery
Battery occurs when a person intentionally makes unauthorized and harmful or offensive contact with another. Quite commonly, battery will be found when someone intentionally strikes another with a fist or an object. A battery can also occur when someone injures another by throwing a rock or firing a gun that causes the other to be struck by the rock or bullet.

Interestingly, for battery to exist, the victim does not need to be aware of the harmful or offensive contact. For instance, an airline passenger can be found guilty of battery if he or she intentionally and offensively touches a fellow passenger who is sleeping.

Assault and battery often occur at the same time. However, that is not always the case. For instance, an aircraft passenger can intentionally hit another passenger on the back of the head. The victim in this case was not assaulted because she or he did not "see it coming." Nonetheless, a battery occurred.

Another interesting facet of battery is the doctrine of transferred intent. Suppose Sam and Dave are airline passengers engaged in a dispute and Sam throws a cup full of water and ice at Dave. If Dave ducks and the water and ice strike Betty, Sam's intent to strike Dave is transferred to Betty and a battery has occurred against Betty.

False imprisonment
False imprisonment involves the intentional confinement or restraint of another person without authority or justification and without that person's consent. The confinement or restraint created by this tort can be achieved through physical barriers, threats of harm, physical force, or any other means that would reasonably lead someone to believe he or she had been confined.

If a flight school manager confines a student to a training room and contacts law enforcement authorities because of a suspicion that the student is a terrorist, the flight school (and manager) may be subject to damages for false imprisonment if it turns out the school had no justification to believe that the student was a terrorist. If the student stayed, but could have walked out to the parking lot via a clearly marked back door, he or she would not be as likely to succeed on a claim of false imprisonment—unless the student reasonably believed his or her safety would be threatened if he or she exited the back door.

Intentional infliction of emotional distress
This tort requires (1) extreme and outrageous conduct that (2) is intended to cause severe emotional distress and (3) causes such severe distress.[1] In some jurisdictions, the

severe emotional distress must be physically manifest (e.g., headaches, ulcers). In other jurisdictions, shame, humiliation, and similar conditions are sufficient damages to justify a finding of intentional infliction of emotional distress.

A finding of intentional infliction of emotional distress does not require any physical contact with the defendant. It is also important to note that the use of foul or rude language will typically not be sufficient to establish this tort.

Intentional infliction of emotional distress has often been associated with debt collection agencies using "over the top" methods to collect bills. Telephone calls in the middle of the night and threats against jobs or the jobs of relatives are all techniques that have been found to constitute intentional infliction of emotional distress.

Defamation

If someone tries to harm another's reputation by stating falsehoods in either written or verbal form, that person may be liable for defamation. Defamation in writing is known as libel. Verbal defamation is known as slander.[2]

In a defamation case, a plaintiff must prove (1) that a defamatory statement regarding the plaintiff (2) was untrue and (3) was communicated to others by the defendant, and (4) that the untrue statements caused harm to the plaintiff.[3]

Merely noting an opinion or an idea about another is not defamation. For instance, a person might say, "I hate flying right seat with Captain Smith, he's such a jerk." However, if that same person made a false statement to airline management that Captain Smith reported to duty with alcohol on his breath, and this false statement caused Captain Smith to be disciplined or discharged, then an action for defamation would likely have merit.

There are several defenses to defamation. However, truth is considered to be a complete defense. Therefore, regardless of the purpose or intent of the person who makes a defamatory statement, if the statement is true, the person has not committed the tort of defamation.

In the context of aviation law, there are several torts that involve intentional wrongs against the property of others. We will take a closer look at the following intentional torts in the discussion below:

Intentional torts against property

- Trespass to land
- Trespass to personal property
- Conversion of personal property

Trespass to land

A landowner has the exclusive legal right to title, possession, and use of his or her property. If another person enters or uses the land without authorization, a tort has been committed.

Trespass to land requires (1) some affirmative action by the defendant, (2) with intention to enter the real property of another that (3) results in the actual entry to the property.[4]

For instance, Eddie is flying over some farm fields when he begins to experience low oil pressure in his Cessna Cardinal. He spots a wheat field below and decides to make an emergency landing at the field. When he lands in the wheat field, he has committed trespass on that farmer's property. First, he made the affirmative decision to land on the wheat field (as opposed to being in a car that is bumped off the road by a truck and winds up in the wheat field). Second, Eddie intended to land in the field (it is important to remember in this regard that Eddie does not have to demonstrate any hostile intent or intent to do harm). Third, Eddie's decision resulted in the actual entry of the farmer's property.

Although a trespass to land might have occurred, there are often no real damages if no property has been found to be injured. In Eddie's case, if he lands and no damage is done, he may have trespassed, but the farmer will be unable to collect damages because there was no physical harm done to the land.

With respect to aircraft, it is noteworthy that a trespass to real property can be committed on, above, or below the surface of the land. However, the law generally regards the altitudes at which aircraft operate above minimum prescribed altitudes for safe flight as being public areas. Therefore, no trespass can be committed by an aircraft operator unless her aircraft enters the airspace below minimum prescribed altitudes and interferes with the landowner's right to the enjoyment and use of his property.

Trespass to personal property

Personal property or "chattel" (as it is sometimes referred to by lawyers) is property other than real property. In the world of aviation, most personal property, such as aircraft and associated equipment, is both tangible and movable, thus making the property susceptible to trespass in the form of intentional dispossession or unauthorized use by others.

Liability for trespass to personal property occurs when a person: [5]

- Dispossesses another of her or his property
- Substantially impairs the condition, quality, or value of that property
- Deprives the possessor of property of its use for a substantial time
- Causes harm to the possessor or to some person or thing in which the possessor has a legally protected interest

Notice that any one of the conditions stated above will suffice for a finding of trespass to personal property. For instance, suppose Amanda flies her Cessna Cardinal to Joy City Airport and parks the airplane at a tie-down location after making arrangements with Lyon FBO. Amanda leaves the airport. Later that day, another aircraft owner, Billy, has Amanda's plane towed to a different parking spot on the other side of the field because Billy typically uses the tie-down location that Amanda had taken. When Amanda returns late that evening, she subsequently looks for her airplane but cannot locate it for several hours. Billy is liable to Amanda for trespass to his personal property.

Conversion of personal property

Conversion occurs when a party (1) intentionally exercises dominion or control (2) over another's personal property (3) such that it substantially interferes with the other's right of control over the personal property and (4) requires payment of the full value of the property.[6] Based on this interpretation of conversion, all conversions are trespasses, but not all trespasses rise to the level of conversion.

As indicated above, conversion can occur if someone either destroys another's property or uses it without authorization. For instance, if Jimmy leaves his aircraft with Sam's FBO for storage in a hangar, and Sam uses the aircraft for 250 hours of flight time while Jimmy is away, Sam is liable to Jimmy for conversion. Further, if Sam intentionally destroys or damages the aircraft by scavenging parts while Jimmy is away, he is also liable to Jimmy for conversion.

NEGLIGENCE

Generally speaking, negligence is an unintentional tort involving carelessness that causes harm. The tort of negligence allows persons to have their interests in safety and economic security protected from the carelessness of others. This also means that each of us has a duty to conduct ourselves in a responsible manner that reflects a concern for others and their interests. As professionals in the aviation environment, you should be particularly sensitive to the fact that lives and expensive equipment are at risk. This section of our chapter on torts will provide an overview of (1) the elements of negligence and (2) legal defenses against negligence.

The following elements are required for a plaintiff to establish the tort of negligence:[7]

1. The defendant owes a *duty of care* to the plaintiff.
2. The defendant *breaches this duty*.
3. The defendant's breach is the *proximate cause* of the plaintiff's injury.
4. The plaintiff suffers *injury* that is protected under the law.

Elements of negligence

Note that all four of these criteria must be met for a defendant to be held liable for negligence. A review of each of these elements follows.

Duty of care
It is sometimes difficult to assess whether a person has a legal duty of care in any given situation. Most courts and legal scholars tend to indicate that the duty of care can be interpreted as a duty that is owed to every person who can be reasonably foreseen to be potentially harmed by that carelessness. For instance, an aircraft pilot would clearly be expected to foresee that her negligence could cause harm to her passengers and persons and/or property on the ground below the aircraft's path of flight. However, how far does this duty extend? Could the pilot be reasonably expected to foresee that her aircraft crash could cause an automobile accident 2 miles from the aircraft crash scene because a speeding ambulance runs a red light on the way to respond to the crash site?

One case illustrating the issue of duty of care involves the tragic collision of two Boeing 747 aircraft in the Canary Islands. In *Burke v. Pan World Airways, Inc.*,[8] the plaintiff sued Pan World Airways for damages she claimed resulted from the crash in the Canary Islands. The interesting twist in this case is that the defendant was in California at the time of the crash. However, her identical twin sister was killed onboard one of the aircraft destroyed in the collision. Burke claimed that at the instant the accident occurred, she felt as though she were "split in two." In support of her claim, Burke was prepared to present evidence that there is a certain "extrasensory empathy" between certain sets of identical twins. The court dismissed her claim, indicating that even if Burke could establish such a phenomenon as extrasensory empathy existed, it was too unusual a theory to allow the court to determine that the defendant had any kind of reasonably foreseeable duty to the plaintiff.

Breach of duty

If a duty to the plaintiff is established, the next element that must be established in a negligence case is the element of breach. The plaintiff must convince a court that the defendant breached a duty owed to the plaintiff.

The test developed by the courts to tackle this question has been dubbed the *reasonable person test*. This test asks whether a defendant failed to exercise the same care as a reasonable person under similar circumstances would have exercised. Just how a reasonable person would react in various circumstances is literally impossible to pin down. Suffice it to say that juries have significant latitude to apply this standard as they deem fit on a case-by-case basis.

Although there is no way to pin down exactly when a duty to others has been breached, certain guidelines have evolved in case law. A review of the reasonable person test reveals that the following factors are typically considered in determining whether a reasonable person would have conducted himself or herself in a certain manner:

- Surrounding circumstances
- Violations of a statute or regulation (negligence per se)
- Special skills or knowledge

Each of these considerations is briefly discussed below.

Surrounding circumstances   In determining whether a defendant acted as a reasonable person, the courts will typically be required to consider the facts and surrounding circumstances. In the context of aviation, it might ordinarily be considered negligent to land your aircraft and run off the paved surface, damaging other aircraft tied down at an adjacent ramp. However, if the cause of the problem was a defective nosewheel steering mechanism that made it impossible for you to maintain directional control, you might not be found to have been negligent.

Violations of a statute or regulation (negligence per se)   Sometimes minimum standards of conduct for the reasonable person can be established by statute or regulation.[9] If a court finds that an applicable statute or regulation was violated by a defendant, it will likely find that the defendant's conduct was negligence per se. In effect, the doctrine of negligence per se creates an inference that a violation of the law indicates a lack of reasonable care.

For instance, a safety regulation requires an aircraft paint shop to operate a certain number of exhaust fans per square foot of floor space. If the shop fails to comply with the regulation and an employee sues for damages caused by the inhalation of paint fumes, the shop (in most jurisdictions) will be found negligent by virtue of the negligence per se doctrine.

Note that the fact that the shop might have met the requirements of the regulations regarding exhaust fans does not mean the shop will not be found negligent. The standards that are set by statute or regulation are typically looked upon as minimum standards, and they do not preclude a finding that a reasonable person would have exercised greater caution than required by statute or regulation under the circumstances.

Special skills or knowledge   A person who has special training or skills may be held to a higher standard of conduct than the reasonable person. Physicians, lawyers, certified public accountants, airline pilots, plumbers, electricians, and other licensed or skilled professionals are expected to perform their duties to the standards of their profession.

In the aviation environment, pilots, aircraft maintenance professionals, and air traffic controllers are held to very high standards of conduct. Almost every task in aviation is subject to one regulation or another. Adherence to the standards is necessary for regulatory compliance and to avoid civil liability.

Res ipsa loquitur   Sometimes, the mere occurrence of an accident or event that causes personal injury or property damage is enough to create a presumption that a duty has been breached. This legal doctrine is known as *res ipsa loquitur*, which means "the thing speaks for itself."

The *Restatements* indicate the following with respect to this doctrine: [10]

1. It may be inferred that the harm suffered by the plaintiff is caused by the negligence of the defendant when:
    a. the event is a kind which ordinarily does not occur in the absence of negligence;
    b. other responsible causes, including the conduct of the plaintiff and third persons, are sufficiently eliminated by the evidence; and
    c. the negligence indicated is within the scope of the defendant's duty to the plaintiff.

This doctrine would apply if a person were injured in an air terminal owing to falling pieces of ceiling tile. If there were no other cause that could reasonably explain the falling ceiling tile (and the defendant were permitted to introduce such evidence), a court could infer that the airport authority was negligent in allowing the condition to exist. As you might imagine, *res ipsa loquitur* is frequently invoked in aircraft accidents where the defendant is unable to cite weather or other reasonable causes to explain a crash. Since aircraft do not routinely just fall from the sky, the only explanation barring weather or mechanical malfunctions is often pilot error or negligence.

Proximate cause

If the plaintiff can prove that the defendant owed her a duty and the duty was breached, the next element that must be established is proximate cause. In its simplest interpretation, this element requires that a plaintiff prove that the defendant's breach of duty was the direct cause of the plaintiff's injury. However, because the courts have faced infinite varieties of facts and circumstances in centuries of building a body of law in this area, the tests that are used to establish proximate cause are a bit more complex.

There is general agreement that two tests must be met to establish proximate cause. The two tests are as follows:

1. Was the defendant's negligence the cause in fact or the actual cause of the plaintiff's injury?
2. Was the harm done to the plaintiff reasonably foreseeable by the defendant?

Each of these tests is discussed below.

Causation in fact   The *causation in fact* or *actual cause* test asks if the defendant's carelessness caused the plaintiff's injury. Courts often apply a rule known as the *but for* rule to test for causation in fact. The but for rule inquires as to whether an event would have happened *but for* a defendant's negligent conduct. In applying this rule, a defendant's breach of duty would not be deemed negligence if the harm caused to the plaintiff would have happened anyway.

For example, an airframe and powerplant (A&P) mechanic is careless and fails to properly install a new autopilot in Jack's aircraft. The installation is signed off and approved in the logbook. Jack takes his aircraft on a cross-country flight and encounters instrument meteorological conditions. Jack is instrument-rated so he is able to press on with his flight. If Jack activates the defective autopilot for an approach and the autopilot failure leads to an accident, the A&P mechanic is likely to be liable for any damage or injury if it is determined that but for the failure of the autopilot, Jack's aircraft would not have crashed. On the other hand, if Jack never activated the defective autopilot, manually flew the approach, and the aircraft crashed, there would not be any liability attributable to the A&P mechanic because the harm to Jack would have occurred regardless of the A&P mechanic's defective autopilot installation.

Foreseeability   The test of foreseeability asks whether the harm done by a defendant's breach of duty could have reasonably been foreseen by the defendant. This question has caused vexing problems to the courts. This test can often be broken down into the following two parts:

- Were the consequences of the defendant's actions foreseeable?
- Was there any superseding cause to the plaintiff's injury?

Each of these inquiries is discussed below.

Foreseeable consequences   In 1928, one of the most well-known tort cases addressed the issue of foreseeable consequences. In *Palsgraf v. Long Island Railroad*,[11] the New York Court of Appeals (New York's highest court) heard a case involving employees of the Long Island Railroad who negligently pushed a passenger who was attempting to board a train. After being pushed, the passenger dropped a package that contained fireworks. The fireworks subsequently exploded under the wheels of the train, and the shock waves from the explosion knocked over a scale at the far end of the train station, injuring the plaintiff, Mrs. Palsgraf. The New York Court of Appeals found that the defendant railroad owed a duty to the passenger and that the subsequent harm done to the passenger's person and property was a foreseeable breach of the railroad's duty of care. However, the court stopped short there. It further ruled that the injury to Mrs. Palsgraf due to the falling scale was unforeseeable, and therefore the railroad's negligence was not the proximate cause of Mrs. Palsgraf's injury. Therefore, the railroad was not liable to Mrs. Palsgraf. The test employed in this case was whether Mrs. Palsgraf was within the reasonably foreseeable *zone of danger*. The court ruled that she was not.

To illustrate this issue, let's look at the following example. Sam is flying a Cessna P-210 from Denver to Colorado Springs. The flight is made largely in instrument conditions, and Sam is instrument-rated and current. Sam makes a navigational error resulting from the erroneous input of a Global Positioning Systems (GPS) coordinate. Regrettably, this causes Sam to fly into the side of a mountain, and he is killed. The aircraft ignites and can be seen from a small town below. The local fire department is dispatched to go to the accident and put out the fire due to drought conditions and the fear of a massive forest fire. Forty minutes later, on the way up the mountainous roads, a fire truck inadvertently runs a passenger car off the road, and the driver of the car, Joe, is killed in the ensuing crash. Joe's widow brings an action against Sam's estate, claiming that it was Sam's negligence that caused Joe's death. In this case it is most likely that Sam's estate will prevail. It is hard to imagine the courts determining that at the time of the crash Joe was within a reasonably foreseeable zone of danger. However, this is the sort of subjective inquiry that is undertaken in determining whether someone is within the requisite zone of danger. Like so much in the law, making a proper determination is often more art than science.

Superseding cause Another possible issue in determining proximate cause is whether a superseding cause relieves the defendant of liability. An *intervening cause* is a happening or event that occurs after the defendant's negligent conduct and that, when combined with the defendant's breach of duty, causes harm to the plaintiff. A mere intervening cause is not enough to relieve a defendant of liability for negligence. However, a *superseding cause* is an intervening cause that is not a normal consequence of the defendant's negligence.

For instance, an airport fails to properly mark an excavation site that it is digging to lay new sewer lines across the airfield for a terminal area on the opposite side. Late one night, a small general aviation aircraft suffers a propeller strike when it runs into unmarked excavated hole. While the low visibility at night may have been an intervening cause, it would be hard to argue that such an accident could be considered unforeseeable. Therefore, this would not qualify as a superseding cause.

However, if Pete and Dave, the airport line workers, were horsing around near the excavation site and Dave purposely threw Pete into the excavated hole, injuring Pete, it is possible that a court would rule that Dave's action was not foreseeable and, therefore, a superseding cause. Therefore, the airport would not be liable to Pete for its failure to mark the excavation site.

Figure 4-1
Determining proximate cause.

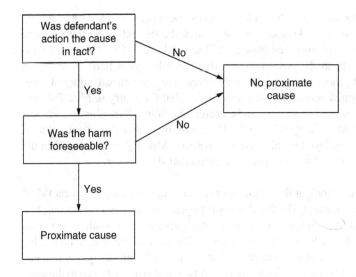

See Fig. 4-1 for a flowchart describing the analysis required to determine probable cause.

Injury

The final element that must be established for a finding of negligence is injury. The court will not find negligence unless it is proved that a legally protected interest of the plaintiff has been harmed. This is relatively easy to establish if someone has suffered serious physical injuries as a result of an airplane accident caused by negligence. However, the issue of injury might become a bit more difficult to sort out if someone were harmed by an offensive touching or accident, but not physically. Typically, recovery of damages in negligence cases is allowed for injury to person or property damage. Recovery of damages that are not manifest physically has not traditionally been embraced by the courts.

Nonetheless, there has been an evolving willingness by the courts to allow damages for negligently inflicted emotional distress. Courts are now willing to find injury when a person suffers bodily harm from emotional distress caused by another's negligence.[12] An example of this situation would be a person who suffered a heart attack during an aircraft crash caused by the defendant pilot's negligence, even if the crash did no direct physical harm to the plaintiff.

Still, in most courts, there is no award of damages when the sole injury is emotional distress.[13] Recently, some courts have allowed damages for negligently inflicted emotional distress with no finding of physical harm to the plaintiff. To recover under this theory, these courts often require that the plaintiff have witnessed injury to a close relative that was caused by the negligence of another.

Defenses to negligence

Even if a plaintiff is able to successfully establish all the elements of negligence, he or she may still be stymied if the defendant is able to counter with one of the recognized defenses to negligence. The three most likely defenses used by defendants in negligence cases are

- Contributory negligence
- Comparative negligence
- Assumption of risk

Each of these defenses is examined below.

Contributory negligence

Contributory negligence exists when a plaintiff is found to have contributed to her or his own injury. Contributory negligence has been defined as "conduct on the part of the plaintiff which falls below the standard to which he should conform for his own protection, and which is a legally contributing cause co-operating with the negligence of the defendant in bringing about the plaintiff's harm."[14] Interestingly, in the states that recognize contributory negligence as a valid defense (and the number of these states continues to shrink), any level of contributory negligence, whether large or small, is enough to prevent a plaintiff from recovering any damages from the defendant. To illustrate the defense of contributory negligence, let's use the following example.

Ralph's Airpark is a small, private, public use airport in a rural, wooded area. Ralph has been making improvements to the airport's taxiway and ramp, using his backhoe. Toward the end of the day, as night approaches, Ralph shuts down the backhoe and leaves it at the intersection of a taxiway and ramp. He does nothing to mark the presence of the backhoe, believing that it is unlikely that anyone will be using the airport that evening. A bit after nightfall, Anna, an experienced airline transport pilot (ATP) with extensive knowledge of the airport and its ramps and taxiways, lands at Ralph's Airpark and immediately turns off on the same taxiway where Ralph's backhoe was left standing. Anna is so comfortable with the airport layout that she just follows the lights of the nearby houses and tie-down area and decides to forgo the use of her taxi lights. Anna runs into the darkness-hidden backhoe, causing substantial damage to her aircraft. Anna sues Ralph for negligence in leaving the backhoe at the intersection of the ramp and taxiway with no marking. If the court finds that Anna was negligent in failing to utilize her taxi lights, she will not be able to recover damages against Ralph for negligence because her negligence contributed to the accident.

Comparative negligence

Because of the "all or nothing" harshness that comes with a contributory negligence approach, many states have opted for a *comparative negligence* defense. This approach is sometimes referred to as *comparative fault* or *comparative responsibility* depending on the state in which it is adopted.

States that have adopted a *pure* comparative negligence approach allocate damages according to a determination of the percentage of blame or fault of each party. In Anna's case, if the court determined that Ralph was 60 percent negligent and Anna was 40 percent negligent, the court would award Anna 60 percent of the damages Anna sustained.

However, in a state that takes a *modified* comparative negligence approach, the plaintiff can recover based on a comparison of fault. However, if it is shown that the plaintiff was "more" negligent than the defendant, the plaintiff cannot recover. To apply this to the example above, if the court finds that Anna was 60 percent negligent for failing to turn on her taxi lights, she will not recover any damages from Ralph.

Assumption of risk

Restatement § 496A indicates that person who knowingly and voluntarily assumes the risk of the negligent or reckless harm from another person may be barred from later making a claim of damages due to a defendant's negligent conduct. Typically, a potential plaintiff might agree to such an assumption of risk in a contract. These types of agreements are explored in greater detail in Chap. 6.

Not too long ago, the law also used to recognize a theory of implied assumption of risk. Under this theory, a person could impliedly assume the risk of harm by engaging in certain conduct. For instance, a person might be deemed to have voluntarily assumed the risk of injury or harm by taking a sightseeing flight in a hot air balloon. However, in recent years, this theory of implied assumption of risk has been abandoned.

WRONGFUL DEATH

You will often hear of wrongful death actions in the context of aviation accidents. A wrongful death action gives close relatives of a deceased the legal authority to sue any person who may have wrongfully caused the death. Under old common law, someone could be held liable for intentionally or negligently causing injury to another. However, someone could escape tort liability if he or she killed another person. To remedy this anomaly, most states have passed wrongful death statutes. These statutes allow for a cause of action when a wrongdoer commits a tort (intentionally or negligently) that causes another person's death.

STRICT PRODUCT LIABILITY

The theory of damage recovery based on strict product liability has had a profound impact on the aviation industry. A review of the basics follows.

Elements

Restatement § 402A lays out strict product liability theory. Section 402A states as follows:

1. One who sells any product in a defective condition unreasonably dangerous to the user or consumer or to his property is subject to liability for physical harm thereby caused to the ultimate user or consumer, or to his property, if (a) the seller is engaged in the business of selling such a product, and (b) it is expected to and does reach the user or consumer without substantial change in the condition in which it is sold.
2. The rule stated in Subsection (1) applies although (a) the seller has exercised all possible care in the preparation and sale of his product, and (b) the user or consumer has not bought the product from or entered into any contractual relation with the seller.

Notice that the liability attaches to a seller or manufacturer even though the seller or manufacturer has exercised all possible care in the preparation and sale of the product and even though the seller may not have any relationship with the consumer who ultimately purchases the goods. As indicated, the term *strict* in *strict product liability* is very fitting—even though a manufacturer makes and sells goods with all due care, that manufacturer can be found strictly liable in tort if the product it sold contains a fault or defect that harms a buyer. Therefore, a plaintiff need not show negligence on the part of the manufacturer. The mere fact that the product was defective and caused injury is sufficient to allow for a claim.

The doctrine of strict liability demonstrates how the law can evolve with the times. Up until modern times, the familiar caveat was "let the buyer beware." In contrast, many have argued that with the emergence of the strict liability doctrine, the new caveat is "let the seller beware."

The philosophy behind this change appears to reflect a desire to shift the burden of injuries from products to the makers or manufacturers of the products. For better or for worse, the thinking is that a manufacturer is in the best position to ensure the safety of products and to absorb the costs of harm to consumers. Therefore, the manufacturer should be ultimately responsible for harm caused by its products.

To better understand the law of strict product liability, we can break down *Restatement* § 402A into four elements. Each of these elements must be established to successfully pursue a product liability claim.

1. *Defendant must be a merchant.* The term merchant as used in this context means someone who regularly deals in goods of the kind in question. Traditionally, this requirement for strict liability has applied only to sellers of newly manufactured goods. However, in some states, the concept has expanded to include merchants selling used goods.
2. *Product is defective at the time it is sold by the defendant.* This element requires only that the plaintiff prove that the product was defective. There is no need to establish why the product is defective. The defect could have been caused during manufacturing or during design—it makes no difference in the ultimate determination of liability. However, it is important that the defect be present at the time the product was sold by defendant.
3. *Product is unreasonably dangerous.* As a general rule, a product is deemed to be unreasonably dangerous if it causes a danger that is beyond what an ordinary purchaser would contemplate.
4. *Plaintiff suffers personal injury or property damage caused by the defective property.* The plaintiff's injuries or property damage must result from the use of the defective property. It must also be established that the plaintiff's injuries were proximately caused by the defective product.

Defenses to strict product liability

There is no question that it is very difficult for a defendant to establish a defense that will create an obstacle to recovery for a plaintiff in a strict product liability case. The following discussion outlines some possible obstacles that a defendant may place before a plaintiff in a strict product liability case.

Plaintiff misconduct

The first line of defenses that might be raised relate to the plaintiff's conduct. These defenses are as follows:

- Product misuse
- Assumption of risk
- Contributory negligence
- Comparative negligence

Product misuse Misuse of a product is a defense available to a defendant if it is established that the plaintiff knew or should have known that she or he was using the product for an inappropriate purpose. For instance, it would clearly be a misuse of an aircraft to tow heavy equipment on the ground. If the aircraft caused property damage because it tipped forward or backward during such an operation, the manufacturer could raise the defense that the product was being misused by the plaintiff—an aircraft is designed and manufactured to transport persons or property in flight, not to tow heavy equipment on the ground.

Assumption of risk The law attempts to protect a manufacturer when a user voluntarily makes use of the goods in an inappropriate manner. To establish this defense, the defendant must show all three of the following:

- The plaintiff knew and appreciated the risk created by his or her inappropriate use of the product (this is a distinguishing feature from product misuse in which the plaintiff might know that he or she is misusing the product but does not appreciate the risk).
- The plaintiff voluntarily engaged in the risky use of the product, knowing the risk.
- The plaintiff's actions/decisions were unreasonable based on the facts and circumstances.

An example of assumption of risk would be the use of normal category aircraft for aerobatic or spin training with knowledge (and notice from the manufacturer) that the aircraft was not certificated for such use.

Contributory negligence As discussed above, contributory negligence is a defense against a plaintiff in a case of ordinary negligence. The defense may be established if the plaintiff's negligent conduct causes, in whole, or in part, his or her injuries. While this defense may be used in certain jurisdictions to defend against ordinary negligence, it cannot be used in the case of strict product liability.

Comparative negligence Comparative negligence is an available defense in a strict liability case. The comparative negligence defense works in precisely the same manner as it would for a negligence case: A comparison of plaintiff and defendant fault is measured by the court, and the percentage of fault attributed to each party sets the amount of damages to be allocated in the case. Pure comparative negligence permits recovery up to the amount of fault assessed against the defendant—however small that percentage of culpability may be. On the other hand, modified comparative negligence permits a successful defense only if the defendant can prove that the plaintiff is responsible for 50 percent or more of the damages incurred.

Notices and disclaimers
An action in strict liability in tort is not subject to the laws of contract.[15] Therefore, notices or disclaimers regarding the potential hazards involved in a product will not always absolve the maker from liability. Nonetheless, some courts will permit notices and/or disclaimers as a defense if the parties involved have roughly equal bargaining powers.

Privity
Typically, to establish privity, a person would be able to pursue an action based on product liability only if he or she used or purchased a product. However, under the doctrine of strict product liability, even an injured third-party bystander can recover.

Therefore, if a bystander is harmed when an aircraft makes a forced landing due to a defective fuel pump, the bystander has the ability to successfully sue the aircraft manufacturer and/or fuel pump manufacturer. The only limit to this "bystander" liability is foreseeability.

There is also no need for a connection between the buyer and seller of a defective product. Again, common law generally required that before you could sue someone because of damages caused by a defective product, you had to have purchased the defective goods from the seller. With strict product liability, the injured party can sue any seller whether the seller is a retailer, wholesaler, or manufacturer. The maker or seller of a defective part can also be held liable as long as the part was incorporated into the whole without alteration.

For example, if an altimeter manufacturer sells a defective altimeter to an aircraft manufacturer, a pilot injured in a crash due to the defective altimeter will have a right to sue both the altimeter manufacturer and the aircraft manufacturer.

Statute of repose
One of the problems faced by manufacturers is the seemingly infinite period of time that they might be exposed to liability for a product. For instance, a manufacturer of an aircraft propeller may have manufactured it to state-of-the-art standards 30 years ago. However, over many years, the propeller may prove defective. Just how long should a manufacturer be on the hook for strict product liability? In certain instances, states have legislated limitations on the period of time for which a manufacturer can be held liable for a defective product. The statute of repose in place for aviation manufacturers is discussed more fully below in conjunction with Case 4-2.

Limits on punitive damages
Punitive damages are damages that are meant to punish the wrongdoer in a tort lawsuit. In recent years, many states have attempted to limit punitive damage awards. To do this, states have used several of the following approaches:

- Placing caps on damage awards
- Specifying ratios that apportion compensation damages with punitive damages (this avoids a finding of very small compensatory damages with very large punitive damages)
- Permitting states to take a portion or a "cut" of punitive damages (this dissuades plaintiffs from seeking large punitive awards)
- Requiring separate hearings to establish liability and then to determine damages
- Increasing the burden of proof from preponderance of the evidence (more likely than not) to "clear and convincing" evidence

All these methods are being employed in the spirit of what is being touted in most states as *tort reform*. The reforms have been deemed necessary by some in industry to reduce high liability insurance and malpractice insurance premiums. Of course, the aviation industry is not immune to these pressures. See the cases below for a discussion of this issue.

CASES AND COMMENTARY

Case 4-1 tackles the issue of strict liability. As you read the case, think about the definition of strict product liability.

CASE 4-1
BROCKELSBY V. UNITED STATES OF AMERICA AND JEPPESEN
767 F.2d 1288 (1985)

OPINION BY: BEEZER

Barbara Brocklesby, Nancy Evans, Frank Evans, Katherine Evans, Norma Chapman, and World Airways, Inc. (collectively "the plaintiffs") brought this action against the United States of America ("the Government") and Jeppesen and Company ("Jeppesen"), seeking damages for wrongful deaths and property damage arising from an airplane crash near Cold Bay, Alaska. In a bifurcated trial, the jury returned a verdict against Jeppesen. Before the district court disposed of claims against the Government, the Government settled with the plaintiffs. An order dismissing the Government was entered. Jeppesen appeals from the jury verdict and seeks indemnification from the Government. We affirm.

I. BACKGROUND

On September 8, 1973, an aircraft owned by World Airways crashed into a mountain near Cold Bay, Alaska, killing all six crew members and destroying the aircraft and its contents. World Airways and the survivors of the deceased crew members brought separate actions against the Government and Jeppesen, alleging, inter alia, that the accident was caused by defects in an instrument approach procedure developed by the Government and published by Jeppesen. The actions were consolidated for trial.

The Federal Aviation Administration ("FAA") designs and publishes standard instrument approach procedures. *See* 14 C.F.R. Pt. 97. The FAA's instrument approach procedures are essentially compilations of data, which are set forth in tabular form. Jeppesen uses the FAA's data to portray the instrument approach procedures in graphic form. One of Jeppesen's charts depicts the instrument approach procedure to the Cold Bay, Alaska airport. On the flight at issue in this case, World Airways' pilot was using that chart.

The...district court entered a final judgment for $12,785,580.81 in favor of the plaintiffs and against Jeppesen.

II. ANALYSIS
[B]. *Liability*
The case was submitted to the jury on three theories of liability: strict liability, breach of warranty, and negligence. The jury returned a general verdict against Jeppesen. Accordingly, the judgment must be reversed if any of the three theories is legally defective. [Citations]. Jeppesen argues (1) that this case does not involve a defective "product" for purposes of strict liability, (2) that the defects in the Government's instrument approach procedure will not support a finding of strict liability against Jeppesen, (3) that the evidence produced at trial does not establish negligence, and (4) that various policy considerations bar any tort liability in this case.

1. *Strict Products Liability*
a. *The Existence of a "Product"*
Jeppesen argues that the Government's instrument approach procedure is not a "product" within the meaning of *Restatement (Second) of Torts* § 402A (1965). While that may be true, it misses the point. The issue is whether Jeppesen's chart is a product, not whether the instrument approach procedure is a product. In [citation], we analyzed the problem as follows:

Jeppesen approach charts depict graphically the instrument approach procedure for the particular airport as that procedure has been promulgated by the Federal Aviation Administration (FAA) after testing and administrative approval. The procedure includes all pertinent aspects of the approach such as directional heading, distances, minimum altitudes, turns, radio frequencies and procedures to be followed if an approach is missed. The specifications prescribed are set forth by the FAA in tabular form. Jeppesen acquires this FAA form and portrays the information therein on a graphic approach chart. This is Jeppesen's "product."

[Citation.] ("Though a 'product' may not include mere provision of architectural design plans or any similar form of data supplied under individually-tailored service arrangements,...the mass production and marketing of these charts requires Jeppesen to bear the costs of accidents that are proximately caused by defects in the charts") (applying Colorado law). We agree with the plaintiffs' position that Jeppesen's chart was a defective product for purposes of analysis under section 402A.

b. *Liability for Defects in the Government's Procedure*
This case is distinguishable from [citations.] because the Jeppesen chart involved in this case accurately portrayed the instrument approach procedure provided by the Government. All of the defects in the Jeppesen chart stem from the Government's alleged failure to establish a safe instrument approach procedure. The district court instructed the jury as follows:

> Now, the fact that Jeppesen's instrument approach charts are based on information which is supplied by the F.A.A. and which Jeppesen cannot deviate from or modify does not relieve Jeppesen from liability if a defect in the Jeppesen chart was a proximate cause of the accident. The manufacturer of a product is strictly liable for defects in that product even though the defect can be traced to a component part supplied by another. Thus if you find that Jeppesen's instrument approach chart is defective and that the defect was a proximate cause of the accident, you must find Jeppesen liable even if the defect exists only because you find that the F.A.A. designed an approach procedure that you find is itself defective. The issue is: was the Jeppesen chart defective?

The validity of that instruction presents a case of first impression.

Jeppesen contends that strict liability is inappropriate because the Government's procedure was beyond its control. In essence, Jeppesen is arguing that it should not be held strictly liable because it is without fault in this case. It is beyond doubt that the concept of fault continues to have a role in strict liability law. [Citation.] Nevertheless, Jeppesen's position is untenable.

Initially, we note that Jeppesen had at least some ability to prevent injuries to users of its charts. Jeppesen's production specifications manual required its employees to research any procedure thoroughly "to determine its validity and completeness." Jeppesen's president, Wayne Rosenkrans, testified that this requirement applied to the construction and composition of charts such as the one involved in this case. Jeppesen's manual requires its employees to contact official sources to resolve apparent discrepancies in information. Jeppesen maintained an office in Washington, D.C. with liaison to the FAA. Rosenkrans testified that Jeppesen's actions had led to changes in Government procedures. Accordingly, Jeppesen had both the ability to detect an error and a mechanism for seeking corrections. Under these circumstances, we reject Jeppesen's argument that the Government's procedure was completely beyond Jeppesen's control.

More fundamentally, however, existing products liability law is contrary to Jeppesen's position. Assuming that the Government's instrument approach procedure was defective, the literal requirements of section 402A are met. Jeppesen's chart was a "product in a defective condition unreasonably dangerous to the user" within the meaning of section 402A(1). Section 402A(2)(a) provides that strict liability is appropriate even though "the seller has exercised all possible care in the preparation and sale of his product." A seller is strictly liable for injuries caused by a defective product even though the defect originated from a component part manufactured by another party. [Citations.] Accordingly, the appropriate focus of inquiry is not whether Jeppesen caused the product to be defective, but whether the product was in fact defective. [Citation.]

The concept of fault has persisted in substantive products liability law primarily in determining whether a defective product is "unreasonably dangerous." [Citation.] We note that the California courts have sought to purge the concept of fault from strict liability law by focusing on the existence of a defect, rather than inquiring whether the product is unreasonably dangerous. [Citations.] To the extent that Jeppesen's product was defective, the case law does not support an attempt to relieve Jeppesen of liability on the ground that the defect was not Jeppesen's fault. Comment *c* to section 402A of the *Restatement (Second) of Torts* states the policies underlying strict products liability as follows:

> The justification for strict liability has been said to be that the seller, by marketing his product for use and consumption, has undertaken and assumed a special responsibility toward any member of the consuming public who may be injured by it; that the public has the right to and does expect, in the case of products which it needs and for which it is forced to rely upon the seller, that reputable sellers will stand behind their goods; that public policy demands that the burden of accidental injuries caused by products intended for consumption be placed upon those who market them, and be treated as a cost of production against which liability insurance can be obtained; and that the consumer of such products is entitled to the maximum of protection at the hands of someone, and the proper person to afford it are those who market the product.

[Citation.] All of those policies apply to the plaintiffs' claim in this case. To the extent that Jeppesen is held liable for the Government's misfeasance, Jeppesen had a right to seek tort law indemnification. The fact that Jeppesen failed to seek tort law indemnification has no bearing on the plaintiffs' substantive rights.

c. *Sufficiency of the Evidence*
Jeppesen contends that the evidence presented at trial did not prove that the chart was defective. The plaintiffs produced evidence regarding several alleged defects. One of the alleged defects was the absence of a transition arc. Jeppesen's

only response to that allegation is that it had no power to alter the Government's procedure. As noted above, that is not a valid defense. Accordingly, we conclude that the plaintiffs introduced sufficient evidence to support a finding that the chart was defective.

2. *Negligence*

Jeppesen contends that it cannot be held liable for negligence on the facts presented at trial. The district court instructed the jury as follows:

Now, one who supplies a product, either directly or through a third person, for another to use, which the supplier knows or has reason to know is dangerous or is likely to be dangerous for the use for which it is supplied, has a duty to use reasonable care to give warning of the dangerous condition of the product or of facts which make it likely to be dangerous to those whom he should expect to use the product or be endangered by its probable use if the supplier has reason to believe that they will not realize its dangerous condition. A failure to fulfill that duty is negligence. This rule applies to a producer of a product.

The producer of a product that is reasonably certain to be dangerous if negligently made, has a duty to exercise reasonable care in the design, production, testing and inspection of the product and in the testing and inspection of any component parts made by another so that the product may be safely used in a manner and for a purpose for which it was made.

A failure to fulfill that duty is negligence.

The district court's instructions correctly stated the law. In light of Jeppesen's internal operating procedures, which required testing for completeness, and its failure to detect the defects in the instrument approach procedure, the jury could reasonably have found that Jeppesen was negligent in failing to warn users of the latent defects in the chart.

Jeppesen argues that it had no ability to change the procedure and that it had no duty to control the actions of the Government. Both arguments miss the point. The negligence theory was submitted to the jury on a "duty to test/duty to warn" theory, rather than on a "duty to control the Government" theory. Jeppesen can be held liable for negligently failing to detect the defect in the product that it marketed. If it had discovered the defect, Jeppesen would have been required either to warn the users of the chart or to refrain from selling the product.

3. *Public Policy Defenses*

Jeppesen argues that it is unfair to hold a chart manufacturer strictly liable for accurately republishing a government regulation. We agree. If, for example, a trade journal had accurately published the government's instrument approach procedure in text form and a pilot had used the procedure as printed in the journal, the journal would be immune from strict liability. This case, however, does not present that situation. Jeppesen's charts are more than just a republication of the text of the government's procedures. Jeppesen converts a government procedure from text into graphic form and represents that the chart contains all necessary information. For example, Jeppesen's catalog contains the following statements:

> When pilots compare approach plates...for information, for readability...they choose Jeppesen. Why? Because the format of Jeppesen charts was designed by pilots, for pilots, and has been time-tested and proven by instrument pilots throughout the world. Every necessary detail is clearly indicated.... Jeppesen approach plates include...EVERYTHING you need for a smooth transition from en route to approach segment of your flight.

It is true that the government's procedures are significant components of Jeppesen's charts. It is apparent, however, that Jeppesen's charts are more than a mere republication of the government's procedures. Indeed, Jeppesen's charts are distinct products. As the manufacturer and marketer of those products, Jeppesen assumed the responsibility for insuring that the charts are not unreasonably dangerous in their intended use.

The judgment of the district court is AFFIRMED.

Do you think the court properly applied strict product liability in this case? Do you think the results are just? Is strict liability an appropriate legal tool to spread the risk of defective products? What suggestions would you make to improve the tort system in the United States?

In response to calls for tort reform, in 1994 the United States passed the General Aviation Revitalization Act (otherwise known as GARA). GARA established a statute of repose that barred all lawsuits against aircraft manufacturers that involved any general aviation aircraft or component part that is more than 18 years old. Case 4-2 was one of the first cases in which a manufacturer invoked the GARA statute of repose.

CASE 4-2
ALTSEIMER V. BELL
919 F. Supp. 340 (1996)

OPINION BY: WILLIAM B. SHUBB

Before the court is defendant Bell Helicopter Textron's motion for summary judgment on all the claims in plaintiffs' complaint.

BACKGROUND

On May 23, 1995, plaintiffs John Altseimer, Horizon Helicopters, Dennis Westerberg, and Sloane Westerberg filed this action for personal injuries, property damage, and economic losses allegedly arising out of a helicopter accident. Bell Helicopter Textron Inc. ("Bell") is the only named defendant. The complaint alleges that Bell designed, manufactured, assembled, tested, fabricated, produced, sold, or otherwise placed in the stream of commerce a defective helicopter and a defective 42 degree gearbox, one of the component parts of the helicopter. The complaint further alleges that Bell failed to provide proper warnings with respect to the negligent and defective design of the helicopter and the 42 degree gearbox.

Bell argues that it is entitled to summary judgment on the grounds that (1) the General Aviation Revitalization Act prohibits lawsuits against aircraft manufacturers arising out of the crash of an aircraft more that 18 years old, and (2) the destruction and rebuild of the "Bell" helicopter, on at least two occasions by unrelated entities, terminated any liability of Bell as manufacturer of the accident aircraft.

STANDARD OF REVIEW

Summary judgment is appropriate if the record, read in the light most favorable to the non-moving party, demonstrates no genuine issue of material fact. [Citation.] Material facts are those necessary to the proof or defense of a claim, and are determined by reference to the substantive law. [Citation.] At the summary judgment stage the question before the court is whether there are genuine issues for trial. The court does not weigh evidence or assess credibility.

DISCUSSION

The General Aviation Revitalization Act of 1994, Pub.L. 103-298, 108 Stat. 1552 (49 U.S.C. § 40101 Note) (1994) ("GARA") is a statute of repose which prohibits all lawsuits against aircraft manufacturers arising out of accidents involving any general aviation aircraft or component part that is more than 18 years old. Section 2(a) of GARA, which sets forth the legislation's basic limitation on civil actions, provides:

(a) Except as provided in subsection (b) no civil action for damages for death or injury to persons or damage to property arising out of an accident involving a general aviation aircraft may be brought against the manufacturer of the aircraft or the manufacturer of any new component, system, subassembly, or other part of the aircraft, in its capacity as a manufacturer if the accident occurred —

(1) After the applicable limitation period beginning on —

    (A) The date of delivery of the aircraft to its first purchaser or lessee, if delivered directly from the manufacturer; or

    (B) The date of first delivery of the aircraft to a person engaged in the business of selling or leasing such aircraft; or

(2) With respect to any new component, system, subassembly, or other part which replaced another component, system, subassembly, or other part originally in, or which was added to, the aircraft, and which is alleged to have caused such death, injury, or damage, after the applicable limitation period beginning on the date of completion of the replacement or addition.

49 U.S.C. § 40101, Note Section 2(a)(1)-(2). Section 3 defines the "limitation period" as 18 years, and section 2(d) provides that GARA supersedes any State law which permits civil actions such as those described in subsection (a) brought after the applicable 18 year limitation period.

Bell has provided undisputed evidence that the helicopter and 42 degree gearbox in question were more than 18 years old at the time of the crash. [Reference to record.] Bell has also produced undisputed evidence that the pinion gear, a component of the gear box, and purportedly the cause of the crash, was more than 18 years old at the time of the crash. [Reference to record.]. Therefore, GARA effectively preempts plaintiffs' action. Although harsh, such a result is consistent with the purpose of GARA to: establish a Federal statute of repose to protect general aviation manufacturers from long-term liability in those instances where a particular aircraft has been in operation for a considerable number of years. A statute of repose is a legal recognition that, after an extended period of time, a product has demonstrated its safety and quality, and that it is not reasonable to hold a manufacturer legally responsible for an accident or injury occurring after that much time has elapsed. [Citation.]

Plaintiffs argue that GARA should not be applied to this action because their claims accrued prior to the enactment of GARA. This contention is without merit. GARA was enacted on August 17, 1994, and plaintiffs did not file their complaint until May 23, 1995. Section 4 of GARA expressly states that "this act shall not apply with respect to civil actions commenced before the date of the enactment of this act." Both Federal and state law provide that a civil action "is commenced by filing a complaint with the court." [Citation.]. Since, plaintiffs' complaint was filed after the enactment of GARA, their claims are unambiguously subjected to GARA's preemptive provisions. Accordingly, Bell is entitled to summary judgment.

IT IS THEREFORE ORDERED that Bell's motion for summary judgment be, and the same is, hereby GRANTED.

Notice that in this case, the accident occurred before GARA became law. However, the court noted that the law itself expressly indicates that any claims filed prior to the enactment of GARA on August 17, 1994, could not be prohibited by GARA. The court also noted that Section 4 of GARA clearly indicated that GARA was applicable to any claims filed after August 17, 1994, and this action was filed on May 23, 1995.

In more recent years, GARA has been challenged on constitutional grounds. In one recent case,[16] a plaintiff filed a lawsuit against Cessna Aircraft Company, Teledyne Continental Motors (engine manufacturer), and Lear Romec (fuel pump manufacturer). The lawsuit arose out of a 1995 fatal accident involving a Cessna 421B aircraft in which the plaintiff's husband died. The plaintiff alleged various defects including fuel pump failure and inadequate single-engine performance. However, the aircraft in question was 22 years old, and the fuel pump was 26 years old. Nonetheless, the plaintiff argued that GARA was unconstitutional, citing the following four arguments:

1. GARA is not a constitutional exercise of Congress' authority under the commerce clause of the U.S. Constitution.
2. GARA offends the Tenth Amendment by foreclosing a plaintiff's ability to pursue a case under state law.
3. The equal protection clause of the U.S. Constitution is violated by GARA because it would deprive her of a case against a manufacturer but would not deprive a claimant when commercial aircraft were involved.
4. GARA violates the due process clause of the U.S. Constitution by depriving the plaintiff of her claim against a general aviation aircraft or part manufacturer.

The court ultimately rejected all four of the arguments above. Do you agree with the court's rejection of the plaintiff's constitutional challenges to GARA? Review the basics of constitutional law as you consider the arguments.

GARA has now been in place since 1994. Do you think this type of statute of repose is good for the aviation industry? Why or why not? If you had the authority, would you repeal GARA, leave it as is, or strengthen the statute of repose?

DISCUSSION CASES

1. The ace of the Base Flight School, nicknamed "Ace," purchases five Acme fire detectors to place in its flight simulator area. The detectors are designed to be powered by alternating current (ac). Four months after purchase of the fire detectors, a fire breaks out in the flight simulator training area. All four of Ace's full-motion simulators are destroyed, and Ace has no insurance for the equipment. The fire detectors did not send an alarm signal when the fire began because the fire was started by a short circuit in one of the flight simulators. The fire detectors were connected to the circuit that had shorted out and that cut off the ac power. Ace sued Acme on the theory that Acme had a duty to warn Ace that the same electrical malfunction that could start a fire in the flight simulator training area might also disable the fire detectors. Will Ace's theory prevail? Discuss.

2. McGoo is a crop duster who became confused while flying over similar-looking farm fields. She flew over Pete's land and sprayed chemicals that killed Pete's crop. McGoo believed that she was spraying the land she had been hired to spray, which was across the street from Pete's land. Pete sues McGoo for trespass. Is Pete's claim likely to prevail? Why or why not?

3. Nina closes the deal on a new XR2 aircraft. After getting the requisite simulator training at the factory, she does a thorough preflight inspection of her new aircraft and departs for her home airport. Approximately halfway through the flight, Nina felt a burning sensation coming from her right arm. She shook her arm and a spider fell to the ground. Nina made it back to her home base safely, but her arm grew redder and the pain got worse. Nina was eventually incapacitated, and it took her 40 days in the hospital to recover. According to Nina's physician, her injury was caused by the bite of a brown recluse spider. Nina seeks to sue the XR2 manufacturer on a theory of negligence and/or strict product liability. Will she succeed in her lawsuit? Why or why not?

4. Late in the evening, a Beech Baron 58P aircraft piloted by Paul crashed shortly after takeoff from Manassas Airport in Virginia, killing Paul, the sole occupant of the aircraft. The Beech Baron 58P is a six-passenger, twin-engine aircraft designed and manufactured by Beech Aircraft. Paul purchased the aircraft from its original owner 1 year earlier. On the evening of the fatal crash, the aircraft was on its first flight following the repair and reinstallation of the actuator trim tabs by a mechanic at Manassas Airport. National Transportation Safety Board (NTSB) investigators determined that the trim tab actuators (an essential component of the aircraft's flight control systems) had been reversed, thus causing the crash. Paul's widow files a lawsuit against Beech Aircraft on a theory of wrongful death. The lawsuit is based, in part, on grounds of negligent design. Evidence during the trial of the case established that the mechanic who did the work on the trim tab actuators failed to follow FAA-mandated compliance with Beech manuals and his training. Paul's widow argued that the trim tab actuators could and should have been designed to prevent the possibility that they could be reversed. Who will prevail in this case? Why?

5. Stan was injured while attempting to take off from the Middle Town Airport in a Piper Super Cub model PA-18-150. The plane was towing a glider that was attached by a tow rope to the tail of the Super Cub. Stan is a cinematographer, and he was planning to film the glider's flight for a television commercial. Prior to the flight, Stan had an FAA-certified mechanic remove the front seat and install a camera that would be used to film the commercial. Stan planned to pilot the Super Cub from the rear seat (which was approved by the Piper Cub manual). A few days before the commercial was to be filmed, the owner of Middle Town Airport learned of the plans for the flight. He closed the airport specifically to prevent the flight. When the airport owner noticed activity at the airport, he parked in van in the runway to prevent takeoffs and landings. A bit later, Stan attempted to take off, and he struck the airport owner's van. Stan's head struck the camera mounted at the front of the aircraft, and he suffered serious head and brain injuries. Stan's family brought an action against Piper Aircraft Company, the manufacturer of the Super Cub. The family argued that Piper negligently designed the aircraft without adequate forward vision from the rear seat of the aircraft and negligently failed

to provide for a rear-seat shoulder harness. What arguments would you make if you were representing Piper? Will those arguments prevail?

6. Dean owns an aircraft paint shop. Dean sends one of his employees, Ricardo, into an industrial tank to clean and inspect the tank. Ricardo (with Dean's knowledge) does not wear the protective gear required by regulation, and he is overcome by the gas. He becomes incoherent and hallucinates. Ricardo is transported to a nearby emergency room where Paula, a nurse, is assigned to his case. While in a state of delirium, Dean strikes Paula, causing her serious injury. Paula sues Dean, charging him with negligence in failing to properly ensure that Ricardo followed required safety procedures when he entered the tank. Will Paula recover? Explain.

7. Collinge takes out a loan on an aircraft from Barracuda Bank. Collinge defaults on the loan, and Barracuda has been unsuccessfully pursuing Collinge for the money owed. Barracuda makes a bogus call to Collinge's mother, Stephanie, telling Stephanie that her grandson (Collinge's son) had been injured and that they needed Collinge's address. When it was learned that the call was falsely made, Stephanie sued Barracuda on a claim of intentional infliction of emotional distress. Will Stephanie prevail? Why or why not?

8. Landy was the pilot in command and sole occupant of a Cessna 210 on a return trip from a business meeting. Landy was a bit tired, and his approach and landing were made in bad weather with lots of rain and high winds. While focusing on weather and communications, Landy did not fasten his shoulder harness during the approach to landing. The aircraft landed normally, but during the landing rollout one of the landing gears experienced a mechanical failure and was sheared off the aircraft. Landy was tossed violently to one side of the aircraft, and he suffered serious injuries. Landy sues Cessna because of the defective landing gear. Will Landy succeed? Discuss the issues presented.

9. National Aircraft Finance Bank sent two of its employees to repossess Charlie's aircraft owing to Charlie's failure to repay his aircraft loan. The two repossessors had a tow bar attached to a tug, and their plan was to tow the aircraft to a secured hangar on the airport. When they arrived at the airport, they spotted the aircraft arriving. After Charlie exited the airplane, the repossessors approached the aircraft with the tug and tow bar. Charlie was concerned and got back into his aircraft. The repossessors requested that Charlie exit the aircraft. However, Charlie refused to leave the aircraft. The repossessors continued their work and towed the aircraft about 350 feet to a hangar. Charlie finally exited the aircraft, but later sued National Aircraft Finance Bank for false imprisonment. Will Charlie prevail? Why or why not?

ENDNOTES

1. *Restatement (Second) of Torts*, § 46.
2. Id. at §§ 568 and 568A.
3. Id. at § 558.

4. Id. at §1158.
5. Id. at § 218.
6. Id. at § 222A.
7. Id. at § 282.
8. 484 F. Supp. 850 (S.D.N.Y. 1980).
9. *Restatement (Second) of Torts*, § 285.
10. Id. at § 328D.
11. 162 N.E. 99 (N.Y. 1928).
12. *Restatement (Second) of Torts*, § 436.
13. Id. at § 436A.
14. Id. at § 463.
15. See comments to Restatement Section 402 A.
16. *Hinkle v. Cessna Aircraft Co., et al.*, 2004 Mich. App. LEXIS 2894, 2004 WL 2413768 (Mich. App. Oct. 28, 2004).

5
Administrative Agencies and Aviation

ADMINISTRATIVE AGENCIES 121
Department of Transportation 122
Federal Aviation Administration 122
National Transportation Safety Board 123

FUNCTIONS OF ADMINISTRATIVE AGENCIES 123
Rulemaking 124
Enforcement 126
Adjudication 131
Special considerations in FAA matters 133

CHECKS ON ADMINISTRATIVE AGENCIES 138
Oversight by traditional branches of government 138
Equal Access to Justice Act 140
Freedom of Information Act 141
Privacy Act 141
Sunshine Act 142

CASES AND COMMENTARY 142

DISCUSSION CASES 153

ENDNOTES 156

Administrative agencies play an increasingly growing role in the modern world. Administrative law is public law created by government agencies that takes the form of rules, regulations, orders, and opinions. Among all the activities regulated by government, aviation is arguably the most profoundly impacted by agency regulation.

In the aviation environment, the administrative agencies having the most direct impact on aviation professionals and the aviation industry are the Federal Aviation Administration (FAA), Department of Transportation (DOT), and the National Transportation Safety Board (NTSB). Other agencies affecting the legal environment of aviation include the National Mediation Board (NMB), Securities and Exchange Commission (SEC), Federal Trade Commission (FTC), Environmental Protection Agency (EPA), and the Department of Labor (DOL).

The increasingly complex nature of our nation's economic and social order has only served to increase the importance of administrative agencies in the United States. In the 1952 case of *FTC v. Ruberoid Co.*, Justice Jackson observes:

> The rise of administrative bodies probably has been the most significant legal trend of the last century and perhaps more values today are affected by their decisions than by those of all the courts.... They also have begun to have important consequences on personal rights.... They have become a veritable fourth branch of the Government which has deranged our three-branch legal theories as much as the concept of a fourth dimension unsettles our three-dimensional thinking.[1]

Beyond the impact of federal agencies, there are also myriad state agencies that impact our daily lives and aviation in particular. State agencies operate airports, provide for workers' compensation benefit rules, regulate pollution control, and oversee the taxation of income and transfers of property.

If you think about it, hardly a move can be made in the world of aviation that is not impacted by one of the many federal and state agencies that make up the "fourth branch" of government cited by Justice Jackson in the *Ruberoid* case. An airline pilot's every move in the cockpit is regulated in one way or another by the FAA. Her employer's routes and economic approval for certification as an air carrier are approved by the DOT. Every paycheck received by the airline pilot is impacted by the Internal Revenue Service (IRS), state and local taxing authorities, and the Social Security Administration. Safety features installed on her aircraft have likely been the result of recommendations from the NTSB. The relationship between the pilot, her union, and her employer is governed by the rules of the DOL and the National Labor Relations Board (NLRB). Her employer's financial statements and ability to solicit credit and shareholder contributions to equity are carefully regulated by the SEC.

There are numerous reasons cited for the emergence of administrative agencies in the United States. However, two reasons tend to be cited most often. The first is the change in the attitude of government toward regulation of business. In the early years of the U.S. history, a more "hands off" approach toward government regulation existed. The prevailing belief was that commerce would thrive as long as government kept its distance and let the marketplace create its own equilibrium. However, abuses of child labor, the emergence of monopolies, and unchecked environmental damage by industry caused a rethinking of the hands-off approach. Citizens and their elected leaders started to believe that there might be a place for at least limited government intervention to protect workers, the environment, and basic levels of competition.

A second reason that is often cited for the emergence of administrative agencies is the inability of traditional government bodies (executives, courts, and legislatures) to deal with the sophisticated problems spawned by new technology and economic complexity. It is not realistic to expect that ordinary legislators and courts can deal with specialized problems in constantly evolving industries such as aviation, telecommunications, and taxation. To regulate these types of activities, traditional government bodies needed to turn to specialized agencies that employed skilled personnel with the ability to understand and develop regulations for targeted industries and activities.

It is important to note that some of the agencies that will be reviewed are part of the executive branch and others are independent of the executive branch. The DOT and FAA are part of the executive branch. The President appoints the head of these agencies and has direct control over the way the agencies are operated. On the other hand, the NTSB is an independent agency. Although the President nominates members of the NTSB, the board members do not answer to the President.

This chapter will focus on the ways that administrative agencies impact the world of aviation. First we will focus on how administrative agencies are created and function. Later we will take an overview of some of the protections built into our legal system to ensure that administrative agencies function as intended. Finally, we will take a look at some aviation-related cases and materials related to administrative law.

ADMINISTRATIVE AGENCIES

As discussed above, one of the primary functions of administrative agencies is to relieve legislators of the seemingly impossible burden of developing laws that will address all the necessary details of particular industries or issues. For example, this allows legislators to pass a law that promotes safety in air transportation without having to address the necessary steps to achieving such safety. Instead, lawmakers can put in place an enabling statute that creates an administrative agency (e.g., the FAA) that will be responsible for tending to the details and promulgating necessary rules for aviation safety. Of course, this also means that the lawmakers must delegate the authority to enact and enforce any necessary rules.

The separate agencies created by an enabling statute can then be staffed by persons with specialized skills and expertise in the industry or environment to be regulated. With this in-house specialization, the agency can devote the necessary time and resources to keep abreast of changes in its field of regulation.

The three agencies that impact aviation activities most directly are the DOT, FAA, and NTSB. Let's turn to a brief overview of the history and laws creating each of these agencies, starting with the DOT.

Department of Transportation

The Department of Transportation was formed when President Lyndon Johnson signed the DOT Act on October 15, 1966. The purpose of the DOT is articulated in 49 U.S.C. §§ 101(a) and (b). That law states that "... development of transportation policies and programs ..." is necessary to meet national "... objectives of general welfare, economic growth and stability, and security of the United States...." The law further specifies that the policies and programs of the DOT should contribute to "... providing fast, safe, efficient, and convenient transportation at the lowest cost consistent with those and other national objectives, including the efficient use and conservation of the resources of the United States."

The head of the DOT is the Secretary of Transportation. The secretary is a member of the President's Cabinet and reports directly to the President. The DOT houses several large agencies including the Federal Highway Administration, Federal Motor Carrier Safety Administration, Federal Railway Administration, Federal Transit Administration, Maritime Administration, National Highway Traffic Safety Administration, and the FAA.

An overview of the FAA follows.

Federal Aviation Administration

The official birth of aviation regulation in the United States came on May 20, 1926, with the passage of the Air Commerce Act of 1926. By 1926 the handwriting was on the wall—commercial aviation was likely to become a major force in national commerce. Under the Air Commerce Act of 1926, the oversight of aviation activities through air traffic rules, airman certification, certification of aircraft, and maintenance of air navigation aids was placed within the purview of the Department of Commerce. The branch within the Department of Commerce that was charged with oversight of aviation was named the Aeronautics Branch.

By 1934, commercial aviation was beginning to play a significant role in the economy of the nation. To reflect its elevated status within the Department of Commerce, the Aeronautics Branch was renamed as the Bureau of Air Commerce. This ushered in a new era of air traffic control and a viable air traffic control system.

The Civil Aeronautics Act of 1938 mandated the transfer of civil aviation responsibilities from the Commerce Department to a newly formed independent agency, the Civil Aeronautics Authority. This legislation also opened the door to government regulation of airfares and routes for air carriers.

The Civil Aeronautics Authority was split into two agencies shortly thereafter in 1940. The first offshoot was the Civil Aeronautics Administration (CAA). The CAA was placed in charge of air traffic control, airman and aircraft certification, enforcement of safety rules, and development of a system of airways. The second offshoot of the Civil Aeronautics Administration was the Civil Aeronautics Board (CAB). The CAB was tasked with safety-related rulemaking, investigating accidents, and economic regulation of air carriers. Both the CAA and the CAB were formally agencies within the Department of Commerce. However, the CAB functioned independently of the Secretary of Commerce.

With greater demand for civil air travel and the advent of civilian jetliners, Congress passed the Federal Aviation Act of 1958. The 1958 act transferred all the CAA's functions to a new Federal Aviation Agency. Further, all the safety rulemaking functions of the CAB were transferred to the new Federal Aviation Agency.

The final step in the evolution of the FAA came when, as noted above, the DOT Act was passed in 1966. Once the DOT came into being, the FAA was placed under the umbrella of the DOT. The FAA's name was also changed at that time from the Federal Aviation Agency to the Federal Aviation Administration. At the same time, the CAB's accident investigation function was transferred to the newly formed National Transportation Safety Board (NTSB).

A discussion of the NTSB follows below.

National Transportation Safety Board

In April 1967, the NTSB became a functioning government agency as mandated by the DOT Act in 1966. In its earlier years, the NTSB was funded by the DOT. However, that changed in 1975 with the Independent Safety Board Act. As a result of that law, all NTSB ties with DOT were severed, and the NTSB became a totally independent federal agency with no affiliation to DOT or any of the agencies housed under DOT.

The NTSB's function is to investigate every civil aviation accident in the United States. It is also responsible for investigating significant accidents related to railway, marine, highway, and pipeline transportation. The NTSB (also commonly referred to as the "Board") is required to determine probable cause for accidents and to make safety recommendations based on its study and investigation of accidents.

The NTSB also serves as an adjudicative body for cases involving FAA certificate actions or civil penalty actions against airmen. The Board also hears appeals in actions by the U.S. Coast Guard against mariners.

FUNCTIONS OF ADMINISTRATIVE AGENCIES

Administrative agencies perform three functions:

- Rulemaking
- Enforcement
- Adjudication

Some administrative bodies perform one or two of these functions. However, many perform all three. When you reflect on these functions, you might consider the parallels between these functions and the constitutional functions performed by the various branches of government. Rulemaking is very much like the legislative process performed by Congress. Enforcement is the same sort of process performed by the executive branch of government. Adjudication is essentially the function performed by the judiciary.

Although the basic functions of administrative agencies track the functions of our constitutional structure of government, the big difference is the lack of separation between the functions in administrative agencies. If an administrative agency makes the rules, enforces the rules, and adjudicates disputes related to the rules, how can the process be fair? There is no easy answer to this question. However, in 1946 Congress passed the Administrative Procedure Act (APA) to address issues of constitutionality within the workings of administrative agencies. All administrative agencies are subject to the APA. As a result, there are great similarities in the way that all administrative agencies make rules, enforce rules, and adjudicate disputes. A review of these processes follows.

Rulemaking

The APA outlines the general requirements that an agency must adhere to when it makes or changes rules.[2] A *rule* is defined as "the whole or part of an agency statement of general or particular applicability and future effect designed to implement, interpret, or process law or policy."[3] One of the major thrusts of the APA is the need to put everyone on notice of the new regulations or changes in regulations.[4] This notice requirement is particularly important since rules created or modified apply to everyone in the United States. The rulemaking process may be best understood when it is broken down into three categories of rulemaking recognized by the APA: (1) regulations, (2) interpretations, and (3) procedural rules.

Regulations

Regulations are often called *legislative rules*. This is so because they are designed to implement legislators' more comprehensive laws. Regulations must be issued by an administrative agency that has the appropriate authority delegated to it by legislators. As long as a regulation does not violate the Constitution or exceed the agency's delegation of authority, it will become law.

For pilots, mechanics, and many others engaged in the aviation industry, the FAA's regulations as found in Title 14 of the Code of Federal Regulations (CFR) carry the force of law. Pilots know that they must adhere to the general operating and flight rules established by 14 CFR Part 91 [otherwise known as Federal Aviation Regulation (FAR) Part 91]. Mechanics must perform their duties in compliance with FAR Part 43, which describes the proper procedures for aircraft maintenance, preventive maintenance, rebuilding, and alteration. These are precisely the type of detailed rules that it would be unrealistic to expect a legislative body such as Congress to promulgate. It is the expectation of the legislators in Congress that the FAA will have the expertise and the experience to put together a coherent set of regulations that implement its desire for safe aircraft and aircraft operations. The details of how that gets accomplished through regulation are delegated to the FAA.

How are regulations created? Section 553 of the APA lays out the framework for most of the rulemaking that goes on at agencies such as the FAA and DOT. This process is generally referred to as informal rulemaking. With informal rulemaking, the APA requires the following:

1. Notice of the proposed regulation must be published—usually in the Federal Register (the Federal Register[5] is a daily publication of the federal government);
2. All interested parties must have an opportunity to comment and participate in the proposed regulation.
3. Publication of the final regulation must occur at least 30 days before the regulation takes effect, along with a statement of the regulation's necessity and purpose.

If an agency is required by an enabling statute, it may need to develop regulations through the formal rulemaking process. The formal rulemaking process is complex and parallels the APA provisions regarding adjudications.[6] The agency must conduct a trial-like hearing with a formal record and introduction of evidence. It must also include a statement of findings and conclusions and reasons for the conclusions reached.

In cases in which informal rulemaking is not permitted, but the rigid process of formal rulemaking is not required, a process of hybrid rulemaking may take place. Hybrid rulemaking is a sort of middle ground, and where applicable, this approach is usually outlined in the agencies' enabling statutes. In some instances, a formal hearing might be required (a necessity for formal rulemaking) while cross-examination is not authorized (much as in informal rulemaking).

Interpretations

Many administrative agencies publish interpretations of their regulations and statutory mandates to assist the public in complying with the law. These interpretations are exempt from the APA's notice and comment requirements.[7]

Interpretations are sometimes necessary and useful to assist affected persons who are attempting to comply with regulations that can often become very technical and complex. One very poignant example of where interpretations can be helpful relates to crew member flight time and duty period limitations.[8] These regulations often generate more questions than answers—this creates a real need for additional agency guidance in explaining how the regulations need to be implemented.

Although agency interpretations are not binding on private parties, they are typically accorded substantial weight. In one early case relating to agency interpretations, the Supreme Court ruled: "The weight of such [an interpretation] in a particular case will depend upon the thoroughness evident in its consideration, the validity of its reasoning, its consistency with earlier and later pronouncement, and all those factors which give it power to persuade ..." *Skidmore v. Swift & Co.*[9]

FAA has a well-established inventory of published interpretations. For instance, the FAA publishes detailed information for pilots in its *Aeronautical Information Manual (AIM)*. Maintenance personnel, pilots, and aircraft owners are provided guidance with FAA's series of Advisory Circulars (otherwise known as ACs). FAA also publishes a multitude of handbooks and opinions from its Office of Chief Counsel. Again, these regulatory interpretations might not be binding on the aviation community; however, they are part of such a comprehensive and well-established scheme of pronouncements that they will often be granted recognition and deference by courts.

Procedural rules

When an agency creates or modifies procedural rules, it does not need to comply with the notice and comment requirements of the APA.[10]

Procedural rules will typically address matters such as qualifications to appear before an agency or commission hearing, business hours and notice of proceedings or hearings, and matters of agency organization.

Enforcement

All administrative agencies have the authority to investigate whether certain conduct has violated statutory or regulations under the jurisdiction of the agency. An agency also needs the authority to gather information and then prosecute violations that it may discover in its investigations. In our discussion of enforcement by administrative agencies, we will use a simplified illustration of an FAA enforcement action to demonstrate how administrative agency enforcement works. We will start with the facts underlying the illustration and then work through how an agency gathers information and then prosecutes once it believes that regulations have been violated.

Facts

Captain Davidson was the nonflying, pilot-in-command of an Arctic Airlines Boeing 747 passenger-carrying flight from Tundra, Alaska, to Capitol International Airport in the Washington, DC, area. Because the runway at Tundra was short and the aircraft was close to maximum takeoff weight, Captain Davidson ordered a static takeoff to be performed in which the brakes are not released until all four engines are producing maximum thrust (this was a routine measure with a fully loaded 747 departing Tundra's airport). As a result of severe frozen weather conditions and the high thrust being created by the 747's engines, considerable portions of the runway behind the 747 started to break apart.

The takeoff proceeded uneventfully, and the experienced cockpit crew and Captain Davidson did not notice anything unusual about the aircraft's handling. However, shortly after takeoff, the airport tower advised Captain Davidson and his crew: "You blew up the asphalt layer in the very south end of the runway, and there was debris coming after you." At least three flight attendants saw the asphalt pieces of the runway being blown up around the back of the aircraft. One reported hearing a "loud bang as the asphalt hit the fuselage." The in-flight manager (IFM) onboard the flight notified the cockpit by intercom that there was considerable dust, stones, and other debris generated behind the aircraft during the static takeoff procedure.

The tower then went on to focus on the question of whether there was any tire damage to the aircraft. Captain Davidson indicated that all the tires appeared to be functioning properly and cockpit indications confirmed that all the tires were intact. However, he did request that the tower check the runway for rubber or tire debris. The tower checked the runway and could not find any sign of tire damage. At that point the tower wished the aircraft a good flight, and the flight proceeded safely to its destination. Throughout the flight, the aircraft showed normal readings and performance appeared to be normal.

However, upon arrival at its destination, it was very apparent that the aircraft had sustained serious damage to its tail section. Asphalt chunks that blew up into the aircraft created three holes in the elevator over 10 inches in diameter and numerous large dents and scratches near the tail section of the aircraft. By any standard, the damage incurred rendered the aircraft unairworthy.

The FAA now begins a process of investigating this incident to determine if Captain Davidson or any of his flight crew violated the FARs. The investigation begins with information gathering.

Gathering information

In pursuing an investigation, agencies have typically been afforded substantial authority in their ability to require disclosure of information by subpoena as long as the following requirements are met:

1. The investigation is related to a legitimate agency purpose.
2. The inquiry is relevant to the legitimate agency purpose.
3. Information requested is not already possessed by the agency.
4. The agency has complied with the administrative steps required by the law.[11]

It is important to note that unlike in a criminal case, an agency does not need to establish that it has "probable cause" to pursue information by subpoena power. Once an agency has established an apparently valid purpose for its investigation, it is very difficult for the target of the investigation to establish that the agency's purpose is not legitimate.

Agencies may also conduct on-site inspections and searches when they are investigating matters within their purview. Health inspectors regularly visit restaurant kitchens on unannounced inspections. Occupational Safety and Health Administration (OSHA) personnel routinely stop by construction sites to ensure compliance with safety standards.

Much of the FAA's investigatory clout is exercised through reviews and inspections of aircraft and airmen. The primary tool utilized by the FAA is inspection of certificates, documents, and logbooks. The FARs require that persons "... must present their pilot certificate, medical certificate, logbook, or any other record required by this part for inspection upon a reasonable request by ..." the administrator, the NTSB, or a federal, state, or local law enforcement officer.[12] The FAA also has the power to subpoena documents that are not covered by the regulations.

Interestingly, another technique used most frequently (and quite successfully) to gather information from airmen is the purely voluntary response to the FAA's letter of investigation (commonly referred to as an LOI). This is often an innocent-looking letter that requests information from an airman related to an incident that the FAA is investigating as a potential violation. For the purposes of our illustration, let's assume that the FAA has photographs of the aircraft damage and now wishes to elicit information directly from Captain Davidson. The inspector in charge of the investigation sends the letter depicted in Fig. 5-1.

Let's assume that Captain Davidson receives the LOI four weeks after the flight in question. His first impulse is to draft a full reply, describing his recollection of the flight in detail. He drafts a letter and is prepared to send it when a colleague suggests that he retain a lawyer to review the letter before sending it to the FAA. Captain Davidson finds a lawyer with some experience in FAA matters and arranges a meeting. The lawyer reviews the situation with him, and by the end of their meeting it is decided that it will be best not to respond to the LOI. The lawyer explains to Captain Davidson that anything he states in the LOI could be used against him if the FAA decided to pursue an enforcement case. The lawyer cites two cases in which the FAA made use of information provided by airmen in LOIs to prove facts against the airmen's interest.[13] The lawyer explains that even though

Figure 5-1
FAA letter of investigation.

```
                FEDERAL AVIATION ADMINISTRATION
                FLIGHT STANDARDS DISTRICT OFFICE
                        1 AIRPORT DRIVE
                     SMITHTOWN, ALASKA 99532


CERTIFIED MAIL, RETURN RECEIPT REQUESTED


April 1, 2006

Captain K. Davidson
555 N. Platte Street
Bethesda, MD 12456

Dear Captain Davidson:

Personnel of this office are investigating an incident which involved the operation of a Boeing
747 aircraft, N321PJ, on March 1, 2006, in the vicinity of Tundra, Alaska.

Information indicates that you operated this aircraft while the aircraft was unairworthy due to
damage sustained to the aircraft during takeoff. Operations of this type are contrary to
Federal Aviation Regulations.

This letter is to inform you that this matter is under investigation by the Federal Aviation
Administration (FAA). We would appreciate receiving any evidence or statements you might
care to make regarding this matter within 10 days of receipt of this letter. Any discussion or
written statements furnished by you will be given consideration in our investigation. If we do
not hear from you within the specified time, our report will be processed without the benefit of
your statement.

Sincerely,


I. M. Kluso
Aviation Safety Inspector
```

Captain Davidson does not think he did anything wrong, it is not known which direction the FAA will take in this case. The lawyer therefore suggests that it might be prudent to refrain from any statements until the full nature of the investigation is better understood. The lawyer does, however, send the FAA a polite note declining further comment until he can review the FAA's investigation report in Captain Davidson's matter.

About four months later Captain Davidson receives correspondence in the mail with a Notice of Proposed Certificate Action (NPCA) from the FAA. The FAA is charging Captain Davidson with a violation of FAR §§ 91.7(a) and (b) (operating an unairworthy aircraft and failure to discontinue the flight once unairworthy conditions occur) and FAR § 91.13(a) (careless or reckless operations). The NPCA calls for a 60-day suspension of Captain Davidson's ATP certificate. Captain Davidson calls his lawyer and asks whether the FAA could still pursue the matter so many months after the flight in question. His lawyer explains that under the Rules of Practice in NTSB hearings, the FAA has up to 6 months from the date of the alleged offense to notify an airman of any allegations.[14] Even then, the lawyer explains, the FAA can still proceed with charges if it can establish good cause for the delay. (For a good discussion of what does and does not represent good cause, see *Administrator v. Ramaprakash*.[15])

A copy of Captain Davidson's NPCA is found in Fig. 5-2.

Figure 5-2
FAA Notice of Proposed Certificate Action.

FEDERAL AVIATION ADMINISTRATION
OFFICE OF REGIONAL COUNSEL
ALASKAN REGION
ANCHORAGE, ALASKA 99513

August 2, 2006

Case No. 2006EA125609

CERTIFIED MAIL, RETURN RECEIPT REQUESTED

Captain K. Davidson
555 N. Platte Street
Bethesda, MD 12456

NOTICE OF PROPOSED CERTIFICATE ACTION

Take notice that upon consideration of a report of investigation, it appears that:

1. You are now, and at all times mentioned herein you were, the holder of Airline Transport Pilot Certificate No. 123456789.
2. On or about March 1, 2006, you served as the pilot in command of civil aircraft N321PJ, a Boeing 747, on a flight in air commerce from Tundra, Alaska, to Capitol Airport in Washington, DC.
3. During its takeoff, N321PJ incurred substantial airframe damage, and the aircraft was rendered unairworthy.
4. Despite the unairworthy condition of the aircraft, you continued the flight to its destination.
5. Your operation of civil aircraft N321PJ was careless or reckless so as to endanger the life or property of others.

Based on the foregoing facts and circumstances, you violated the following Federal Aviation Regulations:

(a) Section 91.7(a) in that you operated a civil aircraft while it was in an unairworthy condition.
(b) Section 91.7(b) in that you failed to discontinue the flight when unairworthy structural conditions occurred.
(c) Section 91.13(a) in that you operated an aircraft in a careless or reckless manner so as to endanger the life or property of another.

Now, therefore, please take notice that by reason of the foregoing facts and circumstances and pursuant to the authority vested in the Administrator by 49 U.S.C. Section 44709(b), we propose to suspend your Airline Transport Pilot Certificate No. 123456789 with all privileges and ratings for a period of sixty (60) days.

(Continued)

Figure 5-2
FAA Notice of Proposed Certificate Action. (*Continued*)

> Unless we receive, in writing, your choice of alternatives listed below within fifteen (15) days after the date you receive this notice, we will issue an order suspending your certificate and all privileges and ratings as proposed above. This notice does not suspend your certificate; however, if you wish to make the suspension effective immediately, you must physically surrender your certificate to this office as provided in Option 1 below.
>
> ---
>
> ABLE LAWYER
> Regional Counsel
>
> Airman's Reply Form
>
> I elect to proceed as follows:
>
> 1. I hereby transmit my certificate. I understand that an Order will be issued as proposed in the Notice of Proposed Certificate Action (NPCA). The Order will be effective on the date I mail this reply. I also understand that I am waiving my right to appeal the Order to the National Transportation Safety Board (NTSB).
> 2. I request that an order be issued so that I may appeal directly to the NTSB.
> 3. I hereby submit my answer to this NPCA Action and request that my answer be considered in connection with the allegations set forth in the NPCA.
> 4. I hereby request an informal conference to discuss this matter with an FAA attorney.
> 5. I hereby claim entitlement to a waiver of penalty under the NASA ASRS and enclose evidence that a timely report was filed.
>
> _____
> Airman Signature

Captain Davidson's lawyer reviews the options offered in the NPCA and explains that the offer of an informal conference is required by law.[16] He adds that it is often a good idea to accept the offer of a meeting with the FAA to see if the matter can be settled without a formal hearing. The lawyer suggests that they prepare for an informal conference with the FAA attorney handling the case. Captain Davidson also learns from his counsel that while awaiting the informal conference he can continue to exercise the privileges of his certificate.

At this point, the lawyer also suggests that it would be a good idea to get a copy of the FAA's enforcement investigative report (otherwise known as the EIR) for this case. This report will include documentation of all the FAA's evidence that is being relied on to charge Captain Davidson with the violations in the NPCA, including tapes of conversations with the air traffic control tower in Norway. Upon receipt of the request, the FAA attorney sends the EIR. Captain Davidson and his lawyer attend an informal conference with the FAA attorney and the inspector in charge of the investigation. Unfortunately for Captain Davidson, the meeting results in the FAA expressing a willingness to reduce sanction, but not drop the charges. Captain Davidson is unwilling to accept any compromise at this point. He instructs his lawyer to tell the FAA that he wants to bring this case to a hearing before an NTSB administrative law judge (ALJ).

Captain Davidson's lawyer informs the FAA's attorney that a compromise cannot be reached based on the settlement offer presented by the FAA at the informal conference.

In view of this, the FAA issues an Order of Suspension, ordering Captain Davidson to surrender his ATP certificate for 60 days as sanction for the violations alleged by FAA. The Order of Suspension looks very much like the NPCA except that it offers the opportunity to stay or put off the suspension of Captain Davidson's ATP certificate if he appeals the FAA's order to the NTSB. Within the required time frame, Captain Davidson's attorney files an appeal with the FAA. A copy of the notice of appeal is found in Fig. 5-3.

The case is now before the NTSB, and the more formal adjudication phase of this matter begins. Once the notice of an appeal is received by the Board, the case will move along very much as in any other hearing or trial before a civil court.

Adjudication

Captain Davidson's lawyer and the FAA's attorney will most likely engage in prehearing discovery. The FAA attorney may want to take depositions of the flight crew, IFM, passengers, and/or flight attendants to get a fuller picture of what occurred during the takeoff. Captain Davidson's lawyer may want to depose or send written interrogatories to the FAA regarding the conversations between Captain Davidson and the tower in Tundra. Both sides will probably request a list of possible witnesses, expert witnesses,

---

UNITED STATES OF AMERICA
NATIONAL TRANSPORTATION SAFETY BOARD
WASHINGTON, D.C.

Marion C. Blakey,   )
Administrator,   )
Federal Aviation Administration,)   NTSB Docket No. SE-_____
)
Complainant,   )
v.   )
)
Captain K. Davidson,   )
)
Respondent.   )
_____)

APPEAL

    Captain K. Davidson, Respondent, by his undersigned counsel, pursuant to Section 821.30 of the Board's Rules of Practice (49 C.F.R. 821), hereby appeals the Administrator's Order of Suspension dated August 2, 2006, FAA Docket No. 2006EA125609, suspending for sixty (60) days Respondent's Airman Certificate Number 123456789. Respondent denies the violations alleged in the Administrator's Order and prays that it be reversed.
    Respondent requests that the hearing for this matter be held in Washington, D.C.

Respectfully submitted,

                                                                       Dated: _____
_____
Pitt and Bull, Attorneys at Law

Figure 5-3
Appeal of FAA order.

and documents that might be presented at a hearing. As discussed in Chap. 1, the primary purpose of this discovery is to allow the parties to properly prepare for the hearing. A secondary purpose is to encourage settlement as the prospects for success or failure at trial become more apparent.

The lawyers for both sides will also do legal research to find previous cases before the NTSB that might have raised similar issues. In doing the research for this case, the parties learn that the central question at hand is whether a reasonable and prudent pilot would have concluded from the information available that there was a possibility that the aircraft was unairworthy. Both parties have noted that the controlling precedent is found in *Administrator v. Dailey* and *Administrator v. Parker*.[17]

Once the trial gets underway, the FAA attempts to prove Captain Davidson should have known about the damage done to his aircraft during the static takeoff run-up of the engines. The FAA brings forward the IFM for the flight, who testifies that he called the cockpit and informed them of the dust, debris, and stones that he observed near the back of the aircraft. However, Captain Davidson's lawyer attempts to establish that Captain Davidson did not have enough information presented to him to know about the damage. The lawyer repeatedly points out that none of the cockpit crew heard the debris strike the aircraft (due to the distance between the cockpit and the separation between the cockpit and back of the plane by four huge engines running at full power); he further points out that the communication from the IFM never indicated that the aircraft had been struck. Finally, Captain Davidson's lawyer emphasizes that all communications with the tower focused on the possibility of damage to the 747's tires. Once the search on the runway for rubber uncovered nothing, Captain Davidson was led to believe that everything was okay.

The ALJ presiding over the hearing makes routine rulings during the course of the proceeding regarding the admissibility of evidence. At the conclusion of the hearing, the ALJ renders a ruling as an Oral Initial Decision.[18] In this case, the ALJ rules in Captain Davidson's favor. Based on the evidence presented, the ALJ concludes that the aircraft was rendered unairworthy after the damage done during the static run-up of the engines. However, the ALJ further concludes that Captain Davidson did not have information put before him that would indicate the aircraft had been damaged. Applying the tests in *Dailey* and *Parker*, the ALJ concludes that Captain Davidson did not violate FAR §§ 91.7(a) and (b) or § 91.13(a).

The FAA reviews the ALJ's opinion and decides that it wishes to appeal the ALJ's decision by filing an appeal to the full five-member NTSB. Once the appeal is filed, a full transcript of the hearing and the evidence presented at the hearing are forwarded to the Board. The Board will not examine witnesses or receive new evidence; therefore, it must rely on the record of the hearing in its review. The FAA files an appeal brief detailing its arguments in an effort to get the Board to reject the ALJ's decision. Captain Davidson's lawyer then files his reply brief in an effort to get the Board to deny the FAA's appeal and sustain the ALJ's decision.

After several months of review and deliberation, the Board issues an Opinion and Order in Captain Davidson's case. This time, the airman prevails. The Board defers to the findings of the ALJ and determines that Captain Davidson did not know and could not have reasonably known about the condition of his aircraft. The Board takes special note that the only communication of any possible problem came from the IFM and that his testimony revealed that he did not speak directly to Captain Davidson and had not relayed to the crew that a flight attendant heard the "loud bang."

After the Board has rendered a decision, an airman who loses can submit a petition for judicial review (an appeal) to the U.S. Court of Appeals. In our case, Captain Davidson was the prevailing party at the Board level, so he will not file an appeal. The FAA can also file a petition with the U.S. Court of Appeals, but only if the FAA decides that the Board's order will have a significant adverse impact on its ability to carry out its regulatory mission.[19] In this case, the FAA decides that it will not appeal because the Board's decision was limited to the facts of Captain Davidson's case. With no appeal from the FAA, the case is closed.

The flowchart in Fig. 5-4 depicts the steps in the course of a typical (nonemergency) enforcement case and appeal to the NTSB. Although our discussion has been limited to the FAA/NTSB process, this model is in many ways similar to the process employed by other agencies.

Special considerations in FAA matters

Beyond the basics of administrative law presented above, there are some unique aspects to FAA matters that warrant further discussion. The four areas that will be addressed here are (1) FAA emergency actions, (2) NASA's Aviation Safety Reporting Program (NASA ASRP), (3) FAA administrative actions, and (4) FAA reinspections and reexaminations of aircraft and airmen.

FAA emergency actions

As discussed above, it can take several months and sometimes more than a year for an FAA enforcement action to be concluded. On occasion the FAA takes the position that an airman's actions indicate that an emergency exists, requiring immediate action by the agency. In those cases, the FAA will issue an emergency order of suspension or revocation (virtually every emergency case involves a revocation action). The emergency order requires immediate surrender of the airman's certificate with no stay of the suspension while the airman might appeal the emergency order.

If the airman who receives an emergency order does not agree with the FAA's determination that an emergency exists, the airman can, within 48 hours of receiving the emergency order, petition the Board for a determination of the existence of an emergency. The Board has 5 days to review the question of whether an emergency exists. If the Board finds that no emergency exists, then the emergency nature of the action is stayed (put off) and the airman can keep his or her certificate(s) throughout the NTSB proceedings. However, if the Board upholds the FAA's determination of an emergency, the airman must surrender his or her certificate to the FAA throughout the NTSB proceedings.

Notice that the big difference between the "normal" and emergency FAA cases is the lack of a notice of proposed certificate action and the option of an informal conference. With an emergency action, an airman who wants to challenge the FAA's findings must

Figure 5-4
Flowchart of FAA enforcement actions, nonemergency cases.

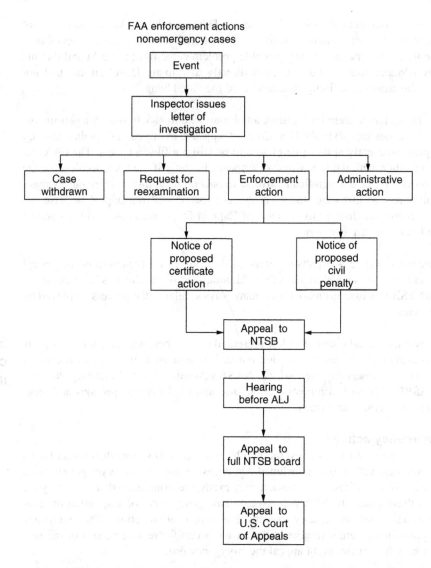

head straight to a hearing before the NTSB (although there is no legal impediment to the airman and the FAA reaching a settlement in the matter prior to the NTSB hearing).

If an airman challenges the FAA's emergency order and the case goes to a hearing, normally the hearing is scheduled just a few weeks after the emergency order has been received. Therefore, the time for the parties to engage in discovery and research is severely limited. The hearing date is set early in the process because if the airman or FAA appeals the findings of the ALJ who presided at the hearing, the case must move quickly to the Board on appeal. In total the NTSB has 60 days to make a determination on an emergency case, starting from the date the airman files a notice of appeal.

If an airman is concerned about the "rushed" nature of the emergency proceedings, she or he may waive the emergency nature of the case. If the airman elects to waive the emergency proceedings, the case will be placed on the regular calendar of the NTSB ALJ. This may mean a wait of several months (without a certificate) before the accused gets to a hearing. Depending on the circumstances of the case, this may or may not be a viable option for an airman faced with an emergency action by the FAA.

Figure 5-5 depicts the steps in an FAA emergency case.

NASA's Aviation Safety Reporting Program

One unique aspect of FAA enforcement matters is the NASA ASRP. This is a program in which the FAA uses the services of the National Aeronautic and Space Administration (NASA) to act as a neutral thirdparty in receiving Aviation Safety Reports from pilots, controllers, flight attendants, mechanics, and others who use the nation's airspace system. The program permits these users to anonymously report unsafe or potentially unsafe conditions as they become aware of such conditions. The reports are often used to report discrepancies and deficiencies, including departures from ATC instructions, near midair collisions, maintenance defects, etc. The system was meant to provide the FAA with honest input from system users while providing the users with a way of noting deficiencies without fear of retaliation or punishment by the FAA. Now we will describe how the NASA ASRP works.

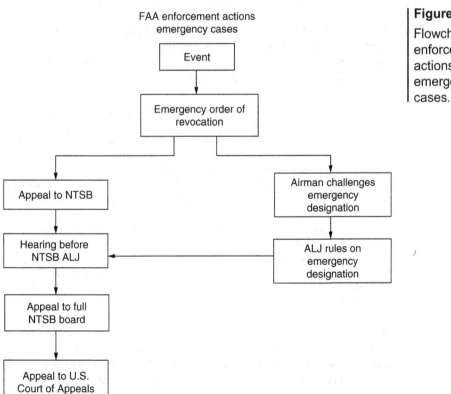

Figure 5-5

Flowchart of FAA enforcement actions, emergency cases.

First, a user of the national airspace system files an Aviation Safety Report. There are four types of reports prepared for users. All the forms are labeled as part of the NASA ARC 277 series. The four types of forms are as follows:

1. ARC 277A for air traffic use
2. ARC 277B for general use including pilots
3. ARC 277C for flight attendants
4. ARC 277D for maintenance personnel

A copy of an ARC Form 277B can be found in App. C.

As you can see on Form 277B, the Aviation Safety Report form has a tear-off portion that is the only place where the reporter is identified. This part of the form is torn off by NASA, date- and time-stamped, and returned to the submitter as a receipt. Except in cases involving accidents or criminal activities, the reporter's identity is not retained in any manner by NASA or the FAA. To further protect anyone submitting a NASA Aviation Safety Report, 14 CFR § 91.25 prohibits the use of any reports submitted to NASA under the program to be used in any action taken against the reporter (with the exception of reporters who submit information regarding accidents and criminal activities). Further ensuring anonymity, NASA will de-identify any reporter names or information that might lead to identification in an Aviation Safety Report within 72 hours after NASA receives the report. (See FAA AC No. 00-46D.)

What is the incentive to file a NASA ASRP? Typically, the reporter is involved in an incident that might lead to an FAA enforcement action. By filing a NASA Aviation Safety Report the reporter may be able to avoid FAA sanction. The relevant excerpts from FAA AC No. 00-46D state as follows:

> The filing of a report with NASA concerning an incident or occurrence involving a violation of [law or the FARs] is considered by FAA to be indicative of a constructive attitude. Such an attitude will tend to prevent future violations. Accordingly, although a finding of violation may be made, neither a civil penalty nor certificate suspension will be imposed if:
> 1. The violation was inadvertent and not deliberate;
> 2. The violation did not involve a criminal offense, or accident, or action under 49 U.S.C. § 44709 which discloses a lack of qualification or competency, which is wholly excluded from this policy.
> 3. The person has not been found in any prior FAA enforcement action to have committed a violation of [aviation related law or FARs] for a period of 5 years prior to the date of the occurrence; and
> 4. The person proves that within 10 days after the violation, he or she completed and delivered or mailed a written report of the incident or occurrence to NASA under ASRS.

To ensure that there is absolute proof of a timely filed NASA Aviation Safety Report, many lawyers will counsel clients to send the form to NASA via certified mail, return receipt requested, or overnight mail. NASA ASRP has proved to be a very useful tool for many airmen and the FAA in improving the system and avoiding penalties for airmen who have made inadvertent errors while attempting to perform their duties in good faith.

ASAP program

Besides the NASA ASRP, the FAA has also enacted a special program known as the Aviation Safety Action Program (ASAP). ASAP is a voluntary reporting program in which employees of Part 121 air carriers and Part 145 repair stations can report inadvertent safety violations "… without fear that the FAA will use reports to take legal enforcement action or that the companies will use such information to take disciplinary action." The program requires that a participating company and the FAA document their participation in ASAP with a memorandum of understanding (MOU).[20]

Administrative actions

If the FAA believes a violation is minor enough, it might decide to forgo formal enforcement action. Just what constitutes a minor violation is often difficult to assess. However, the FAA's internal enforcement guidance indicates that three criteria must be met:

1. The violation must not be deliberate.
2. There was no significantly unsafe condition created by the violation.
3. The violation does not evidence a lack of competency or qualification.[21]

Determining whether a violation will be subject to administrative action is a subjective decision (what is a significantly unsafe condition?) ordinarily made by the FAA inspector investigating an incident.

If the FAA determines that administrative action is the appropriate way to proceed with a case, it has two choices—either a Warning Notice or a Letter of Correction. The Warning Notice takes the form of a letter to the alleged violator. The letter typically recites the facts (as known to the FAA) and indicates that the conduct of the airman was in violation of the FARs. A Warning Notice usually concludes with a statement indicating that the matter does not warrant legal enforcement action but will be made a matter of record in the airman's files for 2 years.

A Letter of Correction is very similar to a Warning Notice. The only difference between the two is that a Letter of Correction will indicate that the alleged violator has taken steps to correct the underlying violation or deficiency that led to the violation. Letters of Correction are usually the end result when an airman is recommended for remedial training by the FAA. Similar to a Warning Letter, a Letter of Correction remains on an airman's file for 2 years.

Although Warning Letters and Letters of Correction are not considered to be violations on an airman's record, many airmen are concerned that the letters remain in their file for a 2-year period.[22] If an airman believes that the FAA was wrong in alleging a violation, can the airman challenge the FAA's administrative action? In *Machado v. Administrator*, the NTSB determined that a Warning Letter cannot be appealed to the Board.[23] In view of this decision, the best thing an airman can do if she or he believes the FAA's administrative action is unwarranted is to respond to the administrative action in writing and to request that the letter be made a part of her or his file while the Warning Letter or Letter of Correction remains in the records. Usually, the FAA's Warning Notice or Letter of Correction will invite the airman to add any explanatory information to the files.

Reexamination

One very important tool available to the FAA to ensure regulatory compliance is the ability to reinspect and reexamine. Applicable law states, "The Administrator of the Federal Aviation Administration may reinspect at any time a civil aircraft, aircraft engine, propeller, appliance, air navigation facility, or air agency, or reexamine an airman holding a certificate issued under section 44703 of this title."[24] As indicated, this law permits the FAA to examine aircraft and airmen at any time.

Despite the literal wording of the law allowing the FAA to examine an airman "at any time," there are several NTSB decisions that shed light on what an airman can expect if the FAA requests a reexamination. First, the request for the reexamination must be reasonable, and it must be limited to those areas in which a lack of knowledge or skill may have been demonstrated by the airman.[25] Second, a refusal to submit to a reexamination will get the airman in a situation in which the FAA takes emergency action to suspend or revoke his or her certificate.[26] Third, a successfully completed reexamination does not stop the FAA from pursuing enforcement through a certificate action.[27] Finally, the NTSB has held that the burdens placed on an airman by a reexamination (time, cost, etc.) cannot supercede the FAA's need to ensure competency, and a reexamination may be warranted even though the airman may not have demonstrated any fault in the underlying incident.[28]

As a practical matter, a vast majority of the time, reexaminations are fairly administered, and they are usually passed on the first try. In very few instances has the FAA appeared to have abused its statutory right to reexamine airmen.

CHECKS ON ADMINISTRATIVE AGENCIES

Because administrative agencies are staffed by unelected officials, there has always been concern that the agencies could act without need for accountability to the public. Traditional branches of government have a duty to oversee the workings of administrative agencies to ensure that they are faithfully performing their statutory duties. Over time, the law has also evolved to create a system that forces administrative agencies to disclose information, open meetings, and generally be accountable to the public whose welfare the agencies were designed to protect. Some of the more significant protections built into the legal system are discussed below.

Oversight by traditional branches of government

The first line of defense against agency abuse of power is the checks and balances provided by the traditional branches of government—the judicial, executive, and legislative branches. With the exception of the judicial branch (which is checked and balanced by the executive and legislative branches), the legislative and executive branches are led by elected officials. This is designed to create an environment of accountability to the public. We will briefly explore the way the traditional branches of government can limit administrative agencies.

Judiciary

As discussed above, a person adversely affected by an action of an administrative agency has the right to appeal the decision of the agency to the courts. In the federal system, appeals go to the U.S. Court of Appeals. The ability to review rules and orders of administrative agencies serves as a check on the agency.[29]

Judicial review is generally available only when there has been a final agency action.[30] This means a party must exhaust all available attempts to adjudicate a matter within agency rules before turning to the courts for relief.

The standards utilized by the courts when examining an agency action vary depending on the agency action being reviewed. If courts are reviewing informal rulemaking, they will ordinarily apply what is known as the "arbitrary or capricious" test. This means that the agency's action will stand as long as it was not arbitrary or capricious. Stated differently, it means that the agency must merely have a rational basis for its action to stand the court's scrutiny.

When an agency (such as the NTSB in FAA enforcement cases) holds a formal hearing, the test applied is the *substantial evidence* test. This essentially means that the courts will uphold the agency's determination as long as it was supported reasonably by adequate evidence. This standard does not allow the court very broad review powers over agency actions. In the case of NTSB proceedings, the Board's power to review is further diminished by federal law which states:

> When conducting a hearing under this subsection, the Board is not bound by findings of fact of the Administrator but is bound by all validly adopted interpretations of laws and regulations the Administrator carries out and of written agency policy guidance available to the public related to sanctions to be imposed under this section unless the Board finds an interpretation is arbitrary, capricious, or otherwise not in accordance to law.[31]

This statutory language requires the Board to exercise deference to the FAA's interpretations of its own regulations. The courts have not fully fleshed out what is meant by the phrase *validly adopted interpretations*. However, from preliminary cases, it appears that the FAA has a relatively low threshold to clear to maintain that its interpretations are validly adopted.

In unusual instances, a reviewing court may be required to apply a standard known as *unwarranted by the facts*. This standard essentially allows the courts to retry the case (sometimes referred to by lawyers as a *de novo* review). This very strict standard of review is put in place only when an enabling statute requires its use.

Executive

The President has substantial ability to exercise control over administrative agencies. Obviously, the President's power is greatest when the agency is part of the executive branch of government. However, even with respect to independent agencies, the President still gets to nominate agency chairs and fill vacancies as they arise. Budgetary control also enables the President to influence agency policies. Finally, the President can significantly alter, combine, or terminate agencies housed within the executive branch unless opposed by Congress. This very potent authority of the President was recently on display with the reorganization of government agencies in the creation of the Homeland Security Department.

Legislative

Congress has the ability to exercise significant control over administrative agencies. The first tool available to Congress is the fundamental tool of modifying an agency's enabling statute. This tool allows Congress to make minor changes to the way an agency does business or to eliminate an agency altogether.

Congress may also exercise control over agencies by enacting guidelines for agency action. In essence, the enactment of the APA was a way for the legislative branch to set standards for all administrative agencies. Congress also creates and maintains oversight committees that regularly review the workings of administrative agencies. To further add to control over agency action, in 1996 Congress enacted the Congressional Review Act. This act gives Congress the ability to review any new rules created by agencies that have significant impact on the economy or on industries. If the new rule is disapproved by Congress, the rule is treated as if it never existed.

Finally, because it holds primary budgetary authority, Congress may withhold or decrease funding to an agency. There may be no more potent tool available than funding to shape an agency's size and scope.

Equal Access to Justice Act

The Equal Access to Justice Act (EAJA) permits eligible individuals and entities to recover attorney fees and related expenses from government agencies.[32] As a general rule, there are four requirements for anyone seeking a recovery of attorney fees under the EAJA:

1. The individual must have been a party to an adversarial adjudication against a government agency.
2. The individual or entity seeking recovery must have been a prevailing party in the adjudication against the government.
3. The government agency's position in the adversarial adjudication must not have been substantially justified (government bears the burden of proof on this issue).
4. Individuals or entities seeking an award under EAJA must establish that they meet certain financial eligibility requirements.

The purpose of the EAJA is an attempt to level the playing field in agency adversarial proceedings. There has always been concern that when an ordinary citizen is faced with government action, she or he starts at a substantial disadvantage. The full weight of government resources and expertise is being faced. In the meantime, retaining counsel and required experts is a substantial expense—often an expense that the person challenged by a government agency cannot afford.

The EAJA can be looked upon as a check on government agencies. The agencies must be mindful of the fact that while they have a duty to enforce agency rules, the position they take must be substantially justified or the agency may be required to pay for the counsel fees and costs of the person it has challenged.

The NTSB has implemented rules for the EAJA.[33] The NTSB rules track the EAJA and clarify what the NTSB requires in an application for an EAJA award related to a case adjudicated before the NTSB.

Freedom of Information Act

The Freedom of Information Act (FOIA) was first enacted in 1967 with substantial amendments made in 1974. It requires that the public be given access to most of the records held by administrative agencies. The purpose behind the FOIA is to make the workings of administrative agencies more transparent, therefore making the agencies more accountable to the public.

The FOIA works in a relatively straightforward manner. Once a request for agency records is received, the agency must notify the individual making the request within 10 working days whether it will comply with the request. If so, the agency must provide the requested records in a reasonable time. An agency can always charge reasonable fees for copying or reproducing records requested.

Nine important categories of records are excluded from the FOIA requirements for disclosure. The nine exceptions to the general rule of disclosure are as follows:

1. Records that have been specifically authorized to be kept secret in the interest of national defense or foreign policy
2. Records relating to solely internal personnel rules and agency practices
3. Records specifically exempted by statute from disclosure
4. Trade secrets and commercial or financial information that are considered confidential
5. Memoranda that are intra- or interagency
6. Personnel and medical information which, if disclosed, would result in a clearly unwarranted invasion of the right of privacy
7. Investigatory records compiled for law enforcement policies
8. Records relating to the regulation of financial institutions
9. Certain geological and geophysical information

If an agency gets a request for information that arguably fits into one of the exclusions written above, the agency must deny disclosure under the FOIA. The requesting party who disputes the agency's denial can always bring a separate action (in court) compelling disclosure of the requested records under FOIA.

In the context of aviation matters, the FOIA comes in handy for airmen who might be involved in an FAA enforcement action. An FOIA request will allow the airmen to request voice tapes or documents related to an alleged violation.

Privacy Act

At the same time the FOIA was being amended in 1974, Congress passed the flip side of the FOIA when it enacted the Privacy Act of 1974.[34] Agencies collect extensive data on individuals to perform their missions (e.g., consider the sheer volume and sensitivity of data collected on every pilot by the FAA's Aeromedical Branch). However, there has always been concern that this information could be misused. The Privacy Act basically prohibits the government from disclosing information relating to an individual without the individual's written authorization. Further, the Privacy Act forbids agencies from providing the names and addresses of citizens without specific authorization by law.

Sunshine Act

The Sunshine Act requires certain agencies to open their meetings to the public. The agencies most affected by the Sunshine Act are the commissions and boards in which the members are appointed by the President. For example, meetings of the NTSB are held in public in accordance with the Sunshine Act. Examples of other agencies that are required to hold public hearings are the Securities and Exchange Commission, the Federal Trade Commission, and the Federal Communications Commission.

As a general rule, agencies may close meetings for the same reasons that they would not disclose information under the FOIA. In addition to the exception provided in the FOIA, an agency may close a meeting under Exemption 10 of the Sunshine Act if it involves discussion regarding future litigation or legal matters pending before the agency.

Similar to the FOIA, the purpose of the Sunshine Act is to ensure that the work of agencies is performed in as transparent an environment as possible. In the end, making meetings public makes agencies more accountable to the public.

CASES AND COMMENTARY

The first case presented, *ATA v. DOT and FAA,* involves the issue of rulemaking.[35] The question presented is whether the FAA should have provided for notice and comment for new civil penalty rules.

To help set the stage, it must be understood that for many years, the FAA had been using its civil penalty authority (issuance of fines) as an enforcement tool. Civil penalties turned out to be a relatively weak enforcement tool for the FAA under the old system. The problem was that if airmen or aircraft owners were fined and they decided not to pay the fines, the case would be referred to the U.S. Attorney's Office. If the U.S. Attorney who got the case wanted to prosecute on the fine, she'd be forced to litigate in U.S. District Court. As time passed, it became clear that the civil penalty laws had no teeth because the Office of the U.S. Attorney was often too busy to prosecute an airman or aircraft owner on a relatively small fine.

The FAA consulted with Congress on this problem. Eventually, the FAA persuaded Congress to pass a 2-year demonstration law whereby the FAA would get a trial run at running the prosecution and adjudication of civil penalty cases involving amounts less than $50,000. Armed with the mandate from Congress, the FAA developed hearing procedures for civil penalty cases and decided that "notice and public comment procedures are impractical, unnecessary, and contrary to the public interest."

A number of organizations, including Air Transport Association, Aircraft Owners and Pilots Association, and various other aviation interests, filed suit in the U.S. Court of Appeals for the District of Columbia Circuit. These aviation organizations argued that the rules being enacted by the FAA required notice and comment. This is the issue the circuit court wrestles with in Case 5-1.

CASE 5-1
AIR TRANSPORT ASSOCIATION OF AMERICA v. DOT AND FAA
900 F. 2d 369 (1990)

OPINION: EDWARDS, Circuit Judge:

* * *

I. BACKGROUND

In December of 1987, Congress enacted a series of amendments to the Federal Aviation Act relating to civil penalties. [Citation.] Among other things, these amendments raised to $10,000 the maximum penalty for a single violation of aviation safety standards, [Citation], and established a "demonstration program" authorizing the FAA to prosecute and adjudicate administrative penalty actions involving less than $50,000, [Citation]. Under the terms of the demonstration program, the FAA was granted the authority to assess administrative penalties for a two-year period beginning on December 30, 1987, [Citation] and was to report to Congress on the effectiveness of the program within eighteen months, [Citation].

Congress' goal in enacting this legislation was to strengthen the enforcement powers of the Federal Aviation Administration. Before the 1987 amendments, the FAA could propose a maximum civil penalty of only $1,000 per violation and had no enforcement authority of its own. When an alleged violator disputed a penalty, the FAA was obliged to refer the case to the United States Attorney's office for prosecution in federal district court; relatively few such cases were prosecuted, however, because of competing work obligations facing U.S. Attorneys. [Citation.] Understandably, Congress did not view this as a particularly effective system for assuring compliance with aviation safety standards. By raising the maximum penalty and giving the FAA the power to prosecute penalty actions administratively, Congress sought to "close the holes in the FAA's safety net" and thereby "provide an incentive for airlines to ensure that [their safety] systems are maintained at the highest of standards." [Citation.]

At the same time, however, Congress remained attentive to the adjudicative rights of civil penalty defendants. Congress provided that the FAA could assess a civil penalty "only after notice and opportunity for a hearing on the record in accordance with section 554 of [the APA]." [Citation.] As the conference report accompanying section 1475 explained, the express incorporation of the APA's procedural protections was designed to achieve two purposes:

First, the requirement is intended to advise the FAA of the appropriate level of procedural formality and attention to the rights of those assessed civil penalties under this demonstration program. Secondly, this requirement is intended to provide reasonable assurance to the potential subjects of such civil penalties that their due process rights are not compromised. [Citations.]

Approximately nine months after enactment of section 1475, the FAA promulgated the Penalty Rules. [Citation.] Effective immediately upon their issuance, the Penalty Rules established a schedule of civil penalties, including fines of up to $10,000 for violations of the safety standards of the Federal Aviation Act and related regulations. [Citation.] The Penalty Rules also established a comprehensive adjudicatory scheme providing for formal notice, settlement procedures, discovery, an adversary hearing before an ALJ and an administrative appeal. [Citation.] In explaining why it dispensed with prepromulgation notice and comment, the FAA emphasized the procedural character of the Penalty Rules and the time constraints of section 1475. [Citation.] The FAA did respond to *post* promulgation comments but declined to make any amendments to the Rules. [Citation.]

* * *

In its petition for review, Air Transport raises two challenges to the Penalty Rules. First, it attacks the procedural adequacy of the Rules, arguing that the FAA was obliged by section 553 of the APA to permit notice and comment *before* the Rules became effective. Second, Air Transport attacks the substantive adequacy of the Rules on the ground that they establish adjudicatory procedures inconsistent with section 554 of the APA. In a previous order, we deferred consideration of the FAA's motion to dismiss the petition on ripeness grounds and directed the parties to address this issue in their briefs on the merits. We now find that Air Transport's *procedural* challenge to the Penalty Rules is ripe for review and grant the petition on that ground.

II. ANALYSIS

* * *

B. *The Merits*

Section 553 of the APA obliges an agency to provide notice and an opportunity to comment before promulgating a final rule. No question exists that the Penalty Rules fall within the scope of the APA's rulemaking provisions. [Citation.] Nonetheless, the FAA maintains that the Penalty Rules were exempt from the notice and comment requirements for two, independent reasons: first, because they are "rules of agency organization, procedure, or practice," [APA] § 553(b)(3)(A); and second, because the time constraints of section 1475 gave the FAA "good cause" to find that prepromulgation notice and comment would be "impracticable, unnecessary, or contrary to the public interest," [APA] § 553(b)(3)(B). The FAA also argues that its entertainment of *post* promulgation comments cured any violation of section 553.

Section 553's notice and comment requirements are essential to the scheme of administrative governance established by the APA. These procedures reflect Congress' "judgment that ... informed administrative decisionmaking require[s] that agency decisions be made only after affording interested persons" an opportunity to communicate their views to the agency. [Citation.] Equally important, by mandating "openness, explanation, and participatory democracy" in the rulemaking process, these procedures assure the legitimacy of administrative norms. [Citation.] For these reasons, we have consistently afforded a narrow cast to the exceptions to section 553, permitting an agency to forgo notice and comment only when the subject matter or the circumstances of the rulemaking divest the public of any legitimate stake in influencing the outcome.[Citations.] In the instant case, because the Penalty Rules substantially affected civil penalty defendants' right to avail themselves of an administrative adjudication, we cannot accept the FAA's contention that the Rules could be promulgated without notice and comment.

* * *

C. *Remedy*

Having determined that the FAA promulgated the Penalty Rules in violation of the APA's notice and comment requirements, we must next consider the appropriate remedy. Ordinarily, when agency rules have been invalidated, the agency may not rely on those rules until they have been repromulgated in accordance with the APA. [Citation.] We find such a disposition appropriate in this case; therefore, we hold that the FAA may not initiate new prosecutions under the Penalty Rules unless and until they are repromulgated. Insofar as the FAA's pending notice of proposed rulemaking seeks public comment on the individual Rules that the agency intends to amend, the agency may rely on the outcome of that rulemaking as a partial fulfillment of this mandate.

* * *

III. CONCLUSION

An agency may dispense with the notice and comment requirements of section 553 "only where the need for public participation is overcome by good cause to suspend it, or where the need is too small to warrant it." [Citation.] Neither of these conditions obtains in this case. The FAA did not have good cause to forgo notice and comment procedures, for nothing in section 1475 either excused or mandated noncompliance with section 553. Nor was this a case in which the need for public participation was "too small to warrant it"; civil penalty defendants have a legitimate interest in influencing agency action affecting their statutory and constitutional "right to avail [themselves] of an administrative adjudication."

[Citation.] We therefore grant the petition for review and order the FAA not to initiate further prosecutions under the Penalty Rules until the agency has engaged in further rulemaking in accord with section 553. Nonetheless, pursuant to our remedial powers, we hold that the FAA is free to hold pending cases in abeyance and resume prosecution upon the repromulgation of a scheme for adjudicating administrative civil penalty actions under section 1475.

It is so ordered.

Do you agree with the circuit court's assessment in this case? Were the rules drafted by the FAA substantive or procedural? What lessons can an administrative agency learn from a case like this?

This next case, Case 5-2, is a relatively recent case of great importance to the aviation community. As discussed in this chapter, the NTSB and the courts are required by law to extend deference to the FAA's validly adopted interpretations of its own regulations. One of the critical questions presented is, What constitutes a "validly adopted" interpretation? The decision contains a rich discussion of FAA enforcement matters. See if you agree with the way the circuit court handles an airline pilot's deviation from a clearance in Case 5-2.

CASE 5-2
GARVEY V. NTSB AND MERRELL
190 F. 3d 571 (1999)

OPINION BY: GARLAND

* * *

The facts of the case are undisputed. On June 19, 1994, Merrell was the pilot-in-command of a commercial passenger plane, Northwest Flight 1024. After Flight 1024 took off in the heavily trafficked Los Angeles area, air traffic control (ATC) instructed it to climb to and maintain an altitude of 17,000 feet. Merrell correctly repeated, or "read back," this instruction to ATC. About a minute later, ATC transmitted an altitude clearance to another aircraft, American Airlines Flight 94, directing it to climb to and maintain an altitude of 23,000 feet. The American flight promptly and correctly acknowledged this clearance with its own "readback."

Merrell, however, mistakenly thought that the instruction to American was intended for his aircraft, so he also read the instruction back to ATC. Unfortunately, because Merrell made his readback at the same time as the American pilot, his transmission was blocked, or "stepped on." The ATC radio system can handle only one transmission at a time on any given frequency; when two transmissions overlap, both may become blocked or garbled, or the stronger signal alone may be heard (i.e., it may "step on" the weaker signal). ATC can often detect that a transmission has been stepped on because, unless the signals overlap completely, ATC will receive a portion of the stepped-on message, and because a loud buzzing noise usually accompanies the period of overlap. On rare occasions, however, two transmissions will overlap completely without creating an identifiable buzz. This appears to have happened in Merrell's case. His readback apparently coincided precisely with that of American Flight 94, and as a result his transmission was entirely blocked. ATC heard neither Merrell's readback nor any indication that it had occurred. And because ATC did not hear the erroneous readback, it could not correct Merrell's mistake.

Meanwhile Merrell, unaware that ATC had not received his transmission, proceeded to ascend toward 23,000 feet. As the Northwest flight rose from its assigned altitude, the ATC controller noticed the deviation and directed the aircraft to return to 17,000 feet. Before Merrell could comply, he had ascended to 18,200 feet and lost the standard safety separation required between commercial flights.

On November 3, 1995, the FAA issued an enforcement order against Merrell. The order alleged that Merrell had violated FAA safety regulations by, *inter alia*, (1) "operating an aircraft contrary to an ATC instruction in an area in which air traffic control is exercised," in violation of 14 C.F.R. § 91.123(b); and (2) "operating an aircraft according to a clearance or instruction that had been issued to the pilot of another aircraft for radar air traffic control purposes," in violation of 14 C.F.R. § 91.123(e).

Merrell appealed the FAA's order to the NTSB. At the outset of the proceedings, the FAA agreed that because Merrell had filed a timely incident report pursuant to the FAA Aviation Safety Reporting Program, it would waive any sanction for the alleged violations. It sought affirmance of its enforcement order, however, arguing that Merrell had deviated from clearly transmitted ATC instructions, that this mistake was due to his own carelessness rather than to ATC error, and that the deviation therefore constituted a regulatory violation. The Administrative Law Judge (ALJ) agreed and affirmed the order. The ALJ found, based on both the recording and the transcript of the radio communications, that the ATC transmission to American Flight 94 had been clear and that the instruction to climb to 23,000 feet had plainly not been intended for Merrell's aircraft. Indeed, after Merrell listened to the tape, he conceded that he had simply "misheard" the instruction. The ALJ concluded that the fact that Merrell's readback was stepped on did not absolve "Captain Merrell of his responsibility to hear that [the] initial clearance" was for another flight. He explained that: "Aviation is ... particularly unforgiving of carelessness or neglect. And in this particular case, the initial mistake was made by Captain Merrell, and he's going to have to be responsible for it." Accordingly, the ALJ held that Merrell "was in regulatory violation as alleged."

Merrell appealed the ALJ's decision to the Board. He argued that under NTSB precedent, a pilot cannot be held responsible for an inadvertent deviation caused by ATC error. His had been such a deviation, he contended, because he had taken actions which, but for ATC, would have kept him from leaving his assigned altitude. He reasoned that because ATC controllers are required to correct erroneous readbacks, his construction of ATC's silence as tacit confirmation had been reasonable and justified. In response, the FAA again argued that because the primary cause of the deviation had been Merrell's misperception of a clear instruction, his actions had violated the safety regulations. The FAA maintained that this outcome was consistent with Board precedent which, it contended, absolves pilots only when "ATC error is the initiating or primary cause of the deviation."

The NTSB accepted Merrell's arguments and dismissed the enforcement order. It found that Merrell had made only "an error of perception," and that there was "no evidence in the record ... that [he] ... was performing his duties in a careless or otherwise unprofessional manner." A "perception mistake," the Board said, does not always result from "a failure of attention," and therefore "careless inattention ... will not be automatically assumed in every case" in which a pilot mishears ATC instructions. Moreover, there was no "failure of procedure" on Merrell's part, as he had "made a full readback so that the opportunity was there, absent the squelched transmission, for ATC to correct his error." The FAA then petitioned the Board for reconsideration of its decision. The agency argued that the Federal Aviation Act requires the Board to defer to the FAA's reasonable interpretation of its own safety regulations. In the FAA's view, 14 C.F.R. § 91.123 obligates pilots "to listen, hear, and comply with all ATC instructions except in an emergency." "Inattention, carelessness, or an unexplained misunderstanding," it said, "do not excuse a deviation from a clearly transmitted clearance or instruction." "When there is an 'error of perception' resulting in a deviation, inattentiveness or carelessness are imputed in the absence of some reasonable explanation for the failure to comply with the ATC clearance." According to the FAA, reasonable explanations include events such as "radio malfunction" or a controller error that precipitates a misunderstanding, but "to excuse [Merrell's] deviation in these circumstances as an acceptable, though unexplained, 'error of perception' " would be inconsistent with the agency's construction of § 91.123. Moreover, the FAA argued that the Board's decision would have a "profound" negative effect on air safety: "Under the decision, airmen can claim, without further proof, that they did not hear or that they misperceived safety crucial instructions as a means to avoid responsibility for noncompliance or erroneous compliance with ATC clearances and instructions." The Board denied the petition for reconsideration. Although it acknowledged its "general obligation to defer to the FAA's validly adopted interpretation of its regulations," the Board considered itself under no such obligation in this case because "the FAA cites no rule *it* has adopted that stands for the proposition the FAA urges here." The Board further noted that the FAA offered "no evidence of *any* policy guidance written by the FAA, validly adopted or otherwise," to support its interpretation, and instead offered only "counsel's litigation statements."

Because the Board determined that it was not required to defer to the FAA's interpretation, it followed its own view of appropriate aviation policy. It stated:

> We ... disagree with the FAA's underlying belief that our policy threatens aviation safety. The premise of our approach is this—human beings make mistakes, and there is no regulatory action, remedial or otherwise, that can eliminate all mistakes... Where an inevitable error of perception does occur, the pilot should not face sanction if he has acted responsibly and prudently thereafter...

Adhering to this principle, the NTSB announced the following rule:

> If a pilot makes a mistake and mishears a clearance or ATC direction, follows all prudent procedures that would expose the mistake (e.g., reads back the clearance), and then acts on that mistaken understanding having heard no correction from ATC, the regulatory violation will be excused if that mistake is not shown to be a result of carelessness or purposeful failure of some sort.

The FAA then petitioned for review in this court.

Under the Federal Aviation Act's split-enforcement regime, Congress has delegated rulemaking authority to the FAA: "The Administrator of the Federal Aviation Administration shall promote safe flight of civil aircraft in air commerce" by prescribing, among other things, "regulations and minimum standards for ... practices, methods, and procedure the [FAA] finds necessary for safety in air commerce and national security." 49 U.S.C. § 44701(a). Pursuant to that authority, the FAA promulgated the safety regulations at issue here, 49 C.F.R. §§ 91.123(b), (e). Congress has also given the FAA authority to enforce its regulations through a number of methods, including the issuance of "an order amending, modifying, suspending, or revoking" a pilot's certificate if the public interest so requires. 49 U.S.C. § 44709(b). The FAA exercised that authority in issuing its enforcement order to Captain Merrell.

Congress has assigned adjudicatory authority under this regime to the NTSB. [Citation.] A pilot whose certificate is adversely affected by an FAA enforcement order may appeal the order to the NTSB. [Citation.] Such an appeal is initially heard by an ALJ, [Citation] whose final decision may be appealed to the full Board, [Citation]. The Board's decision, in turn, may be reconsidered upon the petition of either party. [Citation.] In reviewing an FAA order, "the Board is not bound by findings of fact of the [FAA] Administrator." [Citation.] It is, however, "bound by all validly adopted interpretations of laws and regulations the Administrator carries out ... unless the Board finds an interpretation is arbitrary, capricious, or otherwise not according to law."

If dissatisfied with a final order of the Board, either the FAA Administrator or any "person substantially affected" may petition for review in this court. [Citation.] On judicial review, the "findings of fact of the Board are conclusive if supported by substantial evidence." [Citation.] We must, however, set aside Board decisions if they are "arbitrary, capricious, an abuse of discretion, or otherwise not in accordance with law." [Citation.] And, like the NTSB, we must defer to the FAA's interpretations of its own aviation regulations. [Citation.]

As we have just described, Congress has "unambiguously directed the NTSB to defer to the FAA's interpretations of its own regulations." [Citation.] Here, however, the NTSB explicitly declined to defer to the agency's interpretation of 14 C.F.R. § 91.123. In this Part, we consider the argument that deference to the FAA was not required, either because its interpretation was not validly adopted or because that interpretation was really a factual finding in disguise.

The NTSB declined to defer to the FAA primarily because the agency had offered "no evidence of *any* policy guidance written by the FAA, validly adopted or otherwise," to support its interpretation. Instead, the agency had merely offered the "litigation statements" of FAA counsel, as well as citations to the Board's own case law. *See id.* The NTSB believed the former insufficient to qualify for Board deference under section 44709(d)(3). Accordingly, it rejected the FAA's interpretation and expressly adopted its own policy to govern cases like that of Captain Merrell.

The NTSB's refusal to defer to the FAA on this question of regulatory interpretation and air safety policy was error. The FAA is not required to promulgate interpretations through rulemaking or the issuance of policy guidances, but may instead do so through litigation before the NTSB. We have said as much before, and the Supreme Court so held in [Citation] with respect to the similar split-enforcement regime of the Occupational Safety & Health Act. Indeed, the NTSB itself has repeatedly made the same point. The fact that this mode of regulatory interpretation necessarily is advanced through the "litigation statements" of counsel does not relieve the NTSB of its statutory obligation to accord it due deference.

* * *

Deference, of course, does not mean blind obedience. The agency's interpretation still must not be "plainly erroneous or inconsistent with the regulation" it is interpreting. [Citations.] And even if the interpretation meets this standard, the NTSB need not follow it if it "is arbitrary, capricious, or otherwise not according to law."

* * *

Because the NTSB failed to defer to the FAA's reasonable interpretation of its own regulations, we conclude that the Board's ruling was not in accordance with law. We therefore grant the petition for review, reverse the Board's decision, and remand the case for further proceedings consistent with this opinion.

If anything, this case should draw your attention to the fact that the aviation world is very unforgiving of error, even good faith error. Do you think Captain Merrell was treated justly in this case? Did the court go too far in according deference to the FAA attorney's interpretation?

By way of background, the case heavily relied in by the circuit court in *Merrell* is the Supreme Court decision in *Martin v. OSHRC*.[36] One commentator has characterized deference to an agency's litigation position as "an extreme manifestation of the tendency to extend judicial acceptance to nonlegislatively-issued agency positions."[37] This same commentator indicates that the Supreme Court in *Martin* "just didn't get it." Do you agree that there might be hazard in deferring to an agency attorney's litigation position during the course of a hearing before an administrative agency? Are there any potential Constitutional concerns in granting such deference? Consider whether any due process concerns might arise.

The final "case" for this chapter is not really a judicial case, it is an NTSB Accident Report with findings of probable cause. As indicated earlier in the chapter, the NTSB is responsible for investigating aviation accidents and determining probable cause for the accidents. The Board maintains Accident Investigation Dockets for each accident investigated. The dockets contain preliminary reports, factual reports, and briefs.

The following report, Case 5-3, involves an all too familiar situation in which a non-instrument-rated pilot gets in over his head. Unfortunately, the results are often fatal, as in this case. The NTSB's brief report with a finding of probable cause is found at the beginning. The full narrative follows to illustrate the level of detail that the investigations take on.

CASE 5-3
NTSB IDENTIFICATION: NYC05FA001

14 CFR Part 91: General Aviation

Accident occurred Sunday, October 10, 2004 in Germantown, NY

Probable Cause Approval Date: 6/8/2005

Aircraft: Cessna 172N, registration: N2771J

Injuries: 2 Fatal.

Prior to departing on a round-trip flight, the non-instrument rated pilot telephoned a flight service station and received a standard weather briefing. The briefing included information about marginal VFR conditions for the proposed return trip; including increased cloud cover and scattered showers. VFR flight was not recommended for the return trip. After completing the first leg of the trip, the pilot again telephoned the flight service station. The second briefing included information about a slow moving cool front with poorer weather conditions north and west of the planned route, with the possibility of a shower or two along the route. At the time, one of the airports near the route was reporting marginal VFR conditions. The return flight was conducted at night, and proceeded uneventfully for approximately two-thirds of the planned flight. The pilot was receiving flight following, and asked the controller how high the clouds were, so that he could get out of them. The airplane then descended rapidly in a left turn and struck trees.

The National Transportation Safety Board determines the probable cause(s) of this accident as follows:

The pilot's inadequate in-flight planning/decision which led to VFR flight into IMC and his loss of aircraft control. Factors were night and cloud conditions.

NYC05FA001
HISTORY OF FLIGHT

On October 10, 2004, at 0035 eastern daylight time, a Cessna 172N, N2771J, was substantially damaged during a collision with trees, following a loss of control in cruise flight near Germantown, New York. The certificated private pilot and passenger were fatally injured. Night visual meteorological conditions prevailed for the flight that departed Long Island MacArthur Airport (ISP), Islip, New York; destined for Fulton County Airport (NY0), Johnstown, New York. No flight plan was filed for the personal flight conducted under 14 CFR Part 91.

During the day prior to the accident, the pilot and passenger flew uneventfully from NY0 to ISP. The airplane was fueled before departing NY0, and again after arriving at ISP. The pilot and passenger attended a meeting, and then departed ISP about 2337, for a return trip to NY0.

According to Federal Aviation Administration (FAA) communication and radar data, the pilot made radio contact with the Albany, New York Terminal Radar Approach Control (TRACON), about 0020, and requested flight following. The air traffic controller acknowledged the transmissions, verified radar contact about 0023, and provided a current altimeter setting. The pilot then acknowledged the altimeter setting and verified his altitude. At the time, the airplane was approximately 3,100 feet msl, and 50 miles southeast of Albany, New York.

About 0034, the pilot contacted the Albany TRACON controller and stated, "could you tell me how high these clouds are [unintelligible] get out of them." No further transmissions were received from the accident airplane, and radar contact was lost. Prior to radar contact being lost, a radar target indicated the airplane was at an altitude of 1,700 feet, about 1/3-mile northwest of the accident site. Prior to that, the airplane made a left turn at an altitude of approximately 3,000 feet.

Review of data from the pilot's handheld global positioning system (GPS) receiver revealed the airplane proceeded on a northwesterly track. Prior to the end of the data, the airplane traveled in a 360-degree left turn. The last data point recorded a position of approximately 1/2-mile northeast of the accident site, at an altitude of 2,429 feet.

Three witnesses reported that about 0030, an airplane descended through a cloud layer with an increase in engine noise. The noise was continuous, with no sputtering. Two of the witnesses reported that the airplane was initially on a northwesterly heading, and then made a turn toward a southeasterly direction, before descending rapidly into terrain. The third witness reported that the airplane was heading southeast, and "took a sharp fall-nose-down."

PILOT INFORMATION

The pilot held a private pilot certificate, with a rating for airplane single engine land. The pilot was not instrument rated.

The pilot's most recent FAA third class medical certificate was issued on January 28, 2004.

According to his logbook, the pilot had accumulated approximately 1,231 hours of total flight experience. However, the logbook did not provide a current record of total night experience, total recent experience, or total simulated instrument experience.

AIRCRAFT INFORMATION

The airplane's most recent annual inspection was performed on July 4, 2004. At that time the airplane had accumulated 4,194 hours of operation.

METEOROLOGICAL INFORMATION

Review of recordings from the Burlington, Vermont flight service station (FSS) revealed that pilot telephoned the FSS about 1200 on October 9, 2004. The pilot advised that he was planning a visual flight rules (VFR) flight from NY0 to ISP, departing within the hour, and returning that evening about 2100.

The FSS specialist provided a standard weather briefing. The specialist stated several times that marginal VFR conditions were forecast for the return flight, with thickening cloud cover and scattered showers, and he would not recommend VFR flight in those conditions at night. The specialist further stated that if the pilot waited until the following morning, the weather was forecast to improve.

The pilot again telephoned the FSS about 2100, and advised he was planning a VFR flight from ISP to NY0, departing about 2300.

The FSS specialist stated that there was only one precaution listed for the route; which was for turbulence. However, he did advise of a slow moving cool front in central New York, with most of the weather north of the planned route, but a possibility of a shower or two along the route. The specialist further stated that the current temperature- dew point spread at Schenectady, New York, was 18 degrees C, and 16 degrees C, respectively; and remarked that it was "getting a little close."

The FSS specialist stated that current conditions along the route included: Bridgeport, Connecticut reporting few clouds at 1,200 feet and visibility 10 miles; White Plains, New York reporting marginal VFR conditions with scattered clouds at 1,200 feet, a broken ceiling at 4,000 feet, a broken ceiling at 10,000 feet, and visibility 3 miles in mist; Poughkeepsie, New York reporting sky clear with visibility 9 miles; and Newburgh, New York reporting scattered clouds at 8,000 feet with visibility 7 miles.

The weather briefing contained a forecast for Albany, New York, valid after 2300. The forecast included visibility greater than 6 miles and an overcast ceiling at 6,000 feet.

Columbia County Airport (1B1), Hudson, New York, was located about 10 miles northeast of the accident site. The recorded weather at 1B1, at 0021, was: wind from 350 degrees at 7 knots; visibility 10 miles; scattered clouds at 2,100 feet, ceiling broken at 2,600 feet, ceiling overcast at 3,400 feet; temperature 64 degrees F; dew point 55 degrees F; altimeter 30.05 inches Hg.

WRECKAGE INFORMATION

The wreckage was located in a field about 0700, and examined at the accident site on October 10 and 11, 2004. All major components of the airplane were accounted for at the scene. A debris path was observed, which originated from a row of trees. The debris path extended approximately 330 feet, on a heading of 070 degrees, to the main wreckage. Discolored vegetation was noted along the debris path. A large impact crater was observed about 20 feet along the debris path. The left wingtip was located to the left of the crater. The propeller was located about 75 feet along the debris path, and the right wing was located about 110 feet along the debris path. The left main gear and engine cowling were located about 180 feet along the debris path. The carburetor and cockpit panel were located about 275 feet along the debris path. The magnetos and emergency locator transmitter were found approximately 50 feet beyond the main wreckage.

The right wing was resting inverted in bushes, and a portion of the right aileron was found in the vicinity of the wing. The right wing sustained impact damage to the leading edge, and the damage was greater near the wing root. Flight control continuity was confirmed from the right aileron and flap to the wing root. The right wing fuel tank was compromised.

The left wing was resting inverted on the main wreckage, and remained attached by control cables. Flight control continuity was confirmed from the left flap and aileron to the wing root. The flap was found in the retracted position. The leading edge of the left wing sustained greater impact damage than that of the right wing. The left wing fuel tank was compromised.

The main wreckage was oriented about a 030-degree heading, and the cockpit area was crushed and folded underneath the empennage. The nose wheel and right landing gear were found with the main wreckage. Flight control continuity was confirmed from the elevator and rudder to the forward cabin area. A measurement of the elevator trim jackscrew corresponded to an approximate neutral trim setting. The fuel selector had separated from the cockpit, and was found positioned to the right tank.

Some flight and engine instruments were recovered along the debris path. The airspeed indicator was found with the needle positioned near 150 knots. The attitude indicator was found in a left bank position. The attitude indicator was disassembled for inspection, and scoring was noted on the gyro housing. The altimeter needle had separated, and 29.99 was displayed in the Kollsman window. The directional gyro was positioned near 130 degrees, and the tachometer needle indicated approximately 1,700 rpm. The electric gyro for the turn coordinator was recovered and disassembled for inspection. Scoring was observed on the gyro housing.

The engine had separated from the airframe, and was found inverted on the left side of the main wreckage. The propeller had separated from the engine. The propeller blades exhibited chordwise scratching, s-bending, and leading edge gouging. All spark plugs were removed from the engine for inspection. Their electrodes were intact, and light gray in color. The valve covers were removed, and oil was noted in all cylinders. The vacuum pump was removed, and the drive shaft was intact. The oil filter was examined, and no contamination was observed. The oil suction screen was also absent of contamination. When the single-drive dual magneto was rotated by hand, a spark was produced at all towers.

The crankshaft was rotated through an accessory gear drive. Except for the number two intake pushrod that had separated, and the number four exhaust pushrod that was bent, valve train continuity was confirmed. Thumb compression was obtained on the number two, three, and four cylinders. Thumb compression could not be obtained on the number one cylinder, which had sustained impact damage near the exhaust valve.

MEDICAL AND PATHOLOGICAL INFORMATION

An autopsy was performed on the pilot by the Columbia County Coroner's Office, Hudson, New York.

Toxicological testing was conducted on the pilot at the FAA Toxicology Accident Research Laboratory, Oklahoma City, Oklahoma.

ADDITIONAL INFORMATION

The wreckage was released to a representative of the owner's insurance company on October 11, 2004.

One question that has arisen in judicial proceedings is whether the NTSB's accident reports are admissible as evidence in civil court actions related to aviation accidents. Congress has acted on this question, and 49 U.S.C. § 1154(b) states that "No part of a report of the Board, related to an accident or an investigation of an accident, may be admitted into evidence or used in a civil action for damages resulting from a matter mentioned in the report." A relatively recent judicial interpretation of this statute affirming the absolute nature of the prohibition against using Board reports as evidence is found in *Chiron v. NTSB*, 198 F. 3d 935 (1999). In Chiron, the U.S. Court of Appeals for the District of Columbia Circuit reiterates, "The simple truth here is that NTSB investigatory procedures are not designed to facilitate litigation, and Congress has made it clear that the Board and its reports should not be used to the advantage or disadvantage of any party to a civil lawsuit." Why do you think Congress does not want NTSB accident reports to be used as evidence in civil lawsuits related to the accident NTSB investigated? What harm might come from admitting the reports as evidence?

DISCUSSION CASES

1. Ferguson, a 12,000-hour ATP with no FAA violations ever, was the pilot-in-command of Western Airlines Flight 44 from Los Angeles to seven locations. By the time Flight 44 departed from its Denver, Colorado, to Sheridan, Wyoming, leg, it was already 35

minutes behind schedule. During the flight from Denver to Sheridan, ATC offered Flight 44 a direct clearance to Sheridan. Flight 44 accepted the direct routing to save time and fuel. Ferguson handled radio communications, and the first officer, Bastiani, was flying the aircraft. Neither Ferguson nor Bastiani had ever flown to Sheridan before, but each believed that the other had. Ferguson reviewed navigational charts, but he failed to note that Buffalo Airport was directly under Flight 44's flight path. Sometime around 10:00 p.m., Ferguson and Bastiani noticed runway lights, and they began a visual approach to what they believed to be Sheridan Airport. Ferguson failed to use available radio aids to confirm the destination airport. During the approach, Flight 44 maintained contact with Sheridan's FSS. ATC notified Flight 44 that another aircraft was on final approach. However, neither Ferguson nor Bastiani could spot the other aircraft. The VASI lights Ferguson expected to see at Sheridan were also absent. It was only after the aircraft landed that Ferguson realized that Flight 44 had landed at Buffalo Airport, not Sheridan Airport. Ferguson filed a timely NASA ASRP report outlining the incident. The FAA pursued enforcement action, and Ferguson sought NASA immunity from sanction. The FAA refuses to waive sanction, claiming that Ferguson's actions were not inadvertent. Should Ferguson be able to claim NASA immunity? Why or why not?

2. Money is involved in a fuel exhaustion incident. She files a timely NASA report and on the identification strip where it asks for "Type of Event/Situation" Money writes, "Emergency landing due to fuel exhaustion." The FAA pursues an enforcement action against Money, citing fuel mismanagement and careless operations. The FAA seeks a 90-day suspension of Money's commercial pilot certificate. Money presents the FAA with a copy of the identification strip when she claims immunity from FAA sanction. Money disputes the charges, and the case goes before an NTSB administrative law judge (ALJ). At Money's hearing, the FAA agrees with Money that she is entitled to waiver of sanction from FAA's proposed suspension. However, the FAA seeks to present the copy of Money's NASA ASRP identification strip as an admission that Money ran out of fuel. The NTSB ALJ admitted the identification strip into evidence. Was it correct for the ALJ to admit the NASA identification strip as evidence of Money's alleged violation? Why or why not?

3. In the late 1970s, National Airport in Washington, D.C. (now Ronald Reagan International Airport), was one of the busiest airfields in the United States. Due to air traffic congestion, the FAA limited the number of commercial airline arrival and departure slots to 40 per hour. Up until October 1980, the allocation of 40 slots per hour was worked out by voluntary agreement by an airline scheduling committee (ASC) made up of airlines serving National Airport. In October 1980 the ASC ran into an impasse and could not decide on slot allocations when New York Air insisted on 20 slots at peak hours. To help break the impasse, the FAA issued a notice and request for comments on how slots should be allocated. On October 16, the secretary for the Department of Transportation issued a Notice and Request for Comments. The notice requested public comments on the mechanism that should be used for the temporary allocation of IFR reservations or slots using National Airport. The notice gave one week for comments and described no mechanism for the distribution of slots. Thirty-nine comments were received, and the secretary issued Special Federal Aviation Regulation 43 (SFAR 43),

effective immediately on October 29, 1980. SFAR 43 decreased the number of slots held by current carriers, shifting those slots to less desirable times. Northwest Airlines claimed that SFAR 43 was unlawful and should be set aside because (1) SFAR 43 was not a product of "reasoned decision making" and is "arbitrary and capricious" and (2) the rulemaking did not comply with the usual time allowed for comments and effectiveness of regulations. How should the court of appeals rule in this case? Explain.

4. Krueger was the nonflying pilot-in-command (PIC) of Midwest Express Airlines' flight from Boston to Milwaukee. The experienced copilot on board the DC-9 was Phillips, and FAA inspector Polak was along to perform an en route inspection. During the preflight checklist procedures, Krueger and Phillips properly went through all items in the checklist. However, the copilot mistakenly notified Krueger that he had placed the auxiliary hydraulic pump switch in the on position. In fact, the switch was in the off position throughout the takeoff when it should have been on. Polak noticed the switch in the off position and after the flight notified Krueger of the incident. The FAA sought to suspend Krueger's ATP certificate for 30 days. Krueger argued that he had a right to reasonably rely on his copilot to properly comply with his duties. The switch in question was approximately 3 feet from Krueger to the right and 12 inches from the bottommost engine indicators. There were no annunciator lights that would indicate the switch was off unless an engine failed. As PIC, should Krueger be held responsible for this incident? Why or why not?

5. Thomas is the owner and director of operations for Aero Charter, Inc., a Part 135 air carrier. She is also qualified as a pilot. On October 30, 1991, respondent and an employee pilot, Guin, took the company's Merlin-2 (an aircraft certificated for single-pilot operations) for a sales flight (a Part 91 flight). Thomas flew various portions of the trip. On the return flight to the company's base in St. Louis' Lambert Field, Guin was the pilot flying the aircraft; Thomas worked the radios almost all the time, called out the checklists and altitudes, searched for and called out the runway environment, and deployed the flaps and propeller position. During the first attempt to land, Guin lowered the gear handle but failed to confirm that the landing gear was down and locked. As the aircraft was about to land, Thomas noticed that the gear-down lights were not illuminated and notified Guin. Guin executed an immediate go-around with only minor damage to the aircraft's propellers and an antenna. The FAA pursued both Thomas and Guin with requests for reexamination and enforcement actions. Guin agreed to a reexamination and negotiated a 5-day suspension with the FAA. Thomas was subject to a reexamination by the FAA which he successfully passed. The FAA also initiated an enforcement action against Thomas, ordering a 15-day suspension. Thomas appealed the FAA's order. She argued that (1) she was successfully reexamined and therefore enforcement action was not appropriate and (2) she was not a required crewmember in the aircraft and therefore should not be held responsible for the incident in question. Do you agree with Thomas' arguments? Explain.

6. Roach is the president of Roach Aircraft Company. He has been in the business of selling aircraft for over 20 years, and he has accumulated more than 10,000 flight hours. He has no record of any FAA violations. While demonstrating a Piper Aerostar, Roach

made several low passes and an aileron roll at low altitude. The FAA investigated the matter and issued an order suspending Roach's commercial pilot certificate for 120 days. During the course of the NTSB hearing on Roach's case, the FAA attorney called Roach as an adverse witness. Roach's attorney objected and argued that the case involved a semicriminal case and therefore Roach should not be compelled to testify. The ALJ overruled the objection and Roach went on to testify. Was the ALJ correct in compelling Roach to testify? Explain.

7. Atkins and Richards were pilot-in-command and second-in-command, respectively, of Piedmont Flight 1756. Flight 1756 had been cleared by Atlanta Air Route Control Center (ATC) to climb and maintain flight level 220. Atkins and Richards acknowledged the clearance and had previously expressed their desire to climb higher than flight level 220 because of turbulence. A short while later ATC cleared Piedmont Flight 1258 to climb to flight level 240. Atkins and Richards mistakenly believed that the transmission from ATC was meant for them. Atkins read back the clearance, transmitting the following to ATC: "Piedmont seventeen fifty-six up to two four zero thousand now. Thanks." Neither the controller assigned to the relevant position nor the controller monitoring the communications of the controller (the controller was undergoing recertification) heard the readback by Atkins. When Flight 1756 reached flight level 226, loss of separation occurred with another aircraft and Atkins was instructed to make an immediate left turn. The FAA charged Atkins and Richards with violating FAR §§ 91.75 and 91.9 (currently 91.13). The ALJ hearing the case found that Atkins and Richards were careless because they mistook an ATC communication to another aircraft. Atkins and Richards appeal to the Board. Who should prevail? Why?

ENDNOTES

1. 343 U.S. 470 (1952).
2. See 5 U.S.C. § 553.
3. Id. at 551(4).
4. Id. at § 553.
5. See www.gpoaccess.gov/fr/ for current Federal Registers.
6. 5 U.S.C. § 557(c).
7. Id. at § 553(b)(3)(A).
8. See 14 C.F.R. § 135.261 et seq.
9. 323 U.S. 134 (1944).
10. 5 U.S.C. § 553(b)(3)(A).
11. See *United States v. Powell,* 379 U.S. 48 (1964).
12. See 14 C.F.R. § 61.51(i)(1). See also 14 C.F.R. § 91.417 for aircraft maintenance records.

13. See *Administrator v. Salkind,* 1 NTSB 714 (1970) and *Administrator v. Funk,* 6 NTSB 1016 (1989).
14. See 49 C.F.R. § 821.33.
15. 346 F. 3d 1121 (2003).
16. See 5 U.S.C. § 554(c)(1) and 49 U.S.C. § 44709(c).
17. 3 NTSB 1319 (1978) and 3 NTSB 2997 (1980).
18. Typically, an ALJ will issue an oral initial decision at the conclusion of a hearing. However, in some cases, the ALJ may deliberate further and issue a written decision as per 49 CFR § 821.42(a).
19. See 49 U.S.C. § 44709(f).
20. See FAA AC No. 120-66B.
21. See FAA Order No. 2150.3A.
22. While administrative actions remain on an airman's record for 2 years, a violation resulting in a suspension of pilot privileges will only be expunged after 5 years. See FAA Enforcement Records; Expunction Policy, 56 Fed. Reg. 55,788 (1991).
23. EA-4116 (1994).
24. 49 U.S.C. § 44709(a).
25. See *Administrator v. Hinman,* 2 NTSB 2496 (1976) and *Administrator v. Gamble,* EA-4789 (1999).
26. See *Administrator v. Gamble,* EA-4789 (1999).
27. See *Administrator v. Thomas,* EA-4309 (1994).
28. See *Administrator v. Carson,* EA-3905 (1993).
29. See 5 U.S.C. § 701.
30. See Id. § 704.
31. 49 U.S.C. § 44709(d)(3).
32. See 5 U.S.C. § 504.
33. See 49 CFR § 821.1 et seq.
34. See 5 U.S.C. § 552(a).
35. 900 F. 2d 369 (1990).
36. 499 U.S. 144 (1991).
37. See Robert A. Anthony, "Symposium on the 50th Anniversary of the APA: The Supreme Court and the APA: Sometimes They Just Don't Get It," 10 Admin. L .J. Am. U. 1 (Spring 1996).

6
Commercial Law Applications to Aviation-Related Transactions

CONTRACTS 160
Mutual agreement 160
Consideration 161
Capacity 162
Lawfulness 162
Defenses to contracts 164
Rights, duties, and remedies for breach 167

SALES LAW 170
Sales law versus contract law 171
Warranties 171
Transfer of title and risk of loss 173

DEBTOR-CREDITOR LEGAL ISSUES 174
Credit—unsecured and secured 175
Bankruptcy 177

CASES AND COMMENTARY 179

DISCUSSION CASES 187

ENDNOTES 189

Every day aviation businesses and professionals are engaged in commercial transactions that are aviation-related. The function of this chapter is to provide you with a working knowledge of basic commercial law as it relates to the aviation environment. For the purposes of this chapter, we will loosely define commercial law as the law of contracts, sales law, and debtor-creditor relationships.

CONTRACTS

Have you entered into any contracts today? You may not think you have, but it is likely that you have engaged in one or more contracts by the time any given day is over. Every time you make a purchase at a retail store, whenever you eat out at a restaurant or cafeteria, as you register for classes at your university—during any of these routine occasions you are entering into a contract.

What is a contract? A contract is defined by the American Law Institute (ALI) as "… a promise or a set of promises for the breach of which the law gives a remedy or the performance of which the law in some way recognizes a duty."[1] Boiled down to its essence, the definition indicates that a contract involves promises that the law will enforce. Does this mean that there are promises that the law will not enforce? Indeed, the answer to this is yes—there are many promises the law will not enforce. For example, courts will not enforce social promises; for instance, a promise that you might make with a girlfriend or boyfriend to go on a date this coming Saturday may be morally binding, but it is not legally binding.

What will make one promise legally enforceable while another promise is not? The answer to this question has been developed over many centuries of cases establishing a formidable body of common law for contracts. It essentially distills to the following: For a contract to be legally enforceable, it must meet all these requirements:

1. There must be a mutual agreement or meeting of the minds between all parties.[2]
2. The agreement must be supported by legally recognized consideration.[3]
3. Parties to the contract must have the intellectual capacity to enter into a contract.[4]
4. The agreement must be for a lawful objective.[5]
5. There must not be any legal defense to the contract.[6]

We will briefly go through each of these requirements, along with some other basic contract law issues.

Mutual agreement

To have a legally enforceable contract, both parties must agree on the substance of the agreement. Many times this requirement is referred to as a "meeting of the minds."[7]

When courts check to see if there is a mutual agreement between the parties, one of the first things they might look for is whether there was (1) an offer and (2) an acceptance. An *offer* is defined as "the manifestation of willingness to enter into a bargain, so made as to justify another person in understanding that his assent to that bargain is invited to and will conclude it."[8] If someone says, "I will buy your airplane for $100,000," it will probably be considered a valid offer because it demonstrates the current intent to make a deal. On the other hand, if someone says, "Would you consider selling your airplane for $200,000?" it is probably not an offer—most likely the courts would view this kind of statement as an invitation for negotiations. In the end, to create a valid offer, the terms of the offer must be definite and communicated with clarity.

An *acceptance* is an action or a promise by the party receiving the offer that indicates a willingness to abide by the offer as presented. One issue that comes up from time to time is the offer that is accepted with a condition. Suppose you were selling an airplane and your offer was "as is" with a $200,000 sales price. A few days later a potential buyer comes to you and states, "I accept your offer to sell the airplane for $200,000 as long as you repaint the aircraft before I take delivery." That sort of exchange does not represent an offer and acceptance. The law views this exchange as an offer and a counteroffer. Therefore you, as the seller, have the legal ability to either accept or reject the counteroffer presented by the prospective purchaser.

Consideration

Consideration is the second requirement for an enforceable agreement. The term *consideration* means something of legal value that is given in exchange for a promise or an act.[9] Consideration is typically presented in the form of money. However, it can be other property, performance of services, or the giving up of a legal right (e.g., agreeing to drop a lawsuit in exchange for a monetary settlement). In essence, consideration means giving up or doing something you otherwise would not have to give up or do.

The requirement for consideration makes many promises legally unenforceable. For instance, a promise to give a friend a new Global Positioning System (GPS) receiver for her birthday is not an enforceable contract—your friend does not give up anything in exchange for your promise. This situation is often referred to as a gift or gratuitous promise. If, however, you promised to give your friend a new GPS if she tutored you for your aviation law class—and she did—the promise would be legally enforceable.

Sometimes determining whether consideration exists is a tricky question. Suppose your company sent one of your corporate aircraft for a new paint job. You were placed in charge of seeing the job through and given a budget of $30,000. The paint shop agreed on a fixed price of $25,000, and an estimated completion date was set. About one week before the expected completion date, the paint shop calls your office to inform you that the price of the paint you wanted to use on the aircraft recently increased by 50 percent, and the company would need an extra $5000 to complete the job in a timely manner. You agree to pay the extra $5000; the aircraft is painted, the job looks good, and it is done on time. When your boss sees the bill for $30,000, he discusses the deviation from the original agreed upon price with you. Although the job was still within your budget, your boss refuses to pay the extra $5000 requested by the paint shop. He sends a check for $25,000, and the paint shop sues for the remaining $5000. Who will win?

The initial instinct of most nonlawyers is to say that the paint shop will win. After all, you promised to pay the additional $5000. You should have to live up to your promise.

However, most courts would look at this situation differently. The fact is that the paint shop agreed to paint the airplane for $25,000. That was the deal. When the paint shop called you to ask for $5000, the company offered no additional consideration for the extra money. In other words, the paint shop had a preexisting contractual duty to finish the job for $25,000. The paint shop will not be able to change the deal just because the cost of the job increased. However, the courts would respond differently to this situation if the paint shop simply offered something in the way of additional consideration, for example, some fancy decals or a faster delivery schedule. If that were the case, the courts would uphold the need to pay the extra $5000 because of the additional consideration.[10]

Capacity

For an agreement to rise to the level of a legally enforceable contract, another requirement that must be met is capacity. *Capacity* means the ability to knowingly enter into and understand the obligations that a contract entails. The law presumes that all parties to a contract have the capacity to enter into the contract. However, there are certain instances in which the law recognizes that a person may not be contractually capable. The three most common instances of incapacity to contract are

- Minority age[11]
- Mental incompetence[12]
- Intoxication[13]

In the aviation environment, you seldom run across cases involving the issue of incapacity. However, if a person entering into a contract is deemed to have lacked contractual capacity at the time of the contract, the contract will be deemed to be voidable by the party with the incapacity. This means, for instance, that if an aircraft was sold to a16-year-old, the 16-year-old can return the aircraft (at his option) after the purchase and receive back the entire purchase price (even if the aircraft was used for several hours of flight time).

Lawfulness

Beside the requirements already discussed, an agreement has to have a legal objective in order to be enforceable under the law. If an agreement involves illegal objectives, it will be deemed void by the courts—this means that it cannot be enforced by any of the parties involved.

Obviously, contracts that involve the commission of crimes are illegal and not enforceable by the law. A contract employing a "hit man" to dispose of a nasty boss or curmudgeonly college professor is clearly not going to be enforceable by either the "hit man" or the party paying the hit man. However, often the question of illegality in a contract is much more subtle.

Case law is replete with examples of contracts the courts deem "illegal" and thus unenforceable. Examples of "illegal" contracts include, but are not limited to, (1) contracts violating usury laws (laws that cap interest rates on loans), (2) contracts violating gambling or lottery statutes, (3) contracts contrary to public policy, (4) covenants not to compete, and (5) exculpatory or *hold harmless* contracts or clauses within contracts.

Within the context of the aviation industry, we will take a look at covenants not to compete and exculpatory or hold harmless contract clauses in greater detail here. These are two situations that often arise in an aviation context.

Covenants not to compete

Covenants not to compete often arise in the context of the sale of a business and/or employment contracts.

When someone sells a business, the buyer will often want to ensure that the seller cannot open up a similar business that will have the ability to compete with the very business the seller just sold. For example, State Municipal Airport has one fixed base operator (FBO), State Aero. that sells fuel, provides flight instruction, rents aircraft, and provides maintenance and repair services for aircraft. The owners of State Aero enter into negotiations with you to sell their business to you via a sale of all assets. You and your counsel discuss a potential agreement, and your counsel suggests that to protect your interests, you should require that State Aero sign an agreement that includes a covenant not to compete. Your counsel drafts a provision in the final agreement that prohibits State Aero from operating a similar business on State Municipal Airport or any other airport within 50 miles for a period of 2 years. Will this provision be valid and enforceable against State Aero?

Under general common law principles, this agreement might have a problem. The courts generally disfavored any contracts that restricted trade. However, over time, the courts have recognized that covenants not to compete can be valid if (1) they are ancillary to a legitimate sale of a business; (2) the line of business protected is clearly defined; (3) the geographic area protected is reasonable within the context of the agreement and the type of business to be protected; and (4) the duration of the restriction is reasonable. Obviously, the most common disagreements come about when one is attempting to determine just what a "reasonable" restriction might be; but for the purposes of this example, if 2 years and 50 miles are considered fair, the provision will likely stand up to any legal challenge.

In the context of employment agreements, the test is also one of "reasonableness." What if you are presented with an employment agreement from a regional air carrier that prohibits you from working for a competitor for 1 year after you leave your employment? What sort of restrictions might be deemed reasonable with respect to geographic location or time?

Hold harmless clauses

Exculpatory or hold harmless clauses are contract provisions that relieve parties from tort liability. These types of clauses arise from time to time in an aviation context. Properly drafted and in appropriate circumstances, exculpatory clauses may work to relieve a party of liability for ordinary negligence. However, they will not excuse liability for willful misconduct or gross negligence.

As a general rule, courts will not favor an exculpatory clause unless the parties involved are deemed to be of roughly equal bargaining power. When such parity exists, the courts will usually permit the parties to negotiate their risk for liability. However, in all cases, the courts will ensure that the language of the clause in question is unambiguous and fitting for the circumstances in which it is being evoked.

Defenses to contracts

Even if it appears that all the elements to create a valid contract are present, there may still be defenses available that will cause the contract to be unenforceable. The first set of defenses can be generally classified as challenging whether the mutual agreement between the parties was real. Lawyers often refer to this as genuineness of assent. The second defense deals with a more technical issue: Should the contract be in writing? Most nonlawyers believe that all contracts must be in writing. You'll learn that is hardly the case.

Is the mutual agreement real?

There are several possible grounds to challenge the genuine nature of the "meeting of the minds" between the parties to an agreement. Some of the more likely grounds are the following:

- Mistake
- Fraud
- Undue influence
- Duress

Mistake Can a mistake cause an agreement to be unenforceable? The answer to this question depends on the type of mistake made.

If the mistake was a misunderstanding by only one party, often referred to by lawyers as a *unilateral mistake*, the contract will still be enforceable. For example, you approach an aircraft dealer for a new corporate jet for the corporate flight department you manage. You would like an aircraft with a JXY-2 GPS. You do not say anything to the broker about this preference; however, you have assumed that the aircraft you have settled on has the JXY-2. You sign a contract for the aircraft and soon learn that the aircraft does not have a JXY-2. Your unilateral mistake will not cause the contract to be invalidated. One exception to this rule exists if the nonmistaken party knows of the other party's error and uses it to his or her advantage.

Another type of mistake is the *mutual mistake* of material fact. If both parties are mistaken as to an important fact that forms the basis of their bargain, the contract may be rescinded by either party. The basic theory of rescission in the case of a mutual mistake of fact is that there was never a true "meeting of the minds" between the parties—therefore, there was never genuine assent. To illustrate, assume that you purchase an aircraft from a dealer with the understanding that it had a total of 5000 hours and another 1100 hours until an engine overhaul was required. The dealer had the same understanding when he purchased the aircraft from the original owner. However, it turns out that the aircraft had more than 10,000 hours of total time and just 60 hours until its next overhaul was required. In this case, both you and the dealer made a mutual mistake of material fact. It is therefore likely that you would be able to rescind the contract.[14]

One final type of mistake worth noting is mutual mistake of value. This is a mistake in which the parties to the agreement properly identify the object of the agreement. However, both parties are mistaken as to its value. For instance, suppose you spot an advertisement on the Internet posted by someone selling a vintage Piper Cub that has been taken apart and boxed up in parts. Some of the parts are damaged, and it is unknown whether all necessary parts are available to reconstruct the original airplane. The seller is selling the boxed parts for $1000. You are interested and willing to take a chance. You pay the $1000 and take the boxed parts home to your hangar. Once you begin construction of the aircraft, it is discovered that this was the Piper Cub first soloed by Neil Armstrong, the first astronaut to walk on the moon. The airplane can be sold for $100,000. This is a mistake of value, and the seller cannot rescind the contract. If the rule were any different, all contracts could be subject to rescission by the party who did not get the best of the deal—the courts are not responsible to protect parties from mistakes in contracts.

Fraud   Fraud is a defense to a contract. The terms *fraud* and *fraudulent misrepresentation* refer to situations in which a person deliberately misleads another with regard to material facts that are relied upon in forming a contract. If the fraud induces someone to enter into a contract, the law holds that there was never genuine assent to the contract and, therefore, no meeting of the minds. If this is the case, the innocent party harmed by the fraud may be able to void the contract if she or he can prove the following elements of fraud:

- False representation of material fact
- Intention to deceive
- Justifiable reliance on the part of the innocent party
- Injury to the innocent party

Note that the elements of fraud do not permit a party to void a contract simply because another party made a false statement. There must be proof that the false statement was made with the intention to deceive and was reasonably relied upon by the innocent party. There must also be actual damage incurred.

For example, Sam Seller intentionally misleads Betty Buyer into thinking that his aircraft engines had a recent overhaul, when in fact the overhaul was done many years earlier and was never properly documented by Sam in the aircraft maintenance logs. If Sam's fraudulent statement induced Betty to purchase the aircraft and caused her damage because of the need to overhaul the engines sooner than expected, she can void the contract with Sam.

Undue influence   The law may permit a party to rescind a contract if the injured party can show that the other party took advantage of his or her physical or mental infirmity by unduly influencing that person to enter into a contract. In some cases, undue influence is presumed when someone in a dominant position or a position of trust, such as a physician or lawyer, exerts pressure on a patient or client to enter into a contract that benefits the dominant party.

To illustrate, Charlie is 80 years old and suffers partial paralysis from a stroke. He requires constant nursing care. His children hire Valerie, aged 28, as his full-time, live-in nurse. Valerie is an aviation aficionado, and she and Charlie often converse about airplanes and the private airstrip at Charlie's waterfront house in Connecticut, which is now worth several million dollars. When Charlie passes on several months later, his children learn that Charlie sold the beach house and airstrip to Valcrie for $25.00. They sue to have the contract rescinded based on a theory of undue influence. Will they succeed?

Duress Having a contract rescinded due to duress is a very rare occurrence. However, duress is a ground of defense against a contract. Duress occurs when one person forces another to enter into a contract under threat of bodily harm (to the innocent person or another).

Does the contract need to be in writing?

Nonlawyers are often surprised to learn that most types of contracts do not need to be in writing. In fact, when our current system of contract law was first developing in England, there was no requirement that any contracts be in writing. However, the English courts became increasingly concerned about perjured testimony being used to establish contracts that never existed. In response to these concerns, the English developed a "Statute of Frauds" designed to reduce fraud by requiring that certain contracts be reduced to written form.

Most states have adopted the Statute of Frauds in one form or another. Contracts required to be in writing include (but are not limited to) the following:

- Contracts transferring or involving interests in land[15]
- Contracts that cannot be performed within 1 year[16]
- Contracts for the sale of goods greater than $500[17]

The requirement for contracts involving interests in land to be in writing is expansive. This requirement includes land, land improvements (such as buildings or houses), minerals, timber, crops, or anything else attached to the land. Mortgages representing indebtedness secured by land must also be in writing, as must contracts with real estate agents who assist in selling land and improvements to land.

The "1-year rule" under the Statute of Frauds requires that if a contract cannot be performed by its own terms within 1 year, it must be reduced to written form to be enforceable. The motivation behind this rule is to protect against misunderstandings and disputes that might occur because of the passage of time and difficulty in remembering all the details of a distant transaction. Under this rule, a contract to hire Jim as a chief pilot for 6 months need not be in writing. However, if Jim is hired as chief pilot with a 2-year contract, the contract will need to be in writing.

Perhaps one of the most important requirements of the Statute of Frauds is that any contract for the sale of goods (goods are usually defined as things that can be touched and moved) greater than $500 be in writing. This requirement has been codified in most state laws adopting the Uniform Commercial Code (to be discussed later in this chapter). This rule is especially important in the aviation environment where many sales of goods include equipment or machinery of very high value—for these contracts to be enforceable, they must, as a general rule, be in writing.

Rights, duties, and remedies for breach

Once all the elements of a contract are in place, and no defenses exist, the contract is a legally enforceable promise or set of promises. Each party has rights and duties as they have been laid out in either a written or an oral agreement. But what happens if a party wants to transfer her or his contract rights or duties to a third party? And what happens if a party breaches the contract by failing to live up to his or her obligations under the agreement? We will briefly discuss these issues below.

Assigning contract rights to a third party

In most cases, a party to a contract can assign or transfer rights in the contract to another party. Most times this is referred to as an *assignment* or an *assignment of rights*.[18]

To illustrate, let's assume that Sam's aviation repair station ("Sam's Mufflers") overhauls and remanufactures airplane mufflers. Most of its business comes from repeat customers who buy the mufflers on credit from the repair station. From time to time, Sam needs cash before his credit accounts come due or can pay. When this happens, Sam assigns his rights to the payment from his customers to a finance company ("Aero Finance") that gives him cash (90 percent of the total amount of the credit accounts due) and then collects the full value of the credit account from Sam's customers. The only requirement for this arrangement to work is that any of Sam's credit customers who are involved must be given notice of the assignment and to whom they should make payment. See Fig. 6-1 for an illustration of how this type of assignment works.

In certain instances, the parties may agree that an assignment may not take place or that it may take place only if the other party approves. These provisions are generally enforceable. However, the courts tend to side on the ability to freely transfer or assign contract rights.

Delegation of contract duties

In concert with the right to assign contract rights, the law also permits the delegation of contractual duties. When a party delegates a duty to a third party, and the third party does not perform in accordance with the underlying contract, the person who delegated his or her duties remains liable.[19]

Although most contract duties can be delegated, some are of such a personal nature or require significant skills and experience that they cannot be transferred. To illustrate,

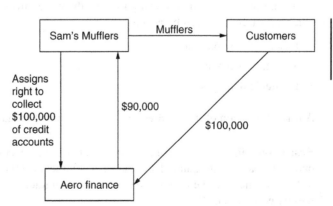

Figure 6-1 Illustration of contract assignment.

suppose a military aircraft manufacturer, Grummor, signed a contract with Charleen, a veteran test pilot and engineer who has flown several thousand hours in aircraft similar to the one she contracted to test for Grummor. If Charleen attempted to delegate her duties to another test pilot, there is a reasonable chance that Grummor could invalidate the designation by arguing that the company specifically wanted Charleen because of her particular experience and skills.

Third-party beneficiaries

Sometimes a contract is designed to benefit someone who is not a party to a contract. Someone designated to benefit under such a contract is referred to as a third-party beneficiary. The most common example of a third-party beneficiary is a designated beneficiary under a life insurance contract.[20]

There are also times when a contract is not designed to intentionally benefit third parties, but it will clearly have the effect of such benefit. For instance, Parker County contracts with a builder to construct a large, modern airport capable of handling large corporate aircraft. Seeing the potential for growth in the area, a real estate developer begins work on developing commercial sites around the area where the expected airport will be located. The developer invests millions of dollars in the project. However, the project is delayed more than 5 years owing to breaches by the builder. The delay causes the real estate developer to cancel the project and declare bankruptcy. Even though the real estate developer would have benefited if the project went through as planned, the developer has no cause of action against either Parker County or the builder—in this case the developer was a mere incidental beneficiary. Incidental beneficiaries have no cause of action against any parties to an agreement.

Remedies for breach of contract

Most contracts, whether verbal or written, come to a successful conclusion with all parties substantially performing their duties in accordance with the contract terms. However, there are contracts that do not have such happy endings. When a party fails to materially adhere to the terms of a contract, the contract has been breached.

So what happens when a contract is breached? What remedies are available to the non-breaching party? For all practical purposes, there are three types of damages available when a contract has been breached:

- Rescission of contract
- Monetary damages
- Equitable damages

A brief discussion of these different forms of contract damages follows.

Rescission Rescinding a contract means undoing it. A contract may be rescinded if one of the parties has materially breached a contract or if the contract is invalid because it does not meet all the required elements of a contract (e.g., the contract was entered into based on fraud).[21]

To undo the contract and restore the parties to the position that they were in prior to the contract, both parties must be prepared to return any consideration transferred by the other party. This return of consideration is often referred to as *restitution*.

Consider this example. Swifty's Pilot Shop enters into a contract with Sonics, a supplier of low-noise headsets, for $200,000. Swifty's makes an initial good faith payment of $20,000, and the first order of approximately $15,000 worth of headsets arrives. It is soon discovered that the headsets are materially defective and the defects cannot be cured. Swifty's can rescind the contract and get back its $20,000. Sonics is entitled to the return of its headsets.

Monetary damages In some cases rescission is not a viable option. In many of these cases, the nonbreaching party may be able to pursue money or monetary damages to remedy a breach. Monetary damages usually come in one of three forms:[22]

- Compensatory damages
- Consequential damages
- Liquidated damages

We will take an overview of each of these different forms of monetary damages.

Compensatory damages With compensatory damages, the law attempts to make the nonbreaching party whole. It does so by compensating the nonbreaching party so she stands in the same position she would have been in if the contract were fully performed.

Consider the following example. An airport owner, Crosswinds County, contracts with Salgado Bros., Inc., a builder, to construct new T-hangars on the airport. Salgado and Crosswinds enter into an agreement requiring Salgado to prepare the land and erect 25 T-hangars within 6 months from the contract date. The contract calls for a $1.5 million fee to be paid to Salgado in exchange for its work on the project. Crosswinds breaches the contract and informs the builder a few days after work is started that it does not wish to continue the project at that time. If Salgado's reasonable expectation was a $1.5 million contract with $1.2 million in costs (labor, materials, etc.), Salgado should be able to collect compensatory damages equal to the amount of its profit if the contract had been fully executed ($300,000) plus any amounts already expended on the work started.

Consequential damages Consequential damages are allowed in cases in which compensatory damages for lost profits or costs paid are not sufficient to bring a party back to where the party would have been if a contract were not breached. Suppose in the example above that just days before Salgado started work on the T-hangars, another airport owner requested Salgado's services, but required that the work start immediately. Salgado declined the offer to do the work because of the commitments to Crosswinds. When Crosswinds breaches the contract, it may be liable to Salgado for the lost profits from the job that Salgado had to forgo because of its expectation that Crosswinds would follow through on its contractual obligations.

Liquidated damages Sometimes, the parties to a contract will agree to damages at the time they come to agreement on the contract. There are two requirements for liquidated damages to be enforceable. First, the circumstances must be such that actual damages would be difficult to determine. Second, the amount of liquidated damages called for in the contract must be reasonable—if the damages are excessive, they will be deemed a penalty and therefore would be unenforceable.[23]

To illustrate, assume that Joan contracts with Pete's A&P Shop to perform an annual checkup on her aircraft. The agreed upon minimum charge for the annual checkup is $4000. Joan tells Pete that it is imperative that the aircraft be ready in 6 weeks on July 1 so that Joan can use the aircraft for some marketing and sightseeing flights she would like to take her customers on during a marketing convention day. Joan and Pete agree that if the aircraft is not ready by that date, she will be entitled to liquidated damages in the sum of $2000. Without good cause, Pete does not have Joan's aircraft ready by July 1, and Joan misses the opportunity to use her aircraft at the marketing convention. In this case, the courts would likely enforce the liquidated damages clause. It is very difficult to assess Joan's damages due to the lost opportunity to schmooze her clients with the flights. Under the circumstances, it also appears that the amount imposed as liquidated damages ($2000) is not excessive or unreasonable, given that the minimum cost of the annual checkup is $4000.

Equitable damages If monetary damages cannot remedy a contract breach, the injured party may have to resort to equitable remedies. The three most common equitable remedies are (1) specific performance, (2) reformation, and (3) injunction.[24]

Specific performance means that the courts will require the breaching party to perform in accordance with the contract. This remedy is very rarely granted because it is permitted only when the subject matter of the contract is so unique that it cannot be replicated. Specific performance is usually available in cases involving land, antiques, heirlooms, etc.

Reformation is also rarely utilized by the courts. However, when necessary, the courts will "reform" a contract to better reflect the parties' intentions. This may be appropriate when typographical or clerical errors need to be corrected in a contract.

Injunctions are granted to prohibit a party from performing an act in violation of a contract. For example, an airport owner signs a 5-year lease for a hangar and office space with a tenant who will use the space for commercial operations. The tenant complies with the terms of the lease in all respects. However, about 18 months into the lease, the airport owner gets a better offer for the space and takes action to have the tenant removed, with an offer to refund the rent paid. The tenant refuses to leave and argues that he has spent a lot of time and money to establish a business at the airport. In a case like this, the tenant will likely file for an injunction ordering the airport owner to abide by the terms of the lease because monetary damages may not be sufficient to compensate the tenant if he is forced to vacate the hangar and office space.

SALES LAW

Sales are transactions involving the transfer of goods from one party to another. Goods are understood to be personal property (as opposed to real property—land or buildings affixed to land). Personal property has the characteristic of being both tangible and movable.

For many centuries, sales of goods were treated just as any other contracts. However, during the early 1900s there was a push to standardize the laws among the various states for transactions involving commerce, including sales transactions. These efforts at standardization culminated in late 1930s with the development of the Uniform Commercial Code (U.C.C.). The U.C.C. dealt with sales law, negotiable instruments, secured transactions, and other commercial law issues. The U.C.C. is now adopted as state law in virtually every state in the United States.

In many respects, U.C.C. sales law is a subset of contract law and property law. For the purposes of our discussion, we will focus on some of the significant aspects of sales law as it relates to transactions in an aviation context. In that regard, first we will look at a few examples of how sales law differs from contract law. Next we will look at how warranties can be created and disclaimed in sales contracts—a significant topic in the aviation environment. Finally, we will review the basics of transfer of title and risk of loss.

Sales law versus contract law

There are several instances in which contracts (driven by common law) and sales (driven by statutory law) differ. Some critical points of departure between the two relate to the issue of definiteness. Under common law, contracts had to be very definite and complete. The U.C.C. loosens these requirements. Specifically, the U.C.C. allows sales contracts to contain open price, delivery, and quantity provisions. This means that a sales contract, unlike a common law contract, can be enforced with no reference to price, mode of delivery, and/or quantity of goods to be sold. The U.C.C. addresses these issues by requiring that parties act in good faith and that any amounts unknown at the time of contract be "filled in" using reasonable valuations, times, and amounts.[25]

Another significant difference between common law contracts and the U.C.C. involves the use of option contracts. Under the common law version of contracts, an offer could be revoked at any time before an acceptance. For instance, if Bob was selling a used GPS unit for $2000 and told Juanita that he'd hold his offer open to her for 10 days, Bob could still sell the unit to a third party within the 10 days because Juanita provided no consideration in exchange for Bob's promise to hold the offer open for 10 days. However, under the U.C.C., Bob's promise to keep the offer open would be treated as an "option" and would be enforceable, even with no consideration from Juanita.[26]

There are numerous additional differences between the U.C.C. treatment of sales contracts and common law contracts. Suffice it to say that the U.C.C. reduced the need for rigid formality required under the common law. The rigidity is replaced with standards of good faith and reasonableness that the courts must enforce, if necessary. In the end, the law of sales under the U.C.C. was designed to make it less cumbersome for parties to engage in commerce.

Warranties

In the aviation environment, the issue of warranties is frequently encountered. Under the U.C.C. warranties are assurances from a seller to a buyer that the goods being purchased meet certain standards. If the goods fail to meet the standards set in the warranty, the buyer has a claim against the seller for breach of warranty.

The U.C.C. recognizes two types of warranties: express and implied. A quick review of the legal essentials for both types of warranties follows.

Express warranties

Express warranties are the most recognizable type of warranties. They are created when a seller affirms that the goods being sold meet certain quality standards. Contrary to what most lay people believe, express warranties do not have to be created in writing. They can be generated by oral statements or inferred by the conduct of the seller.[27] Express warranties may also be created by descriptions of goods that become a part of the bargain[28] and samples of models of the goods to be sold.[29]

It can be confusing at times to sort out the difference between express warranties and salesmanship involving statements of opinion. If an aircraft dealer tells you that the aircraft you are looking at is the "sweetest flying aircraft this side of the Mississippi," the dealer is stating an opinion, not creating a warranty. On the other hand, if the dealer tells you the engine on the aircraft has 1000 hours before the next required overhaul, the seller creates an express warranty.

Implied warranties

Three types of implied warranties are important in the aviation environment: (1) implied warranty of title, (2) implied warranty of merchantability, and (3) implied warranty of fitness for a particular purpose.

The implied warranty of title is relatively straightforward. This implied warranty means that whether stated or not, all sellers warrant that (1) they hold title to the asset they are selling and have the legal authority to sell the asset and (2) there are no claims by third parties (liens) against the asset. With aircraft, a buyer does have the ability to ensure (by searching FAA records) that the aircraft being purchased belongs to the person who is selling the aircraft. However, with many aircraft parts and accessories, such an inquiry may be difficult, if not impossible. Therefore, the buyer gets the protection of this implied warranty of title.[30]

An implied warranty of merchantability is created when a merchant (someone who regularly deals in goods of a certain type or who has special knowledge of such goods) sells goods. This warranty implies that the goods sold are fit for the ordinary purposes for which they are used. The first thing to note with this type of warranty is that it applies only to sales made by merchants. It does not apply to casual sales made by individuals selling their personal-use goods second hand. The second thing to note is that the implied warranty of merchantability does not require that the goods sold be of the highest quality—only that the goods be usable for their ordinary function.[31]

An implied warranty of fitness for a particular purpose applies to both merchant and nonmerchant sellers. This warranty applies if the following conditions are met: (1) the seller has reason to know that the buyer will make a particular use of the goods to be purchased; (2) the seller indicates that the goods will function appropriately for the contemplated use; and (3) the buyer is dependent on the seller's skill or judgment in making the purchase. To illustrate, Joel considers the purchase of new seats for his Cessna Caravan, an aircraft he uses in his cargo operations. He contacts Torgeson Aircraft

Refurbishment to see if it can help. After Joel describes his needs, the salesperson at Torgeson leads Joel to purchase a set of reconditioned aircraft seats for the Caravan. After a few uses, it becomes apparent to Joel that the seats are fine, but they are not the right fit for the Caravan—they are hard to adjust and too tight against the side panels of the cockpit. Joel would have a cause of action against Torgeson for breach of warranty of fitness for a particular purpose.[32]

Implied warranties may be disclaimed by sellers. The U.C.C. provides for several different approaches to legally disclaiming implied warranties. Perhaps the most widely used approach to disclaiming implied warranties is the use of language clearly and conspicuously indicating that a sale is being made "as is" and/or "with all faults."[33] Implied warranties may also be inapplicable when a buyer has been offered an opportunity to inspect goods but has refused the opportunity to examine.[34] Both of these disclaimers and exclusions come about frequently in cases of aircraft sales in which used aircraft are sold and when a buyer is offered the opportunity for a prepurchase inspection of an aircraft but fails to do so.

Transfer of title and risk of loss

Rules related to transfer of title and risk of loss are particularly important in the aviation environment. Aircraft and aircraft parts and accessories are expensive, and it is imperative that the parties know just when they own these types of goods and when they bear the risk that they will need to insure for losses. While the discussion below is by no means exhaustive, it is meant to acquaint you with the fundamentals of transfer of title and risk of loss issues as they relate to aviation transactions.

Transfer of title

Once a buyer identifies the goods that she or he will be purchasing and a contract for purchase is in place, either the goods can be physically delivered to the buyer or the buyer can pick up the goods at a predetermined place. The determination of when title is transferred depends, in part, on whether the goods are delivered to the buyer or not.

If the goods have to be transported to the buyer, the next question concerns whether a shipment contract or a destination contract is in place. In a shipment contract, the seller's responsibility is to get the goods to a common carrier (e.g., UPS, FedEx, Yellow Line). In a destination contract, the seller has the responsibility to transport the goods all the way to a place designated by the buyer. In a shipment contract, title passes once the goods are delivered to a common carrier. In a destination contract, title passes once the goods arrive at the buyer's designated destination. Therefore if Earhart Avionics enters into a contract with Air Transport in which Earhart will sell and deliver to Air Transport 5 new weather radar units, via a shipment contract (using UPS as the common carrier), then title to the radar units will pass once the goods reach UPS. If the contract is a destination contract and the designated destination is Air Transport's hangar at Fort Lauderdale, Florida, then title to the radar units does not transfer to Air Transport until they are received by Air Transport at its Fort Lauderdale hangar.[35]

If the goods are delivered to the buyer without the need for transportation, title transfers at the moment of contract. For instance, if Air Transport simply contracted and simultaneously picked up the 5 radar units directly from Earhart, the transfer of title would be complete at the time and place of the contract.[36]

Risk of loss

Article 2 of the U.C.C. establishes a rather intricate set of rules for when risk of loss passes from a seller to a buyer. Determination of risk of loss is important because sometimes goods are destroyed, lost, or damaged before delivery to a buyer (not due to any fault by either the buyer or the seller). The question is, Who will bear such a loss? Interestingly, risk of loss does not always pass to a buyer at the same time as the title does. A review of some of the more commonly encountered risk of loss issues in aviation follows.

The first thing to note is that the U.C.C. allows the parties to determine who bears the risk of loss. Therefore, by agreement, the parties can assign risk of loss, and the law will respect any such agreement by the parties.[37] If the parties do not specifically agree on how risk of loss will be assigned, then the determination of who bears the risk depends on whether the goods are to be transported or not (see example above for Earhart Avionics).

Risk of loss rules when goods are transported to the buyer closely parallel the rules for transfer of title. If the agreement is a shipment contract, then title and risk of loss pass at the time the goods are delivered to the common carrier. If the agreement is a destination contract, then title and risk of loss pass at the time the goods are delivered to the designated destination agreed to by both seller and buyer.[38] Therefore, in transportation cases, the risk of loss will transfer at the same time as the title.

In most other situations (in which no transportation is required to deliver goods to the buyer), risk of loss will depend on whether the seller is a merchant or a non-merchant. If the seller is a merchant, risk of loss will only pass upon the buyer's receipt of the goods. If the seller is a non-merchant, risk of loss will pass only upon the seller's offer to the buyer to pick up the goods. For example, suppose Betty goes to Sidney's pilot aircraft supplies shop and purchases a new transponder for $2000 upon Sidney's agreement to install the transponder in Betty's plane the next day. That evening, the transponder is accidentally destroyed by a fire. Although the parties had a contract for the purchase of the transponder, and although title had passed to Betty, the risk of loss is borne by Sidney because he is the seller. On the other hand, assume that Betty purchased a used transponder from Samantha, a bank executive selling some used parts from an old aircraft she owned. In this situation, the risk of loss would pass immediately to Betty upon Samantha's notification that the transponder was available for pick up.

DEBTOR-CREDITOR LEGAL ISSUES

In *Hamlet* (1600), Shakespeare wrote: "Neither a borrower nor a lender be: For loan oft loses both itself and friend. And borrowing dulls the edge of husbandry." Despite his admonitions regarding debt, there is little doubt that the U.S. economy is driven in large part by borrowing and lending activities. There is even less doubt that in the aviation industry, debtor-creditor transactions have become an integral part of doing business. In the first part of this section, we will take a brief look at the legal aspects of creditor-debtor relationships. In the second part of this section, we take a broad overview of bankruptcy laws—a topic that comes up all too frequently in recent years in the aviation environment.

When creditors lend money or sell goods on credit, they can do so on a secured or an unsecured basis. Unsecured credit is credit that does not require the debtor to provide collateral (security) for the loan. This type of credit carries the highest risk for creditors and often the highest rates for debtors (e.g., credit card debt). Often if a debtor defaults on amounts due to a creditor, the creditor may not be able to recover any of the debt due if the debtor has no assets or income (this type of debtor is often referred to as *judgment-proof*).

Credit—unsecured and secured

On the other hand, secured credit is supported by collateral that the debtor must put up to get the extension of credit. Of course, this type of credit gives the creditor a stronger measure of protection. For this reason, many transactions in the world of aviation are secured transactions. These transactions are governed by Article 9 of the U.C.C. A discussion of the basics of Article 9 follows.

Nature of secured transactions

The two most common forms of secured transactions are two-party and three-party transactions. In two-party transactions, a seller sells goods (such as an airplane) to a buyer and retains a security interest in the goods. For example, Campbell wants to sell his airplane to Palmer, but Palmer does not have the money to pay for the purchase price and cannot get credit elsewhere. Campbell may agree that Palmer can pay over time, but with Campbell retaining a security interest in the aircraft until such time as Palmer has fully paid. A two-party secured transaction is illustrated in Fig. 6-2.

In three-party transactions, the seller sells goods to a buyer who has obtained financing from a third-party lender. The third-party lender then takes a security interest in the goods purchased by buyer. To illustrate, assume in the example above that Campbell is unwilling to sell his aircraft unless Palmer can pay the full purchase price. Palmer is able to find a bank that will finance a loan for the purchase price. Campbell is paid in full by funds from the bank, and Palmer now owes the bank the amount borrowed to purchase the aircraft. The bank retains a security interest in the aircraft until Palmer pays back the loan. A three-party secured transaction is illustrated in Fig. 6-3.

Figure 6-2
Two-party secured transaction.

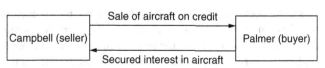

Figure 6-3
Three-party secured transaction.

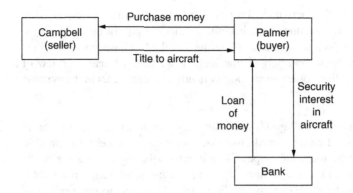

Protecting security interests

While the bank in the example above may have a security interest in the aircraft purchased by Palmer, what would happen if Palmer defaulted on an earlier (and unrelated) loan by another bank and that bank sought to obtain the aircraft in a judgment against Palmer? This scenario raises the question of how a secured party (such as Palmer's bank) can protect its security interest against the interests that other creditors may have in a debtor's assets.

In U.C.C. Article 9, the law provides for three ways to protect a security interest. The U.C.C. refers to this process as *perfecting* a security interest.

The first and most common method of perfection is to file a financing statement or other approved document with the appropriate government recording office. This filing puts the world on notice that the creditor has a security interest in the debtor's property.[39] In many aviation transactions involving aircraft, engines, and propellers, the place to send any such documents is the FAA's Civil Registry Branch in Oklahoma City, Oklahoma. This allows anyone who extends credit to review an aircraft's files to determine whether there are any prior creditors with an interest in the same collateral. Once the debt is satisfied by the debtor, the creditor must file a termination statement indicating to anyone reviewing government files that the debt has been discharged and the collateral released from the security agreement.

A second method, which is very rarely used, is *perfection by possession*. This method requires the creditor to physically hold an item as collateral until the loan or extended credit is paid off. As you might guess, this approach is somewhat impractical in most circumstances, especially in an aviation setting in which an aircraft or equipment must be used by the debtor. When this method is used, the collateral must be returned upon the payment of any debt owed.[40]

The third method is also a rarity in aviation, but not so uncommon in other transactions. It is known as *perfection by purchase* of the money security interest. This method allows for automatic protection or perfection in the case of consumer goods (e.g., home appliances, furniture, televisions) when the seller extends the credit.[41]

Other types of secured transactions

Artisans' and mechanics' liens are liens that artisans or mechanics may place on property they have worked on in order to secure payment for their services. In the aviation environment, if a mechanic has performed work on an aircraft or related equipment, most state laws will permit the mechanic to hold a lien against the aircraft or equipment regardless of whether the mechanic has possession of the aircraft or equipment. In other states, the lien can be created without the need for possession. The FAA will record mechanics' liens from 32 states. In the remaining states, the FAA requires a court judgment to record the lien.

Guarantees and surety arrangements

From time to time, a creditor may refuse to grant any credit to a party unless a third person either guarantees or serves as a surety. In both cases, the guarantor or surety holds out his or her credit as the security for the transaction.

A guarantor agrees to pay the debt of a debtor if the debtor defaults. In this type of arrangement, if Joe's parents guaranteed his debt on an airplane he purchased for his new flight school, and Joe defaults, his parents will have to pay the full amount due on Joe's debt.

On the other hand, a surety is primarily liable. This means that the creditor does not have to go after the debtor first. The creditor can simply pursue the surety. Sometimes sureties are also referred to as cosigners. In the previous example, if Joe's parents were cosigners, the creditor who loaned the money would not have to pursue Joe first in order to collect on the debt. Instead the creditor could simply turn to Joe's parents as soon as the debt came due.

In the case of a guaranty or a surety, the debtor will be liable to the guarantor if she or he fails to pay the obligation. If Joe failed to pay his loan, his parents would have a claim against him if they were required to pay as guarantors or sureties.

Bankruptcy

Sometimes the relationship between a creditor and a debtor does not end on a happy note. Debtors are sometimes unable to make good on their loans, and creditors are sometimes harmed as a result.

The framers of our Constitution were well aware of this problem. They provided in the Constitution that "the Congress shall have power ... to establish ... uniform Laws on the subject of Bankruptcies throughout the United States."[42] Because of this very explicit mandate to the federal government and subsequent laws, federal bankruptcy law has superceded state laws on issues of debtor insolvency or bankruptcy. Consistent with the need for a uniform body of bankruptcy law, all bankruptcy courts are federal courts.

Bankruptcy laws are very complex—we have no intention here of providing anything but a brief overview with a focus on the most relevant bankruptcy issues affecting the aviation industry. The discussion will begin with an introduction to the purpose of bankruptcy laws, a description of the different types of bankruptcy available, and a closer look at reorganizations under Chapter 11 of the Bankruptcy Code—an all too familiar journey for many prominent air carriers in recent years.

Purpose of bankruptcy laws

The need for uniform bankruptcy laws resonated with the framers of the Constitution. They understood that to create a society with an economic vibrancy, you had to acknowledge that progress required risk taking. Of course, risk taking always carries the chance of failure—the greater the risk, the greater the chance for failure. Rather than stifle the motivation to take risks, the framers thought it would be best to provide failed debtors with a chance at a fresh start. For better or for worse, that is what bankruptcy laws are designed to do—allow debtors a chance to get out from under burdensome debt and to move forward.

Specifically, bankruptcy laws achieve their general purpose by

- Creating an orderly process for creditors to get at a debtor's remaining assets without taking unfair advantage of other creditors
- Protecting creditors from improper or illegal attempts by a debtor to hide assets
- Shielding debtors from abusive collection methods
- Providing a chance to maintain existing business relationships

Different types of bankruptcy

There are several different provisions for bankruptcy. The Bankruptcy Code contains nine chapters that deal with different types of bankruptcies and procedural issues in bankruptcy courts. The most common types of bankruptcy are *liquidation* (Chapter 7), *reorganization* (Chapter 11), and *consumer debt adjustment* (Chapter 13).

In a Chapter 7 or liquidation bankruptcy, a debtor files for protection with a bankruptcy court, and the filing automatically suspends or stays all creditors' actions against the debtor. All of a debtor's property, with the exception of certain property exempted by federal and/or state law (e.g., a residence or retirement funds), is sold for cash. The cash remaining after subtraction of the expenses of the bankruptcy (e.g., lawyer and accountant fees) is used to pay off creditors to the extent possible. This discharges the debtor from having to make any further payments on the debt involved in the bankruptcy proceeding. In most modern-day liquidations, creditors receive only pennies for each dollar they are owed. Sometimes a Chapter 7 bankruptcy is referred to as a "straight" bankruptcy.

Chapter 11 bankruptcy is also referred to as reorganization. In a Chapter 11 proceeding, the debtor can continue to operate its business, but must do so under the supervision of the bankruptcy court. This is typically the approach employed by a business that is viable, but unable to continue presently under the strain of current liabilities. One important feature of a Chapter 11 bankruptcy is the right of the debtor to retain possession of its assets during the bankruptcy proceedings. This gives the debtor the ability to continue its operations as it works its way through bankruptcy.

A Chapter 13 bankruptcy is designed to allow consumers to readjust their debt. The readjustments can take the form of a plan authorized by the bankruptcy court in which the debtor stretches out payments (over no longer than a 5-year period) and/or reduces the total payments to be made. Once the plan is executed, the court can discharge the debtor from any unpaid debts under the plan of readjustment.

Airlines and Chapter 11 bankruptcy

In the recent past, several airlines in the United States have filed for bankruptcy protection under the provisions of Chapter 11. Under Chapter 11 proceedings, the airline is permitted to continue operations during the process of reorganizing and restructuring its debts. Creditors' committees—consisting of the airline's largest creditors—are formed and participate in the reorganization plan. During the bankruptcy proceedings, the debtor airline gets a stay from certain legal actions by creditors. Once a plan is approved by the airline's creditors and the bankruptcy court, the debtor airline is granted a discharge of all debts not included in the approved reorganization plan—this terminates the discharged airline's legal obligation to pay these debts.

Some large air carriers in the United States have filed for Chapter 11 bankruptcy once or even multiple times. Some have survived the process, and others such as Eastern, TWA, and Pan Am have not. Beyond the legalities involved in the bankruptcy process, you should consider, in conjunction with Case 6-3 (found below in the "Cases and Commentary" section), the following questions:

1. Does seeking Chapter 11 bankruptcy protection give an airline an unfair advantage over its competitors?
2. Should there be a limit on the number of times an airline can seek the shelter of Chapter 11 bankruptcy (some have been back to the courts two or three times)?
3. What industries and companies are most likely to suffer (as creditors) when an airline gets Chapter 11 protection from the courts?
4. Who wins (financially) when an airline seeks Chapter 11 reorganization?

Delta and Northwest Airlines recently filed for bankruptcy protection under Chapter 11. At the same time, tighter measures for Chapter 11 protection go into place and will be in place by the time this text is published. It is anticipated by some legal experts that there will be more filings—maybe even a raft of filings—in the near future to avoid the obstacles imposed by the new bankruptcy laws. Some even believe that airlines that might have gone to Chapter 11 bankruptcy for protection will now chose to file for Chapter 7 and liquidate instead. Only time will tell how the new laws affect the airlines and the industry as a whole.

CASES AND COMMENTARY

Sales of aircraft and related equipment are usually subject to the requirements of the U.C.C. This means that there should be a written agreement laying out the terms of the purchase and sale. It also means that the parties have to pay close attention to the issue of warranties. If sellers do not wish to create warranties (and most sellers of previously used aircraft and parts do not want to create warranties), they must disclaim them in accordance with U.C.C. rules. On the other hand, buyers should be aware of the fact that a lawful disclaimer of a warranty means that they purchase an aircraft or aircraft parts at their own peril—careful inspections and research may be necessary. Case 6-1 makes the point that a disclaimer of warranty, if properly drafted and agreed to by the parties, will be enforced by the courts.

CASE 6-1
DALLAS AEROSPACE, INC. V. CIS AIR CORPORATION
352 F. 3d 775 (2003)

OPINION BY: JOHN M. WALKER, JR.

Plaintiff-appellant Dallas Aerospace, Inc. ("Dallas"), a buyer of a used jet engine from defendant-appellee CIS Air Corporation ("CIS"), appeals from the judgment of the United States District Court for the Southern District of New York (Barbara S. Jones, District Judge), granting summary judgment to CIS.

Dallas and CIS are both corporations in the business of buying, selling, and leasing aircraft and aircraft engines. The various claims at issue in this appeal arose out of CIS's sale to Dallas of a JT8D engine in August 1997 under a written agreement. Months after the purchase, Dallas discovered that the engine had been involved in a hard landing years earlier that rendered the engine not "airworthy." "Airworthy" is a term of art in the aviation industry indicating that an engine is safe and that it comports with FAA requirements. Dallas brought this diversity action alleging various claims under New York law to recover the $1.15 million it paid for the engine.

The district court agreed with CIS that there was no genuine issue of material fact precluding a grant of summary judgment in CIS's favor and concluded that, as a matter of law: (1) for the purposes of Dallas's breach of contract claim, the agreement between Dallas and CIS, which disclaimed all representations about the engine, had not been modified; (2) for the purposes of its fraud claim, Dallas could not show it justifiably relied on any purported misrepresentation under the contract because (a) the contract specifically disclaimed the very representation alleged to be fraudulent, and (b) the truth of the allegedly misrepresented matter was easily discoverable by Dallas; (3) the contract terms were not unconscionable; and (4) no special relationship existed between the parties that would trigger a duty to disclose on CIS's part, for the purposes of Dallas's negligent misrepresentation claim.

I. BACKGROUND

A Japan Air Systems ("JAS") aircraft experienced a hard landing in Japan in April 1993. While the ensuing fire substantially destroyed the aircraft, the engine remained intact and was salvaged from the wreckage. The insurance company that took title to the engine sold it to Charlotte Aircraft Corporation ("Charlotte"). In 1996, American Air Ventures, Inc. ("AAV"), a broker, negotiated the sale of the engine to CIS and took title from Charlotte pursuant to a separate contract before transferring it to CIS for $425,000, which was paid directly by CIS to Charlotte. CIS paid AAV a finder's fee of $10,000.

CIS understood that it would have to overhaul the engine to get it back into "serviceable" condition under guidelines established by the engine's manufacturer, Pratt & Whitney ("P&W"), for returning a used engine to service in compliance with FAA regulations. CIS claims that it had no specific knowledge of the hard landing, however, and undertook a less expensive overhaul that was appropriate for used, but not incident-related, engines. CIS sent the engine for overhaul to ST Aerospace ("ST"), a reputable repair shop authorized by the FAA, with overhaul instructions that had been provided by AAV. CIS asserts that it relied on AAV because CIS has no internal technical staff. CIS paid approximately $350,000 for the overhaul and, in due course, ST returned the engine to CIS, certifying it—mistakenly as it turned out—as airworthy. ST was not aware of the engine's incident-related status, which, under the P&W guidelines, would have necessitated a $500,000 overhaul.

Upon return of the engine to CIS, CIS found Dallas as a willing buyer in August 1997, and the two parties quickly reduced their agreement to a written contract, subject to Dallas's inspection of the engine and its records. While the contract between Charlotte and AAV expressly stated that the engine had been involved in an accident, neither the contract between AAV and CIS, nor the one between CIS and Dallas contained any such provision. The extent of CIS's own knowledge about the hard landing and CIS's corporate relationship with AAV are both disputed. It is undisputed, however, that Dallas was not told prior to its purchase about the hard landing in Japan or that ST's overhaul was other than adequate. Dallas's extensive boroscopic physical inspection of the engine prior to purchase revealed no defects and its month-long "back-to-birth" review of the engine's records did not bring the fact of the JAS accident to light. Accordingly, Dallas consummated its purchase of the engine from CIS.

Dallas's contract with CIS, dated August 26, 1997, (the "Agreement") disclaimed that CIS had made any representations regarding the engine. It specifically disclaimed any representation as to the engine's airworthiness in Paragraph 8 and obligated plaintiff to accept delivery of the engine and its records "as-is, where-is" in Paragraph 7. All of the exclusions and disclaimers in the Agreement were "conspicuous," as required by § 2-316 of the Uniform Commercial Code ("UCC"), and the Agreement contained an integration clause stating that "no warranties, representations or undertakings have been made by either party except as expressly set forth herein." On August 29, 1997, Dallas signed the Agreement as well as an Engine Delivery Receipt, which recited that Dallas "accepted delivery of" the engine and "confirms its acceptance of the [engine], in 'AS IS' 'WHERE IS' condition."

Dallas did not pay for the engine until September 9, 1997 or thereabouts. Dallas claims that at the time it wire-transferred payment, it also delivered to CIS a purchase order that, after stating that "all the terms and conditions of this purchase are stated in the contract dated August 28, 1997," purported to modify the contract by requiring CIS to deliver a serviceable and airworthy engine that had "not been subjected to extreme stress or heat as in a major engine failure[,] accident, incident or fire." The status of the purchase order is a disputed issue and is discussed more fully in connection with Dallas's breach of contract claim below.

Dallas subsequently leased the engine to Sky Trek Airlines, and the engine was flown in daily service for several months without incident. In 1999, Dallas attempted to sell the engine, but the prospective buyer walked away from the negotiations after informing Dallas that the engine was incident-related. P&W confirmed this information to Dallas; P&W had always known about the engine's history because it keeps records on every engine it manufactures, including data related to incidents and accidents. While just who had access to the P&W records at the time of Dallas's negotiations with CIS is disputed, it is not disputed that both JAS and Charlotte knew about the engine's incident-related history and would not have withheld the information if they had been asked about it by Dallas at any time.

After unsuccessfully trying to recover its purchase price from CIS, Dallas filed suit against CIS for breach of contract, fraudulent misrepresentation, and negligent misrepresentation in the district court. Applying New York law, pursuant to the Agreement of the parties, the district court granted summary judgment in favor of CIS on all of Dallas's claims. This appeal followed.

II. DISCUSSION

* * *

B. The Breach of Contract Claim

Dallas alleges that CIS breached their agreement by delivering an engine that was falsely represented as airworthy based on an airworthiness inspection for non-incident-related engines performed by ST. CIS responds that two clear and unambiguous contractual provisions in the Agreement disclaim any representation as to airworthiness. Paragraph 8 of the Agreement states (with emphasis in the original):

> DISCLAIMER OF WARRANTY Seller has not made and does not make, nor shall Seller be deemed to have made or given, and hereby expressly disclaims, any warranty, guaranty or representation, express or implied, as to ... the [engine's] title, airworthiness, design, value, operation, condition, quality, durability, suitability, merchantability or fitness for a particular purpose....

Paragraph 13D of the Agreement states:

> This Agreement contains the entire understanding of the parties with respect to the purchase and sale of the engine, and no warranties, representations or undertakings have been made by either party except as expressly set forth herein. Any other previous oral or written communications, representations, agreements or understanding between the parties are no longer of any force and effect, and are superseded and replaced in their entirety by the provisions of this Agreement.

[The court then examines claims by Dallas that the disclaimer provisions were nullified by subsequent actions by the parties].

Under all of the circumstances, no reasonable jury could find that the parties agreed to modify the Agreement so as to nullify its disclaimer provisions. Accordingly, for all the foregoing reasons, the district court's grant of summary judgment on the appellant's breach of contract claim is affirmed.

III. CONCLUSION

For the foregoing reasons, the district court's grant of summary judgment in favor of appellee CIS is affirmed.

The language in paragraphs 8 and 13D in the *Dallas Aerospace* case is very commonly used to disclaim all warranties and to remind all parties that everything that needs to be said is stated in the agreement. If nothing else, this case provides a lesson: What a contract says is what the courts will enforce.

Case 6-2 illustrates the hazards involved when the parties fail to prepare a well-drafted agreement for a sale. Please note that this case involves the issue of express warranties.

CASE 6-2
EDWARD MILES, RICHARD W. KEENAN AND KENNETH L. "DUSTY" BURROW, APPELLANTS, v. JOHN F. KAVANAUGH, APPELLEE
350 So. 2d 1090 (1977)

OPINION BY: Hubbart, J.

OPINION: This is an action for breach of express warranty and misrepresentation in the sale of an airplane. Judgment was rendered for the plaintiff-buyer and the seller-defendant appeals. Party defendants responsible for repairing the airplane prior to the sale also appeal. We affirm.

In March, 1973, the plaintiff [John Kavanaugh] answered a newspaper ad placed by the defendant [Richard Keenan] advertising the sale of a used 1956 Cessna 172 private airplane. The plaintiff and defendant Keenan met on several occasions to examine the airplane and to discuss the sale. The defendant Keenan stated that the engine in the airplane had recently been completely overhauled during which time a number of new mechanical parts had been placed in the engine. The defendant Keenan gave the plaintiff an engine and propeller logbook detailing the mechanical repair and flight history of the airplane which the plaintiff carefully inspected.

The logbook reflected that on May 16, 1972, the engine had been given a major overhaul in which new mechanical parts were placed in the engine all in conformity with the manufacturer's engine overhaul manual. The repair work had been done by the defendant [Kenneth L. "Dusty" Burrow] whose work was certified in the logbook by the defendant F.A.A. inspector [Edward Miles]. Based on the accuracy of this information, the plaintiff purchased the airplane from the defendant Keenan. The plaintiff specifically testified that he would not have purchased the airplane had he not been able to inspect and rely upon the information contained in the logbook.

The plaintiff flew the airplane without incident for several months. Thereafter, he experienced a harrowing engine malfunction while the airplane was in flight. On December 5, 1973, he took off from a narrow airstrip in the Everglades approximately fifty miles out of Miami. After takeoff, the engine began to lose power, shake violently and emit a loud clanking sound. The plaintiff was barely able to land on the Everglades airstrip without crashing.

Subsequent thereto, the plaintiff had to arrange at considerable expense for the airplane to be transported in parts to an aircraft repair shop and there completely reoverhauled. It was there discovered that the prior overhaul had not included new parts as represented and that the prior overhaul had been performed in a completely defective manner. All parties to this appeal agree that the logbook contained inaccurate, misleading and false information about the prior repair history of the airplane.

The plaintiff paid approximately $350 to transport the airplane from the Everglades for repairs and $5,700 for the re-overhaul job. In addition, the plaintiff estimated his loss of use of the airplane during this repair period to be $600.

The plaintiff sued the defendant Keenan and the defendants Burrow and Miles for breach of express warranty and misrepresentation. After a non-jury trial, the court awarded a judgment in favor of the plaintiff against all defendants in the amount of $5,800. The defendant Keenan appeals questioning his liability on the sale of the airplane as well as the amount of damages awarded. The defendants Burrow and Miles appeal solely on the damages issue.

I

The first issue presented by this appeal is whether a private party, who sells his used airplane to a buyer and to induce the sale shows the buyer an engine and propeller logbook setting forth the repair history of the airplane, expressly warrants the accuracy of the information contained in the logbook within the meaning of Florida's Uniform Commercial Code, Section 672.313, Florida Statutes (1975). We hold that the seller expressly so warrants the accuracy of the information contained in the logbook where it forms part of the basis of the bargain between the parties.

The controlling law in this case is set forth at Section 672.313, Florida Statutes (1975), as follows:

> 672.313 Express warranties by affirmation, promise, description, sample.
>
> (1) Express warranties by the seller are created as follows:
>
> (a) Any affirmation of fact or promise made by the seller to the buyer which relates to the goods and becomes part of the basis of the bargain creates an express warranty that the goods shall conform to the affirmation or promise.
>
> (b) *Any description of the goods which is made part of the basis of the bargain creates an express warranty that the goods shall conform to the description.* [Emphasis added.]
>
> (c) Any sample or model which is made part of the basis of the bargain creates an express warranty that the whole of the goods shall conform to the sample or model.
>
> (2) *It is not necessary to the creation of an express warranty that the seller use formal words such as 'warrant' or 'guarantee' or that he have a specific intention to make a warranty*, but an affirmation merely of the value of the goods or a statement purporting to be merely the seller's opinion or commendation of the goods does not create a warranty. [Emphasis added.]

The official comments of the above provision of Florida's Uniform Commercial Code are instructive on the issue presented in this case and state in part as follows:

> (1)(b) makes specific some of the principles set forth above when a description of the goods is given by the seller.

> *A description need not be by words. Technical specifications, blueprints and the like can afford more exact description than mere language and if made part of the basis of the bargain goods must conform with them.* [Emphasis added.]

In the instant case, the defendant Keenan gave the plaintiff-buyer the engine and propeller logbook which, much like a blue-print, set out in some detail the prior repair and flight history of the airplane. The accuracy of the information contained in the logbook formed the basis of the bargain as the plaintiff relied upon the accuracy of such information and would not have purchased the airplane if he had not been permitted to see the logbook. The logbook thus constituted a description of the goods purchased by the plaintiff and an express warranty of the accuracy of such description.

The defendant Keenan argues that he never in so many words warranted the accuracy of the information contained in the logbook and was in fact ignorant of the admittedly false information on the prior repair history of the airplane. The simple answer to that argument is that an express warranty need not be by words, but can be by conduct as well, such as, the showing of a blueprint or other description of the goods sold to the buyer. Moreover, fraud is not an essential ingredient of an action for breach of express warranty and indeed it is not even necessary that the seller have a specific intention to make an express warranty. It is sufficient that the warranty was made which formed part of the basis of the bargain. We find such an express warranty in this case through Keenan's showing of the logbook to the plaintiff without which this sale would never have been made. For breach of such warranty, the defendant Keenan is liable to the plaintiff. [Citation.]

* * *

Is this a fair result? What if the sellers did not even own the aircraft at the time of the misleading maintenance entries? What does the court say about that situation? Would the language used in the *Dallas Aerospace* disclaimer protect the sellers in this case?

The final case for this chapter involves a recent bankruptcy court decision. By way of background, United Airlines sought protection from creditors under Chapter 11 of the Bankruptcy Code.

One of United's biggest liabilities was its pension obligation to retired employees. In these plans, called *defined benefit plans*, United promised to pay its employees a fixed amount during their retirement years. The pension plans were negotiated and agreed to between the various employee groups and United over the years. However, by the time United Airlines filed bankruptcy, United's benefit promises exceeded its ability to pay on those promises by almost $10 billion.

During bankruptcy proceedings, United sought an agreement with the Pension Benefit Guaranty Corporation (PBGC), a taxpayer-subsidized entity that insures private pension providers. PBGC charges companies a premium to insure their pension plans. If the plans fail, PBGC bails them out. In the agreement, United wanted to terminate its pension plans and hand over the liability for payment to the PBGC. The court addresses this issue in Case 6-3.

CASE 6-3
In re: UAL Corporation, et al., Debtors.
Chapter 11, Case No. 02-B-48191 (Jointly Administered)

2005 Bankr. LEXIS 816
May 11, 2005, Decided

OPINION BY: EUGENE R. WEDOFF

OPINION: ORDER APPROVING DEBTORS' EMERGENCY MOTION TO APPROVE AGREEMENT WITH PBGC

Upon the emergency motion (as amended hereby, the "Motion") by the Debtors to approve the Agreement with PBGC attached hereto as Exhibit 1 (as amended by this Order, the "Agreement") and incorporated herein; all interested parties having been afforded an opportunity to be heard with respect to the Motion and all relief related thereto; it appearing that the relief requested is in the best interests of the Debtors' estates, their creditors and other parties in interest; it appearing that the relief requested is essential to the continued operation of the Debtors' businesses; it appearing that this Court has jurisdiction over this matter pursuant to [Citations]; it appearing that this proceeding is a core proceeding within the meaning of [Citations]; it appearing that venue is proper in this District pursuant to [Citations]; adequate notice having been given; it appearing that no other notice need be given; and after due deliberation and sufficient cause appearing therefore, it is hereby ORDERED THAT:

1. The Motion is granted in all respects. All objections not otherwise resolved or withdrawn are hereby overruled, except to the extent set forth in the Court's statements on the record in open court on May 10, 2005. The Court's statements on the record in open court on May 10, 2005 are incorporated herein by reference, including, without limitation:

(A) Under Section [Citation], PBGC may terminate a pension plan in order to protect the pension benefit guaranty system with the consent of the plan sponsor without a court hearing even though that overrides the provisions of a collective bargaining agreement;

(B) Aggrieved parties have their rights under [Citation], to bring actions against PBGC to challenge the propriety of its actions under ERISA (and PBGC reserves its rights in any such action); and

(C) The Agreement, and United's entry into the Agreement, does not violate the law.

2. Any findings by the Court herein and conclusions of law stated herein shall constitute findings of fact and conclusions of law pursuant to [Citation], made applicable to this proceeding by [Citation]. To the extent any finding of fact shall be determined to be a, conclusion of law, it shall be so deemed, and vice-versa.

3. The Agreement is hereby approved.

* * *

Exhibit I
Agreement
IN RE UAL CORPORATION, ET AL. (Case No. 02-B-8191)
Settlement Agreement By and Among UAL Corporation and all Direct and Indirect Subsidiaries and Pension Benefit Guaranty Corporation

This Settlement Agreement (this "Agreement") is made effective as of the Approval Date (defined below) by and among UAL Corporation, all of its direct and indirect subsidiaries, and all members of its "controlled group" as defined under the Employee Retirement Income Security Act of 1974 (as amended, "ERISA"), and all of its successors and assigns (collectively, "United"), and Pension Benefit Guaranty Corporation ("PBGC") (United and PBGC each shall be referred to herein individually as a "Party" and collectively as the "Parties").

WHEREAS, on December 9, 2002, United filed voluntary petitions for relief (the "Chapter 11 Cases") under Chapter 11 of Title 11 of the United States Code (the "Bankruptcy Code") in the United States Bankruptcy Court for the Northern District of Illinois, Eastern Division (the "Bankruptcy Court"), as Case Numbers 02-B-48191 et. seq. and continue to operate their business as debtors-in-possession pursuant to Sections 1107 and 1108 of the Bankruptcy Code;

WHEREAS, the Parties have reached a settlement and compromise (the "Agreement") with respect to their various disputes and controversies in connection with their defined benefit pension plans, and this Agreement sets forth the principal terms and conditions under which United and PBGC will agree to a settlement and compromise of their various disputes in connection with United's Pilot Plan, the Flight Attendant Plan, the Ground Plan, the Management, Administrative and Public Contact Employee ("MA&PC") Plan, and the Variable Plan (collectively, the "Pension Plans"), all on the terms and conditions hereinafter set forth.

NOW, THEREFORE, in consideration of the above recitals and for other good and valuable consideration, the receipt and adequacy of which are hereby mutually acknowledged, and intending to be legally bound hereby, the Parties do hereby agree as follows.

1. Termination of Agreement. Prior to the Approval Date, the Parties may continue to explore alternatives to the termination of the Pension Plans; and if the Parties agree, in each of their respective sole

> and absolute discretion, to pursue any such alternative, this Agreement will terminate.
>
> * * *
>
> 4. Termination of Pension Plans. Subject to Paragraph 1:
>
> a. Flight Attendant and MA&PC Plans. As soon as practicable after the date that the Bankruptcy Court enters an order approving the Agreement (the "Approval Date"), PBGC staff will initiate termination under [Citation] of the Flight Attendant and MA&PC Plans. If and when PBGC issues Notices of Determination that the Flight Attendant and MA&PC Plans should terminate, then PBGC and United shall execute termination and trusteeship agreements with respect to such Plans;
>
> * * *
>
> c. Pilot Plan. As soon as practicable after the Approval Date, PBGC and United shall execute termination and trusteeship agreements with respect to the Pilot Plan, with a termination date that is either mutually agreed by the Parties or judicially determined (but, in neither event any later than the latest termination date for any of the other Pension Plans); * * *
>
> * * *

The United bankruptcy case raises some interesting legal and ethical issues. From a legal standpoint, the bankruptcy law is clear that the pension plan could be terminated as part of United's bankruptcy. The question that arises here is whether this is good law.

At the time of this ruling, the PBGC was $23.3 billion in the red due to earlier defaults by U.S. Airways, United, and other airlines. The issue is whether U.S. taxpayers should be required to subsidize the pensions of a failed business plan. What do you think?

At the personal level, United's failure took a heavy toll on employees. In a related news piece, one recently retired United pilot stated

> I was required by federal law to retire at age 60. Another federal law made me ineligible for Social Security for some period of time, for a minimum of two years after that. And should the pension be terminated, there is a further concern, agewise, because the Pension Benefit Guaranty Corporation, which would take over these plans, imposes very strict penalties if you are less than 65 years old when you retire. We may, in fact probably will, have to sell our home and start over. And you can't start over as a pilot after the federal government shuts you down at age 60.[43]

What, if anything, should be done to assist the current and former employees of a failed company? Is it more important to try to keep the airline alive by terminating the pension plan or to force an airline to keep its pension plan intact for the benefit of retired employees?

DISCUSSION CASES

1. Byron Construction Company was employed as a general contractor to build a hangar/office at a large municipal airport. Byron hired Donovan Carpentry, Limited, to perform the carpentry work on the building. The contract with Donovan clearly indicated that Donovan was responsible for providing all the labor, materials, tools, scaffolding, and any other items needed to complete the carpentry work. On July 17, 2006, Donovan's employees constructed 18 trusses at the job site and installed them. The next day, 16 of the trusses fell off the structure. There was no explanation for the fact that the trusses fell, and evidence indicated that it was not due to any fault of Byron or Donovan. Byron told Donovan that it would pay Donovan to reerect the trusses and finish the job. When the job was finished, Byron paid Donovan the original contract price, but refused to pay the additional costs of reerecting the trusses. Donovan sues Byron for the added expenses. Who will win this lawsuit? Explain.

2. Blake and Associates purchases and sells antique aircraft. Blake purchased a vintage warplane purportedly flown in a famous air duel during World War I. The purchase price was $60,000. Blake then sold the aircraft to Hinds Antiques, Inc., a vintage aircraft collector, for $70,000. A principal of Hinds examined the aircraft for several hours prior to its purchase. Soon afterward, Hinds received an offer of $100,000 for the aircraft subject to certification of its war heritage by the British War Bird Society. When this organization concluded that the aircraft had never seen any action in World War I, and was therefore not worth more than $40,000, Hinds sued Blake to rescind their contract. Will Hinds be successful in rescinding the contract? Why or why not?

3. Kissick invites Jonsen, LeBlanc, and Schmierer to fly with him to Coghill Lake for a fishing trip. Kissick and his three passengers were killed when their aircraft hit a mountain bordering Burns Glacier. Kissick was a major in the U.S. Air Force and a member of the Elmendorf Aero Club. His passengers were all civilians. The Aero Club is established by the U.S. Air Force and managed in accordance with Air Force regulations. Before they departed for the flight, an Aero Club representative directed the civilian passengers to sign Air Force Form AF-1585 in which they agreed not to bring a claim against "the US Government and/or its officers, agents, or employees, or Aero Club members … for any loss, damage, or injury to my person or my property which may occur from any cause whatsoever." Following the accident, the widows of Jonsen, LeBlanc, and Schmierer filed wrongful death claims against the Kissick estate. Kissick asserted as an affirmative defense that Air Force Form 1585 barred all claims. Will the agreement not to sue bar the claims of the widows in this case? Explain.

4. Glover owned an aircraft avionics shop located at Powell Municipal Airport for more than 25 years. His business was very successful, and aircraft owners flew in from surrounding states to have him paint their aircraft. Glover wanted to retire and sell his business. Portier was interested in buying the business, but only on the condition that Glover agree to a noncompete clause in their buy/sell agreement. Glover agreed and signed an agreement with a noncompete clause that stated, "Seller agrees not to enter into the business of aircraft painting within a 500 mile radius of Powell Municipal Airport for period of ten (10) years from the date of this agreement and will not compete in any manner with the Buyers." Two years later, Glover was financially devastated by his compulsive gambling habit. He needed money and opened up a paint shop at Hutchins Town International Airport, about 400 miles from Powell Municipal Airport. Portier sues Glover, claiming that Glover violated their noncompete clause. Who will win? Discuss.

5. Northwest Airlines leased space from the Port of Portland for its Portland Airport Operations. The Port entered into a contract with Crosetti to provide janitorial services for the building. The contract specifically required that Crosetti keep the floors clean in terminal areas. Crosetti also had to indemnify the Port for any losses resulting from Crosetti's failure to perform and to provide public liability insurance for the Port and Crosetti. A Northwest Airlines passenger sued Northwest after injuring herself in a fall at Northwest's ticket counter. It is undisputed that the passenger slipped on a foreign substance on the floor right in front of the ticket counter. After settling the lawsuit with its passenger, Northwest sued Crosetti to recover the amount of its damages in the lawsuit. Northwest claimed that it was a third-party beneficiary of Crosetti's contract with the Port of Portland. Will Northwest win this case? Why or why not?

6. Prior to relocating at a new airport, Burtner attempts to sell some of the equipment he has been storing in his hangar. One of the items for sale is a gasoline-powered tow bar. Martin makes an offer on the tow bar that Burtner accepts for $250. Burtner offers the tow bar to Martin for pickup whenever he is ready. Martin is delayed a few days. In the meantime, the tow bar is stolen from Burtner's hangar. Will Burtner be able to recover the $250 sales price? Explain.

7. Would your answer to discussion case 6 above be different if Burtner were an aircraft parts and equipment dealer? Why or why not?

ENDNOTES

1. *Restatement (Second) of Contracts*, § 1.
2. Id. § 3.
3. Id. § 71.
4. Id. §§ 13, 14, and 16.
5. Id. §§ 181 and 195.
6. Id. §§ 152, 153, 161, 174, and 175.
7. Id. § 3.
8. Id. § 24.
9. Id. § 71.
10. Id. § 89.
11. Id. § 14.
12. Id. § 13.
13. Id. § 16.
14. Id. § 152.
15. Id. § 125.
16. Id. § 130.
17. U.C.C § 2-201.
18. *Restatement (Second) of Contracts*, § 317.
19. Id. § 318.
20. Id. § 302.
21. Id. § 373.
22. Id. § 346.
23. Id. § 356.

24. Id. §§ 361 and 364.
25. U.C.C. §§ 2-204, 2-305, 2-308, and 2-306.
26. Id. § 2-205.
27. Id. § 2-313(1)(a).
28. Id. § 2-312(1)(b).
29. Id. § 2-313(1)(c).
30. Id. § 2-312.
31. Id. § 2-314.
32. Id. § 2-315.
33. Id. § 2-316(3)(a).
34. Id. § 2-316(3)(b).
35. Id. § 2-401.
36. Id. § 2-401.
37. Id. § 2-303.
38. Id. § 2-509.
39. Id. § 9-402.
40. Id. § 9-207.
41. Id. § 9-302.
42. U.S. Constitution, Article I, § 8, clause 4.
43. See *Religion and Ethics Newsweekly*, June 17, 2005, No. 842.

7
Entity Choice for Aviation Enterprises

THE SOLE PROPRIETORSHIP   192
  Establishing a sole proprietorship   192
  Liability issues for sole proprietors   193
  Taxation of sole proprietorships   193

PARTNERSHIPS   194
  Establishing a partnership   194
  Operation of a partnership   194
  Liability issues for partnerships   195
  Ownership of assets   196
  Taxation of partnerships   196
  Termination of a partnership   197

LIMITED PARTNERSHIPS   197
  Establishing a limited partnership   198
  Liability of general and limited partners   198
  Taxation of limited partnerships   200

LIMITED LIABILITY COMPANY   200
  LLC formation   200
  LLC operating agreement   200
  Liability of LLC members   201
  LLC taxation   202

CORPORATIONS   202
  Formation of a corporation   203
  Operating the corporation   204
  Duties of directors and officers   204
  Liability issues for corporations and shareholders   205
  Taxation of corporations   206

CASES AND COMMENTARY   208

DISCUSSION CASES   212

Whether you are a pilot, an air traffic controller, an aircraft maintenance specialist, or a management executive, you will likely work for, or work with, business entities of all different types and sizes. It is important that you gain an appreciation of the basics of business entity law and the pros and cons of each type of entity.

In this chapter, we will review the fundamentals of business entities under U.S. law. Our approach will include a description of each of the more prominent types of business entities. Specifically, we will take a look at the following entity forms:

- Sole proprietorship
- Partnership
- Limited partnership
- Limited liability company
- Corporation

Beyond the legal basics, this chapter will also allow you to take a look at how each of these entities is affected by current tax laws in the United States. Understanding how the tax code impacts an entity is often a critical step in making business decisions and/or deciding what type of entity will best suit your business plan.

THE SOLE PROPRIETORSHIP

One of the most common forms of operating a business in the United States is the sole proprietorship. Although this form of business is often associated with small business enterprises, a substantial number of large businesses are also sole proprietorships.

In the sole proprietorship, the owner is the business enterprise. There is no distinction between the owner and the business. The business is not a separate entity from the owner.

There are advantages to a sole proprietorship, including the following:

- There is no additional cost to starting a sole proprietorship.
- The owner calls all the shots in the business, including the hiring and firing of employees.
- It is a simple process to transfer ownership of a sole proprietorship with no approvals from partners or shareholders required.
- All profits inure to the benefit of the sole proprietor.

Establishing a sole proprietorship

The sole proprietorship is probably the simplest of all business entities to create. There are no government approvals required, and no legal formalities are necessary. Although some local and state governments may require a business license, this is likely to be required of any business. In many cases, a sole proprietorship is looked upon as being the default form of business. If no other form of business has been established, the law will look to the sole proprietorship as the most likely form of business.

In selecting a name for a sole proprietorship, the owner can use her own name or select a trade name that appeals most. A budding aviation entrepreneur named Arnold Tuddle may wish to name his new flight school "Arnie's Future Aces" or Arnold Tuddle d/b/a (doing business as) "Arnie's Future Aces." In many states, any trade names must be registered with the state under fictitious name statutes. The statutes allow consumers and vendors to be able to track down the owner of a business that uses a trade name who might be otherwise difficult to identify by name.

Liability issues for sole proprietors

Perhaps the most significant disadvantage of the sole proprietorship is the unlimited personal liability borne by the owner. Unlimited personal liability means that creditors of the sole proprietorship may recover on their claims by pursuing the sole proprietor's business or personal assets. This disadvantage is arguably magnified in the context of the aviation industry where liability exposure is a constant hazard.

For example, let us assume that Ray wants to establish an air charter business. He names it Ray's Air Services. If anyone is injured while the business operates an aircraft, or if the business cannot pay its creditors in a timely manner, those with claims related to the business can go directly after Ray's personal assets. See Fig. 7-1 for a depiction of how liability to business creditors works in the context of a sole proprietorship.

Because of pervasive concerns regarding liability exposure in the world of aviation, you will seldom find an aviation business that is operated as a sole proprietorship. One alternative to the sole proprietorship model that is discussed below is the one-member limited liability company (LLC) or corporation. For a fuller discussion of these forms of doing business, see the discussion below.

Taxation of sole proprietorships

The sole proprietorship is not taxed as an entity separate from its owner. The profits of the company are treated directly as the profits of the entity's owner. Federal and Internal Revenue Service (IRS) rules require that sole proprietors prepare an income statement for their businesses on a form called a Schedule C. Schedule C includes a summary of all the corporation's revenue and expenses. Any net income or net loss is then filtered through to the sole proprietor's individual income tax return (IRS Form 1040) and in turn taxed at the sole proprietor's individual tax rate. See Fig. 7-2 for a flowchart depicting the approach to taxation for sole proprietorships, assuming an effective tax rate for the owner (Ray) of 20 percent, revenues of $100,000, and expenses of $40,000.

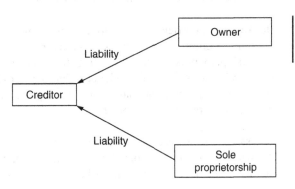

Figure 7-1
Sole proprietorship liability.

Figure 7-2
Sole proprietorship taxation.

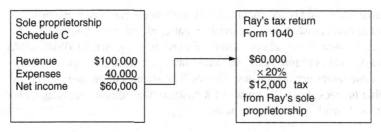

PARTNERSHIPS

Partnerships are a voluntary combination of two or more individuals or entities for the purposes of engaging in a trade or business for profit. Partnerships are also referred to as general partnerships to distinguish them from limited partnerships (discussed below).

Establishing a partnership

A partnership can be created with very little formality or effort. All that needs to be present are two or more persons or entities that voluntarily agree to coown a business with the intention of making and sharing profits. No formal documents or agreements are required for a partnership to be legally established. Interestingly, there is no requirement that partners be active participants in the management of the partnership business. The day-to-day operations of a partnership can be delegated to employees and hired managers without any impact on the legality of the partnership's existence.

Similar to a sole proprietorship, a partnership can function using the name of its partners. A partnership can also use a fictitious trade name, in which case it will be subject to registering the trade name under state law.

One important note is that a partnership is not the same as a coownership. A coownership is evidenced by two or more persons owning an asset. A coownership does not require that the parties have any intention of coowning a trade or business with the intention of making a profit (as is required for a partnership). The terms *coownership* and *partnership* are often confused in the aviation industry. It is important to make the distinction correctly because aircraft can be registered to either a partnership or a coownership. See Chap. 8 for further discussion.

Operation of a partnership

Unlike a sole proprietorship, a partnership has more than one owner. Therefore, there is a bit more complication in the way the partnership will be operated. With a sole proprietorship, there is one person calling the shots, and that person's way of doing things is generally not subject to challenge. With the introduction of two or more owners, a partnership requires greater attention to the rights and duties of each partner in the functioning of the business enterprise.

Unless there is an agreement that states differently, all partners have a right to share in the management of the partnership. This means that each partner will have a single vote, no matter how large or small his or her contribution of capital to the partnership unless an agreement indicates something different. Partners also have the flexibility to designate management of the partnership to one or more partners.

Although it is not required, a well-drafted partnership agreement is strongly recommended for any partnership. In a partnership agreement, the partners can address issues such as capital contributions, management of the partnership, and termination or sale of the partnership.

As indicated above, perhaps the strongest indicator of a partnership is the sharing of profits from the partnership business. The general rule for partnerships is that, unless expressly stated differently, all partners have an equal right to sharing in the partnership's profits. However, most partnership agreements indicate that partners will share in profits in proportion to capital contributions to the partnership by each partner. The same rules also apply to partnership losses. It is assumed, unless expressly stated otherwise, that partnership losses are to be applied equally. Again, most partnership agreements will call for sharing losses in proportion to capital contributions made by partners to the partnership.

Just as with a sole proprietorship, partners share exposure to unlimited personal liability to creditors of the partnership. In addition, there is the added complexity of whether liability can attach to one partner when it was another partner who created the liability in question.

Liability issues for partnerships

With respect to torts, a partner will generally be held liable for the torts of another partner or an employee of the partnership. This means that a partner can be held liable for the negligence or intentional tort(s) of one or more of the other partners. For example, Ray (from the previous example) now wants to expand his business. He takes on a partner and friend named Bob. Bob and Ray form a 50-50 partnership for the purpose of providing on-demand air charter service (B&R Air Services). Both Bob and Ray share duties managing the partnership and piloting the aircraft for charters. One day Ray gets into a landing accident owing to his negligence in providing for proper crosswind corrections, and two passengers are seriously injured. Naturally, Ray is personally liable for the injuries to the passengers. However, Bob will also be liable to the passengers. The passengers may even collect all their damages from Bob, who would then have to pursue Ray for an appropriate contribution of his share of the loss.

Similarly, a partner is liable for the contractual obligations entered into by the other partners on behalf of the partnership. Therefore, all partners are jointly liable for the debt created by the contract. If one partner is required to pay more than her fair share on a judgment, she may be able to seek contributions from other partners to equalize the loss from the judgment. Let's say B&R Air Services also entered into a contract to purchase new Global Positioning System (GPS) equipment for their aircraft. They financed the transaction by taking a note and promising to pay the bank that underwrote the transaction in monthly installments. If Bob and Ray default on the note, the bank can recover damages from either or both of them. Again, if one of them pays an unequal amount of the damages to the bank, that partner can pursue the other partner for contribution to equalize the liability borne by each partner.

See Fig. 7-3 for a chart depicting how partnerships and individual partners share liability in a general partnership.

Figure 7-3
Partnership liability.

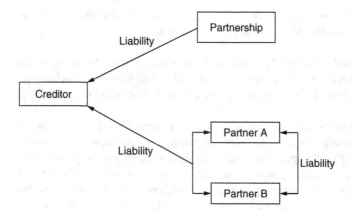

Just as with sole proprietorships, the issue of legal liability, especially in the aviation industry, has greatly diminished the number of pure partnerships that run aviation-related enterprises. Alternatives such as limited liability companies and corporations have become the more likely form of doing business in the aviation environment. These forms of doing business are discussed in greater detail below.

Ownership of assets

Another interesting issue with partnerships is the question of whether individual partners own specific partnership assets. Put another way, this question asks, What do the partners in a partnership actually own? With a sole proprietorship, the question is easily answered. All assets belong to the sole proprietors. However, with a partnership, the answer is a bit more complex. Partners own a right to only their share of the partnership's profits. They do not own any rights to specific partnership assets. Therefore, upon the death of a partner, a deceased partner's heirs cannot seek to recover specific assets such as buildings, aircraft, and computers. Instead, the heirs would be entitled to the deceased partner's share of the partnership profit.

Let's go back to the example of B&R Air Services to further explore this issue. Suppose that B&R Air Services owns two aircraft (both Cessna 310s) of equal fair market value. The partnership owns no other assets (all office space and equipment are leased). If Bob dies, his heirs cannot stake a claim on one of the Cessna 310s based on the fact that Bob was a 50 percent owner of the partnership and is thus entitled to one-half of the partnership's assets. Instead, Bob's heirs can only make a claim to an interest in 50 percent of the profits of B&R Air Services. Further, Bob's heirs do not have any rights to the management of the partnership.

Taxation of partnerships

For taxation purposes, a partnership files a separate information tax return as a business entity on IRS Form 1065. The tax return requires the partnership to report all revenue less expenses. However, the profit is not taxed at the partnership level. Instead, it flows (based on partnership arrangements for sharing profits) to each partner individually. It is at the individual level that partnership profits are taxed at the individual partner's tax rate.

For example, assume that B&R Air Services had revenue of $100,000 and expenses of $40,000 in tax year 2006. The resulting profit of $60,000 would not be taxable at the

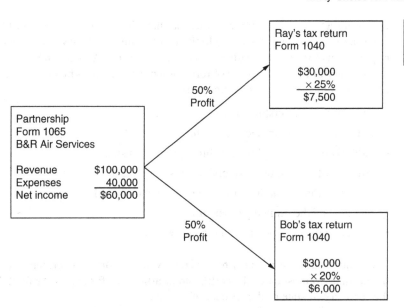

Figure 7-4
Partnership taxation.

partnership level, meaning that B&R Air Services would not be taxed on the $60,000 profit. However, the profit of $60,000 would flow to each partner individually and then be taxed at that partner's individual tax rate. Therefore both Bob and Ray would have to report $30,000 (50 percent of $60,000) as income, and they would be taxed at their respective personal income tax rates (which increase with greater personal income). Figure 7-4 depicts the flow of income from a partnership to its individual partners, assuming an effective tax rate of 20 percent for Bob and 25 percent for Ray.

A partnership is terminated upon the death of a partner or the withdrawal of a partner from the partnership. The termination of a partnership is often referred to as dissolution of the partnership.

Termination of a partnership

However, the remaining partners may still proceed with the business of the dissolved partnership, albeit in a different business form. If Ray decided to withdraw from B&R Air Services, the partnership between Bob and Ray would be dissolved. However, that would not necessarily stop Bob from continuing the business as a sole proprietor or taking on a new partner and starting a new partnership. Of course, it would be best for all parties to have a carefully drafted agreement outlining the procedures to be followed to buy out a withdrawing or deceased partner and continue the business. Such an agreement lessens the possibility of confusion and disputes upon the death or withdrawal of a partner.

LIMITED PARTNERSHIPS

One special type of partnership is the limited partnership. A limited partnership consists of two types of partners: general partners and limited partners. General partners manage the partnership and invest capital in the partnership. Limited partners invest only capital in the partnership and do not participate in the management of the partnership.

Establishing a limited partnership

Unlike a regular partnership, a limited partnership requires that certain formalities be met prior to the recognition of the limited partnership by the law. Each state has its own requirements for a limited partnership to be established. As a general matter, most states require the filing of a certificate of formation or articles of formation. Typically, the certificate will identify the following items:

- The limited partnership name
- Purpose(s) of the limited partnership
- Address of the limited partnership's place of business
- Address of the limited partnership's resident agent
- Identification and address of each general and limited partner
- Duration of the limited partnership
- The amount and type of capital contributions

The limited partnership is only recognized by the law once the written filing is accepted by the state. In most cases the certificate or articles of formation are filed directly with the secretary of state or department of revenue.

The general and limited partners are generally not required to enter into a written agreement outlining the operation of the limited partnership and the rights and duties of each of the parties. However, it is prudent for the parties to draft such an agreement to avoid confusion and/or misunderstanding.

If Bob and Ray of B&R Air Services wished to draw more capital from investors, one approach they might consider is the limited partnership. Bob and Ray might be attracted to the idea that they could run the day-to-day business as general partners while still being able to add to their capital through investors who will not participate in the management of the business. Let us assume they took that option and found four new investors to help them expand their business: Joy, Dean, Joel, and Rose. The parties would prepare a certificate of limited partnership and submit it to the state for approval. The new entity would have to be identified as a limited partnership. One possible name would be B&R Air Services, L.P. Once approved, Bob and Ray would run the business as general partners. Joy, Dean, Joel, and Rose would invest in the partnership but not play an active role in the daily business of the enterprise.

Liability of general and limited partners

One of the unique features of a limited partnership is that it limits the liability of limited partners to the amounts of capital contributions they have made to the partnership. On the other hand, general partners remain personally liable to creditors of the partnership.

As indicated above, limited partners do not participate in the management of the partnership. Limited partners may not contractually obligate the partnership or do business on behalf of the partnership. Once a limited partner gets too actively involved in partnership affairs and "crosses the line" by taking on the role of a general partner, that limited partner steps into the shoes of a general partner and becomes personally liable for the debts of the partnership.

However, the laws for limited partnerships in most states permit some involvement by limited partners before any such loss of limited liability status. The laws that set the bar for how far limited partners can go are often referred to as *safe harbor* provisions. Some of the activities that a limited partner can engage in without "crossing the line" include

- Serving as an employee of the partnership
- Consulting and advising with general partners
- Requesting meetings or attending meetings of general partners
- Voting on major issues such as admission or termination of a partner

These rules were instituted to allow a limited partner some authority to intervene in important partnership matters as necessary.

Following the example of B&R Air Services, L.P., it would not be advisable for Joy, Dean, Joel, or Rose to get involved in the daily operations of the business. If they did, they would risk the possibility of personal liability to partnership creditors. Just how far they could go is largely governed by state law and the safe harbor provisions outlined above. It is also fair to say that the limited partners do not have to sit around as bystanders if the partnership encounters serious enough problems that their investment is in jeopardy. It is often a tough judgment call, but in such situations, there may be room to argue that the active involvement of a limited partner would not necessarily place him or her in peril of personal liability.

See Fig. 7-5 for an illustration of how liability exposure is distributed among general and limited partners in a limited partnership.

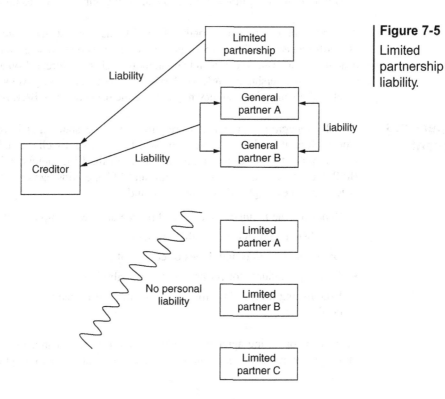

Figure 7-5
Limited partnership liability.

Taxation of limited partnerships

Limited partnerships get the same tax treatment as general partnerships. The partnership is not taxed as an entity. Instead, each partner's share of income or loss is passed through to that partner and taxed at his or her individual tax rates. See the section "Partnerships" above for a fuller discussion of the taxation of partnerships.

LIMITED LIABILITY COMPANY

In many ways, the limited liability company, or LLC, has rendered the sole proprietorship, partnership, and limited partnership forms of doing business obsolete. LLCs are relatively new as recognized business entities. They embody most of the favorable features of corporations (discussed below), partnerships, and limited partnerships. An LLC is an unincorporated business entity. In most states, owners who contribute capital are referred to as *members*. The LLC is recognized as an entity separate and apart from its members.

LLC formation

An LLC requires the approval of the state. Typically, the entity must file articles of formation or articles of organization with the secretary of state in which the LLC is to be formed. Usually, the LLC articles must include the following information:

- Name of LLC
- Address of company and resident agent
- Purpose of LLC
- Duration of LLC (although in many states the duration is presumed to be perpetual)

It is important to note that the name of an LLC must include the abbreviation *LLC* or *LC* or words *limited liability company* in most states. This puts persons on notice that they are dealing with a business entity and not an individual. Therefore, if Bob and Ray decided to convert the partnership of B&R Air Services to an LLC, then they would be required to file articles of organization and they might name the new company B&R Air Services, LLC.

LLC operating agreement

An LLC operating agreement is required in some states and is optional in others. Nonetheless, it is always prudent for an LLC to have a well-drafted operating agreement in place to lay out the rights and duties of the LLC members. If Bob and Ray of B&R Air Services, LLC, wished to draft an LLC operating agreement, some of the specific issues they might address would include

- Reporting the members' names and percentage ownership of each member
- Establishing a residence for the business
- Outlining the rights and duties of certain officers
- Noting procedures for further capital contributions
- Establishing a procedure for the termination of a member's interest or a member's death

Some LLC operating agreements will provide for the issuance of certificates that evidence the ownership of each member. Such certificates are not required in most states.

Typically, an LLC operating agreement that notes percentage ownership by each member is sufficient evidence of each member's share of the LLC.

One other important issue that should be tackled in an LLC operating agreement is whether the LLC will be member-managed or manager-managed. In some cases, all the members wish to act as agents of the LLC and manage the LLC in accordance with their percentage of ownership or some other agreed upon formula for sharing management authority. In other cases, the members will appoint a manager to manage the LLC. In a manager-managed LLC, the nonmanager members do not have a right to manage the LLC. The LLC managers can be members or nonmembers of the LLC.

Liability of LLC members

One big advantage to an LLC over any form of sole proprietorship or partnership is the ability to shelter the members of the LLC from personal liability to creditors of the LLC. As indicated earlier, an LLC is treated as a separate entity from its owners.

With respect to contract claims against the LLC, creditors may not pursue the personal assets of the LLC members. For example, suppose B&R Air Services, LLC, has a claim filed against it by a company that overhauls aircraft engines. If the claim relates to the LLC's failure to pay for services performed, the claim cannot be pursued against Bob's or Ray's personal assets.

When it comes to tort claims, the same basic rules apply. If an employee of the LLC operates an aircraft negligently and damages someone else's property, neither Bob nor Ray will be personally liable to the injured party. The injured party can go after the LLC, but not after Bob or Ray. Of course if Bob or Ray were operating the aircraft, that person could be personally liable for his own negligence.

It is also worth noting that LLCs are available to single-member entities. Therefore, a sole proprietor can readily convert his or her sole proprietorship to an LLC to take advantage of the protection from personal liability.

See Fig. 7-6 for an overview of the liability exposure of LLCs and LLC members.

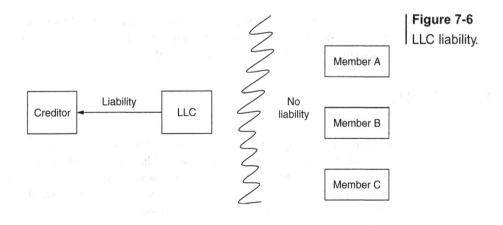

Figure 7-6
LLC liability.

Figure 7-7
LLC taxation.

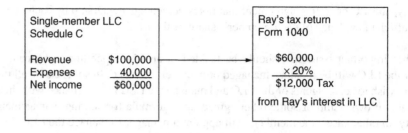

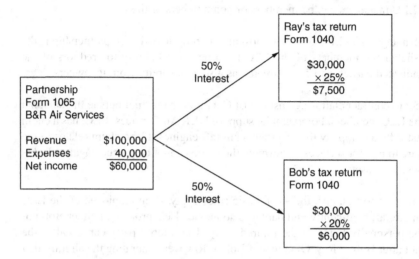

LLC taxation

Another great advantage of LLCs is the ability to choose the approach to taxation for the LLC. Members can choose to have the LLC taxed as a sole proprietorship (for single-member LLCs), a partnership, or a corporation. Most LLCs choose to be taxed as sole proprietorships or partnerships because that allows the individual members to be taxed at their individual tax rates. Under current tax law, individual tax rates are somewhat lower, and thus more favorable, than corporate rates.

See Fig. 7-7 for an illustration of how single-member and multiple-member LLCs will be taxed, using the same figures as we employed in Figs. 7-2 and 7-4 for the examples related to sole proprietorships and partnerships.

CORPORATIONS

Corporations are legal entities that are created by virtue of state law (and in some special cases federal law). Just as with the limited liability company, a corporation is a separate legal entity—often referred to as an *artificial person*—that can do the same things that a natural person can do, such as doing business, making legal claims and having claims made against it, and engaging in contracts.

Formation of a corporation

To form a corporation, certain legal requirements must be met. The first thing to do is to file articles of incorporation. Typically, state law requires that articles of incorporation include the following (at a minimum):

- Corporate name
- The number of stock shares that the corporation can issue
- The name of the person acting as the incorporator
- The address of the corporation's principal office
- The name and address of a resident agent for the corporation
- The purpose of the corporation

The articles of incorporation can be simple or very complex, depending on the needs of the corporation involved. Usually, the document is prepared and filed with appropriate state agencies by attorneys.

The next formality that must be complied with in forming a corporation is the establishment of bylaws. Bylaws are the internal rules of the corporation. They detail how the corporation will be operated and the processes involved in selecting management of the corporation.

Once the bylaws are in place, the next step is the corporation's organizational meeting. This meeting usually involves the adoption or approval of bylaws, the selection of a board of directors, the selection of a bank, the issuance of stock shares, and any other provisions deemed necessary to get the new corporate entity started as a viable business entity.

Following our earlier example, if Bob and Ray decided to incorporate B&R Air Services, they might decide to name the corporation B&R Air Services, Inc. (most states require the corporate name to include the term *Incorporated*, *Inc.*, or *Corporation* after the name of the corporation). Once the name is established, Ray and Bob will prepare and submit articles of incorporation for the new entity. When the articles are accepted by the state, the corporation can finalize its new bylaws and hold an organizational meeting to select the board of directors. In a small corporation such as B&R Air Services, Inc., most likely Bob and Ray will be the only two shareholders and directors in the entity. In a larger corporation, there may be many directors with several of the directors coming from outside the corporation. This type of director is often referred to as an outside director. Many larger corporations are relying more and more heavily on outside directors owing to increased pressure from shareholders and regulators to ensure independence on the board.

Another important action at the organizational meeting will be the issuance of stock certificates for each of the corporation's shareholders—in our case Bob and Ray. Let us assume for the rest of this discussion that the corporation has been authorized by the state to issue up to 1000 shares of stock. Bob and Ray are the only shareholders, so they have decided that the corporation will actually issue only 100 shares, with 50 going to Bob and 50 going to Ray, once Bob and Ray contribute capital to the corporation. The 1000 shares are called *authorized shares* since the corporation was authorized to issue only that many shares. The 100 shares that have been issued to Bob and Ray are known as *outstanding shares* as they have now been issued "outside" the corporation.

Operating the corporation

By law, corporations have formal standards and procedures for governance. Shareholders own shares in the corporation through the issuance of corporate stock. The shareholders, in turn, select the management of the corporation when they vote for the corporation's board of directors. The board of directors is responsible for the management of the corporation. However, the board works through officers that it appoints to manage the day-to-day affairs of the corporation's business.

Naturally, the formality of the typical governance structure for a corporation will vary from entity to entity. In a small corporation such as the one established by Bob and Ray, the two shareholders will likely be the board of the directors and officers of the corporation as well. In a huge, multinational corporation, the formality of corporate governance and the separation between ownership and management of the corporation become more apparent.

In any corporation, large or small, the shareholders must hold at least one annual meeting, typically at a time and place detailed in the corporation's bylaws. Special meetings can also be called to convene shareholders regarding special issues or new developments in a corporation. In a small corporation, shareholder meetings may take place by telephone or at informal settings. Sometimes small corporations waive the requirements for in-person meetings and permit meetings to take place through written resolutions in lieu of meetings. In large corporations, shareholder meetings are usually well planned and orchestrated in a more formal manner. Most of the time, shareholders can vote according to the number of shares of stock they own. In large corporations, most shareholders vote by proxy, permitting another person (often a large shareholder and/or director of the corporation) to cast their votes for them. Before any vote can count, a quorum of members (a majority of shareholders entitled to vote) must be present unless the requirement is modified in the articles of incorporation and/or bylaws.

Once a board of directors is selected, the board will typically meet on a regular basis to discuss corporation business. Just as with shareholder meetings, board meetings are held on a regular basis according to bylaws, and sometimes special meetings may be called. Quorums are typically required for board actions to be considered valid.

One of the primary duties of a board of directors will be the selection of corporate officers. Most corporations employ at least a president, one or more vice presidents, a secretary, and a treasurer for the corporation. Beyond these offices, the corporation bylaws can provide for however many officer positions are deemed necessary. These officers are delegated authority for the day-to-day operations of the corporation. They report directly to the board, and usually they serve at the pleasure of the board. The duties of the corporation's officers are detailed in the corporate bylaws. Sometimes officers are also board members—a board member in this situation is often referred to as an *inside director*.

See Fig. 7-8 for an illustration of the governance structure for the typical corporation.

Duties of directors and officers

Directors and officers of a corporation are placed in a position of high trust in the corporation. The relationship of trust with the corporate entity is often referred to as a *fiduciary duty*.

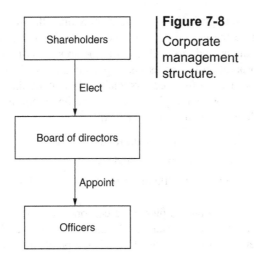

Figure 7-8
Corporate management structure.

One of the primary duties the director or officer has is the duty of loyalty to the corporation. Directors and officers must be ready to place their interests second to the interests of the corporation. In a large corporation, it might mean that directors or officers who might lose their positions due to a merger cannot oppose the merger if it will bring benefits to the majority of shareholders holding stock in the company. In a corporation as small as that of Bob and Ray, it might mean that neither one can take on "side business" without full notification to the other.

Directors and officers are also held to a high standard of care. This is often referred to as a *duty of care.* Many commentators indicate that to measure compliance with this standard, the director or officer must act in good faith, with ordinary care (as compared to someone in a similar position), and for the betterment of the corporation. Although this legal standard sets a high mark, it does not mean that directors or officers must be right all the time. A standard like that would drive anyone away from a directorship or officership of an institution. Instead, this legal standard means that honest mistakes or lapses in judgment cannot be held against directors or officers as long as the directors or officers exercised due diligence in their performance of duties.

Liability issues for corporations and shareholders

Similar to LLCs, corporations are treated as entities separate from their owners. Therefore, the general rule for corporations is that shareholders are not liable to creditors of the corporation beyond the shareholders' levels of capital contribution. The law treats contract and tort liability in a manner consistent with the description above for LLCs. The only difference with a corporation is that the owners are shareholders and not members.

The high level of liability exposure in the aviation industry often persuades business owners to incorporate their enterprises. However, once an entity is incorporated, care must be taken to ensure that all corporate formalities are met and that the corporation's business is treated as an entity separate and apart from its owners.

Failure to maintain the requisite "separateness" may open the door for creditors of the corporation to "pierce the corporate veil" and pursue the assets of the corporation's shareholders. Sometimes creditors will resort to the approach of piercing the corporate veil because the corporation has no assets to pursue and the shareholders have assets that may satisfy all or part of the creditors' claims. In making their argument to pierce the corporate veil, creditors will typically argue that the corporation is the mere "alter ego" of its shareholders. This may be evidenced by

- Commingling of personal and corporate funds
- Failure to follow corporate formalities and records
- Forming the corporation with little or no funding by shareholders

The courts are usually reluctant to pierce the corporate veil. However, it does happen from time to time when the circumstances dictate. Also note that the same issues may be faced by members of an LLC if the LLC fails to operate as an entity separate from its members.

Taxation of corporations

Corporations are looked upon as a separate taxable entity under the tax law. The IRS has established special rates at which it taxes corporate profits. Under current tax law, corporate tax rates are somewhat higher than the tax rates applied to individuals. This creates some disadvantage for corporate taxpayers. The disadvantage becomes even more apparent when the corporation has its profits taxed at the corporate level at corporate rates and then any distributions of profit by way of dividends are taxed at the shareholder level at individual rates for each shareholder receiving dividends. This creates the tax phenomenon known as *double taxation*. First the corporation's profits are taxed at corporate rates; then the profits that are distributed to shareholders as dividends are taxed again at individual rates.

In the case of B&R Air Services, Inc., assume that the corporation enjoys a profit of $100,000 after the first year of operation as a corporation. If the applicable corporate rate is 30 percent, the corporation will have to pay tax of $30,000.[1] If the corporation distributed a dividend of all remaining profits to Bob and Ray in a 50-50 split (since each owns 50 percent of the shares of the corporation, the $70,000.00 distributed would be taxed at the individual rates applicable to both Bob and Ray). If Bob's applicable individual rate were 20 percent, he would have to pay tax of $7000.00 on his dividend of $35,000.00. If Ray's applicable individual rate were 10 percent, he would have to pay tax of $3500.00 on his dividend of $35,000.00. All told, in this situation, the corporation and shareholders combined would have paid a total tax of $40,500.00, consisting of $30,000.00 paid by the corporation, $7000.00 paid by Bob, and $3500.00 paid by Ray.

However, when a corporation meets certain requirements, it may elect to be taxed as a Subchapter S corporation, or an *S corporation*.[2] The rules for S corporations allow any corporate income to pass through directly to shareholders for taxation at their individual rates with no tax paid at the corporate level. In the case of Bob and Ray, they could pass through the entire $100,000 of corporate profit to themselves. Bob would pay tax of $20,000.00 based on his individual tax rate of 20 percent. Ray would pay tax of

$10,000.00 based on his individual tax rate of 10 percent. All told, the S corporation election in this case would save more than $10,000.00 in taxes overall.

Not all corporations can qualify for treatment under S corporation status. To be eligible for an S corporation election, the following conditions must be met:

- The corporation making the election must be a domestic U.S. corporation.
- The corporation cannot be a member of an affiliated group of corporations.
- There can be no more than 75 shareholders in the corporation.
- All shareholders must be individuals, estates, or certain trusts (corporations and partnerships cannot be shareholders).
- All shareholders must be U.S citizens or permanent residents of the United States.
- The corporation cannot have more than one class of stock.

For many small corporations, the S corporation election makes a lot of sense. The election is fairly simple to make, and it can be rescinded. (However, if it is rescinded, the corporation cannot go back to S corporation status for 5 years following the rescission.) An S corporation election makes sense for shareholders who have individual tax rates lower than their corporation's tax rates. It may also make sense for certain corporations that experience losses if the shareholders wish to pass the losses through to their individual tax returns to offset other individual income that will be subject to taxation.

See Fig. 7-9 for a summary of corporate tax procedures for regular corporations and S corporations.

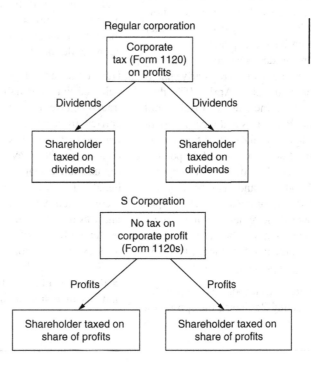

Figure 7-9
Corporate taxation.

CASES AND COMMENTARY

Case 7-1 deals with the corporate opportunity doctrine. As noted in the discussion above, it is critical that directors and officers of corporation act with loyalty to the corporation. The importance of that fiduciary relationship to the corporation is emphasized in Case 7-1.

CASE 7-1
KLINICKI V. LUNDGREN
298 Ore. 662; 695 P. 2d 906 (1985)

OPINION BY: JONES

The factual and legal background of this complicated litigation was succinctly set forth by Chief Judge Joseph in the Court of Appeals opinion as follows:

In January, 1977, plaintiff Klinicki conceived the idea of engaging in the air transportation business in Berlin, West Germany. He discussed the idea with his friend, defendant Lundgren. At that time, both men were furloughed Pan American pilots stationed in West Germany. They decided to enter the air transportation business, planning to begin operations with an air taxi service and later to expand into other service, such as regularly scheduled flights or charter flights. In April, 1977, they incorporated Berlinair, Inc., as a closely held Oregon corporation. Plaintiff was a vice-president and a director. Lundgren was the corporation's president and a director. Each man owned 33 percent of the company stock. Lelco, Inc., a corporation owned by Lundgren and members of his family, owned 33 percent of the stock. The corporation's attorney owned the remaining one percent of the stock. Berlinair obtained the necessary governmental licenses, purchased an aircraft and in November, 1977, began passenger service.

As president, Lundgren was responsible, in part, for developing and promoting Berlinair's transportation business. Plaintiff was in charge of operations and maintenance. In November, 1977, plaintiff and Lundgren, as representatives of Berlinair, met with representatives of the Berliner Flug Ring (BFR), a consortium of Berlin travel agents that contracts for charter flights to take sallow German tourists to sunnier climes. The BFR contract was considered a lucrative business opportunity by those familiar with the air transportation business, and plaintiff and defendant had contemplated pursuing the contract when they formed Berlinair. After the initial meeting, all subsequent contacts with BFR were made by Lundgren or other Berlinair employees acting under his directions.

During the early stages of negotiations, Lundgren believed that Berlinair could not obtain the contract because BFR was then satisfied with its carrier. In early June, 1978, however, Lundgren learned that there was a good chance that the BFR contract might be available. He informed a BFR representative that he would make a proposal on behalf of a new company. On July 7, 1978, he incorporated Air Berlin Charter Company (ABC) and was its sole owner. On August 20, 1978, ABC presented BFR with a contract proposal, and after a series of discussions it was awarded the contract on September 1, 1978. Lundgren effectively concealed from plaintiff his negotiations with BFR and his diversion of the BFR contract to ABC, even though he used Berlinair working time, staff, money and facilities.

Plaintiff, as a minority stockholder in Berlinair, brought a derivative action against ABC for usurping a corporate opportunity of Berlinair. He also brought an individual claim against Lundgren for compensatory and punitive damages based on breach of fiduciary duty.

The trial court found that ABC, acting through Lundgren, had wrongfully diverted the BFR contract, which was a corporate opportunity of Berlinair. The court imposed a constructive trust on ABC in favor of Berlinair, ordered an accounting by ABC and enjoined ABC from transferring its assets. The trial court also found that Lundgren, as an officer and director of Berlinair, had breached his fiduciary duties of good faith, fair dealing and full disclosure owed to plaintiff individually and to Berlinair. The court did not award plaintiff any actual damages on the breach of fiduciary duty claim. All the issues were tried to the court, except that a jury was empaneled to try the punitive damages issue. It returned a verdict in favor of plaintiff and assessed punitive damages against Lundgren in the amount of $750,000. Lundgren then moved to dismiss plaintiff's claim for punitive damages. The court granted the motion to dismiss and, *sua sponte*, entered judgment in favor of Lundgren notwithstanding the verdict on the punitive damages claim.

ABC appealed to the Court of Appeals contending that it did not usurp a corporate opportunity of Berlinair.

ABC petitions for review to this court contending that the concealment and diversion of the BFR contract was not a usurpation of a corporate opportunity, because Berlinair did not have the financial ability to undertake that contract. ABC argues that proof of financial ability is a necessary part of a corporate opportunity case and that plaintiff had the burden of proof on that issue and did not carry that burden.

There is no dispute that the corporate opportunity doctrine precludes corporate fiduciaries from diverting to themselves business opportunities in which the corporation has an expectancy, property interest or right, or which in fairness should otherwise belong to the corporation. The doctrine follows from a corporate fiduciary's duty of undivided loyalty to the corporation. ABC agrees that, unless Berlinair's financial inability to undertake the contract makes a difference, the BFR contract was a corporate opportunity of Berlinair.

Where a director or principal senior executive of a close corporation wishes to take personal advantage of a "corporate opportunity," as defined by the proposed rule, the director or principal senior executive must comply strictly with the following procedure:

(1) The director or principal senior executive must promptly offer the opportunity and disclose all material facts known regarding the opportunity to the disinterested directors or, if there is no disinterested director, to the disinterested shareholders. If the director or principal senior executive learns of other material facts after such disclosure, the director or principal senior executive must disclose these additional facts in a like manner before personally taking the opportunity.

(2) The director or principal senior executive may take advantage of the corporate opportunity only after full disclosure and only if the opportunity is rejected by a majority of the disinterested directors or, if there are no disinterested directors, by a majority of the disinterested shareholders. If, after full disclosure, the disinterested directors or shareholders unreasonably fail to reject the offer, the interested director or principal senior executive may proceed to take the opportunity if he can prove the taking was otherwise "fair" to the corporation. Full disclosure to the appropriate corporate body is, however, an absolute condition precedent to the validity of any forthcoming rejection as well as

to the availability to the director or principal senior executive of the defense of fairness.

(3) An appropriation of a corporate opportunity may be ratified by rejection of the opportunity by a majority of disinterested directors or a majority of disinterested shareholders, after full disclosure subject to the same rules as set out above for prior offer, disclosure and rejection. Where a director or principal senior executive of a close corporation appropriates a corporate opportunity without first fully disclosing the opportunity and offering it to the corporation, absent ratification, that director or principal senior executive holds the opportunity in trust for the corporation.

Applying these rules to the facts in this case, we conclude:

(1) Lundgren, as director and principal executive officer of Berlinair, owed a fiduciary duty to Berlinair.

(2) The BFR contract was a "corporate opportunity" of Berlinair.

(3) Landgren formed ABC for the purpose of usurping the opportunity presented to Berlinair by the BFR contract.

(4) Lundgren did not offer Berlinair the BFR contract.

(5) Lundgren did not attempt to obtain the consent of Berlinair to his taking of the BFR corporate opportunity.

(6) Lundgren did not fully disclose to Berlinair his intent to appropriate the opportunity for himself and ABC.

(7) Berlinair never rejected the opportunity presented by the BFR contract.

(8) Berlinair never ratified the appropriation of the BFR contract.

(9) Lundgren, acting for ABC, misappropriated the BFR contract.

Because of the above, the defendant may not now contend that Berlinair did not have the financial ability to successfully pursue the BFR contract. As stated in [citation] "if the challenging party satisfies the burden of proving that a corporate opportunity was taken without being offered to the corporation, the challenging party will prevail."

[Judgment Affirmed]

In large and small corporations, there may be ample opportunities for directors and officers to make secret profits from their relationship with the corporation. The law is clear on this point—officers and directors must subordinate their own personal interests to the interests of the corporation and of the shareholders who own the corporation.

Beyond issues related to the corporate opportunity doctrine, there have been recent scandals involving publicly held companies in which officers and directors have not acted in the best interests of their constituents—shareholders and employees. In the scandals involving Enron, Worldcom, and Tyco, the shareholders, employees, and creditors suffered huge losses due to failures of corporate governance at these entities. These scandals prompted the government to address the issues of corporate governance.

The government response came in the form of the Sarbanes-Oxley Act of 2002. Sarbanes-Oxley was enacted to meet the following goals:

- Improve corporate governance by forcing directors to be more vigilant in guiding the conduct and business practices of corporate officers.
- Eliminate problems created by conflicts of interest.
- Foster a sense of shareholder confidence in the system of corporate governance.

Sarbanes-Oxley attempted to meet these goals through the following provisions:

- Chief executive officers (CEOs) and chief financial officers (CFOs) must now personally certify that the financial statements of the corporation are free of material misstatements. Any CEO or CFO who violates this provision knowingly and willfully will be subject to severe penalties including prison terms and large fines.
- Any bonuses received by a CEO or CFO must be reimbursed to the corporation if it is later learned that the corporation must restate its financial statements owing to noncompliance with the law related to financial disclosures.
- Corporations are now prohibited from making personal loans to directors or executive officers.

These are just some of the provisions enacted in Sarbanes-Oxley to ensure stronger corporate governance and to protect shareholders. Although the law is applicable to only larger, publicly held corporations, many nonpublic and not-for-profit organizations are adopting the practices required by Sarbanes-Oxley in order to strengthen their own corporate governance. Compliance with Sarbanes-Oxley is now a part of life in publicly held U.S. corporations. The jury is still out on whether the law will have a positive impact on the ethical conduct of directors and officers in U.S. corporations.

Case 7-2 involves an aircraft leasing company operated as an LLC. The case demonstrates the peril of failing to meet formalities that ensure that the line between business and personal matters is not crossed with a separate entity such as an LLC or corporation.

CASE 7-2
NELSEN V. MORRIS
2004 U.S. Dist. LEXIS 6928

OPINION BY: WILLIAM T. HART

Plaintiff Daniel Nelsen, an Illinois resident, alleges that he lost money investing in limited liability companies ("LLC's") that leased aircraft. Plaintiff alleges that defendant Waldo Morris, an Iowa resident, caused these losses by breaching his fiduciary duty as a managing member of the LLC's and by committing fraud in order to induce plaintiff's investment. There is complete diversity of citizenship and the amount in controversy exceeds $75,000. Defendant has moved to dismiss based on lack of personal jurisdiction.

According to the allegations of the complaint, defendant was the managing member of two LLC's known as Interlease IV and Interlease V. Both LLC's were in the business of leasing aircraft that were owned by the LLC's. Plaintiff was a 25% member of Interlease IV and a 5% member of Interlease V. Plaintiff allegedly loaned Interlease V $1,000,000 in return for a promissory note secured by an aircraft, but later relinquished the note for his 5% share of Interlease V. Plaintiff allegedly invested another $1,000,000 and also personally guaranteed a $5,800,000 loan as part of his purchase price for his interest in Interlease IV. Defendant allegedly used the loan proceeds to pay off debts defendant would have otherwise owed. Also, defendant allegedly intermingled funds among various LLC's and used some funds for his own benefit. After the foreclosure on the loan, defendant allegedly improperly purchased one of the aircraft that had been security. Plaintiff lost his $2,000,000 investment and, based on the personal guaranty, was required to pay $1,451,250 in principal plus $92,940 interest and incurred $46,060 in attorney fees.

In response to the motion to dismiss, plaintiff need only make a "minimally viable" showing that the LLC's are a sham. [Citation]. It is enough to show that the allegations are not "patently without merit." The alter ego rules applicable to LLC's are generally the same as those for corporations. [Citation] (Personal liability of LLC member is same as that of a corporate shareholder except that failure to observe formalities as to meetings is not a consideration.) [Citations]. Here, it must be taken as true that funds of the various LLC's were commingled, that Interlease IV became insolvent in that the guarantors had to pay off the outstanding balance of its loan, and that defendant used LLC funds for his own benefit. There is also evidence that defendant managed and/or directed activities of the LLC's. It cannot be held that plaintiff's alter ego theory is not viable. [Citation]. Therefore, the conduct of the LLC's is attributable to defendant for personal jurisdiction purposes and, at this stage of the proceedings, it is appropriate to continue to exercise personal jurisdiction over defendant. Because personal jurisdiction is being upheld on this basis, it is unnecessary to consider whether the fiduciary shield doctrine should not apply because defendant was acting in his own interest instead of the interests of the LLC's and whether, on that basis, it would be appropriate to exercise general and/or specific personal jurisdiction.

Although Case 7-2 is limited to a discussion of whether personal jurisdiction could be exercised, it provides a good discussion of what can trigger the personal liability of a person for his or her actions as a member of the LLC. As discussed earlier, one of the primary reasons for selecting an LLC or a corporate form of doing business is to shelter the owners of the business from personal liability to the creditors of the business. As can be seen in this case, by treating an LLC as an alter ego, a member can potentially expose herself or himself to personal liability.

DISCUSSION CASES

1. Samantha, Joe, Marty, and Charlie formed DataTrax, LLC, a limited liability company located in Detroit, Michigan. The company designs and manufactures custom equipment for airport automated weather systems. Charlie is an engineer, and he has been sent to discuss specifications for a new weather station at an airport outside Reno, Nevada. The business trip is scheduled to require 4 days of out-of-town travel for Charlie. Charlie is

a pilot, and he flies a King Air owned by DataTrax to the business meeting location outside of Reno. After a long day of meetings with airport management, Charlie decides to take the aircraft for a flight to Las Vegas to unwind for the evening. He arrives safely at Las Vegas. However, while he is taxiing to the general aviation parking area, he misjudges his distance from a parked Gulfstream V and damages the wingtips on the Gulfstream. The Gulfstream owners seek to file claims against Samantha, Joe, Marty, and Charlie. Are all four LLC members liable for the damage incurred by the Gulfsteam in this accident? Is the LLC liable for damages related to the accident? Why or why not? Would your answer be any different if DataTrax were a general partnership rather than an LLC?

2. Kirk and Bill form MSM LLC. MSM is a limited liability company that operates a chain of flight schools all along the eastern seaboard. The LLC is a manager-managed LLC. Kirk has been designated as the manager of the LLC in the articles of organization filed with the state. Kirk spots a hangar and shop location in the Pittsburgh, Pennsylvania, area that would, in his opinion, be an excellent location for a new flight school. He enters into a 5-year lease with the hangar owner on behalf of MSM, LLC. While visiting the Cumberland Maryland area, Bill sees a location that he thinks will be the perfect spot for a flight school. Bill signs a lease with the Cumberland Maryland hangar owner. Is MSM, LLC, bound to either of these leases? Please explain.

3. Freemont Aviation, Inc., is an incorporated entity located in the state of Connecticut. The corporation operates an on-demand air taxi service out of Bridgeport Airport. The corporation is owned by two shareholders, Tara and Sandra, in equal 50 percent shares. For tax year 2006, Freemont has net profits of $100,000.00. If Tara and Sandra agree to pay a dividend consisting of all net profits for 2006, what are the tax ramifications to Freemont and to Tara and Sandra? How would your response differ if the corporation made a timely election with the IRS to be treated as a subchapter S corporation? Assume that the applicable corporate tax rate is 30 percent and that the applicable individual tax rates for Tara and Sandra are 15 percent and 25 percent, respectively.

4. Dana was the president and general manager of Air Freight, Inc., a subsidiary of Flying Tiger Cargo Services, Inc. Flying Tiger Cargo is a corporation with national and international offices. Air Freight, Inc., concentrates on air cargo services in the mid-Atlantic region of the United States. In 2004, Dana recruited three other Air Freight executives to join her in starting up a new air freight enterprise. The four remained at Air Freight for the 3 years it took to develop a business plan and put their new business enterprise into place. During this time, the four used their positions at Air Freight to travel to customers, gathering ideas, new customer leads, and working out equipment and aircraft purchases for their new business. At no time did they reveal their plans to Flying Tiger Cargo. In 2006, they all resigned their positions at Air Freight and started their new business, incorporated as Chesapeake Air Cargo, Inc. When the executive officers at Flying Tiger Cargo learned of the new venture formed by the former Air Freight executives, they sued all four for damages. Who will win the lawsuit? Why?

5. Stanton comes to you seeking legal and tax advice. He currently owns seven aircraft that are being leased to various flight schools and air charter companies. All the aircraft are owned by Stanton as a sole proprietor doing business as Stanton Air. During the course of the last several years, Air Stanton has been profitable, with profits ranging from $50,000 to $80,000. Stanton's individual income tax rate is 20 percent. Corporate income tax rates for the income range of Stanton Air are 30 percent. Stanton's primary concern is whether he should continue to operate Stanton Air as a sole proprietorship or some other type of legal entity. What would you advise Stanton to do? Explain fully, describing the positive and negative aspects of possible options.

6. Terry purchased a new aircraft manufactured by Koda Aircraft, Inc., a privately held corporation with 10 shareholders that designs and manufactures light general aviation aircraft. Terry was killed when her Koda aircraft crashed shortly after an engine failure in mountainous terrain. Terry's widowed husband brought a lawsuit against Koda, alleging that the aircraft crash was caused by a defect in Koda's engineering that causes the aircraft engine to seize in certain weather conditions. If Terry's widowed husband is successful in suing Koda, will the shareholders of Koda be held personally liable for any judgment? Explain.

7. Jake and Sally formed a limited liability company that will allow them to own and lease aircraft to flight schools. What sort of provisions would you suggest for an operating agreement between the parties? Please list your suggestions along with a brief explanation of why you would propose certain provisions.

8. Aviation Properties was a limited partnership engaged in buying, selling, and managing corporate aircraft. The limited partnership consisted of Greer as the sole general partner and a group of limited partners. About a year after its formation, the business ran into cash flow problems and was near insolvency. The limited partners called several meetings to address the cash flow crisis. Eventually, the limited partners hired two employees to take over the management of the business from Greer. During this time, all the limited partners also were personally involved in dealing with creditors and vendors with respect to the debts owed by the partnership. Greer was still legally the general partner and was looked upon as such by the creditors and vendors. However, he was no longer treated as a general partner and no longer received compensation from the partnership. Eventually Greer went to work for another aircraft dealer. Creditors of the company now file suit to recover debts owed by Aviation Properties. The creditors file a lawsuit against both the limited partnership and the limited partners of the enterprise. Will the creditors be able to successfully pursue the limited partnership? What about the individual limited partners? Explain your response fully.

9. Steve sets up a corporation for the purpose of aircraft muffler repairs. Steve is the sole shareholder. He receives FAA certification as a repair station and opens his shop for business with the name AvMuffler, Inc. The corporation is profitable, but Steve wants to expand the operation. AvMuffler applies for and receives a commercial loan of $500,000.00 from a bank. In the loan application, the corporation indicates that the loan proceeds are to be used for working capital needs (inventory, increased labor costs, and small tools). Around the same time that he receives the loan proceeds from the bank, Steve and his wife find a home that they are very interested in purchasing. However, they do not have the money available for a down payment. Steve decides that he will use $150,000.00 of the commercial loan proceeds received by AvMuffler for a down

payment on the home. Steve takes the money from AvMuffler's account and puts it in his personal bank account, where it is eventually used for the down payment on the house. There are no efforts made by Steve to repay the $150,000 to AvMuffler. Later, AvMuffler runs into cash flow difficulty and defaults on the bank loan of $500,000.00. The bank learns of the $150,000.00 transfer from AvMuffler to Steve and sues both AvMuffler and Steve. Steve argues that AvMuffler is the only party that may be sued in this case. Is Steve correct? Explain.

8 Property Law Issues for Aircraft Owners and Airport Operators

REAL PROPERTY 218
Ownership rights in real property 219
Multiple ownership 219
Transfers of real property 220
Easements 221
Local zoning issues 221
Leasing 222

AIRPORTS 223

PERSONAL PROPERTY 224
Tangible property 224
Intangible property 230

INSURANCE 230
Hull insurance and liability insurance 231
The insurance contract 232
Other types of insurance 234

CASES AND COMMENTARY 236

DISCUSSION CASES 247

ENDNOTES 251

The rules of law relating to the ownership of property are among the most fundamental in our system of law. Acquisition and use of property are an important part of life in most societies. Without some stable and enforceable set of laws regarding property ownership and rights, ownership and use of property would be dependent on the physical or military strength of the possessor.

In modern property law there are basically two types of property recognized—real property and personal property. Real property is sometimes referred to as real estate or realty. This category of property consists of land and things that are annexed or affixed to the land.

Personal property is in many ways a default category—anything that does not fit the definition of real property is categorized as personal property. Tangible property that is movable is considered personal property and often referred to as goods. Personal property can also be intangible and can take the form of copyrights, patents, trademarks, or other intellectual property.

Property law is a vast field of study. In this chapter, we will endeavor to highlight some of the major features of both real and personal property law and will provide relevant examples of applications of these principles to businesses and individuals in the aviation environment.

REAL PROPERTY

Some of the different types of real property are

- Land and buildings
- Rights to minerals or subsurface rights
- Plants and vegetation (although plant life severed from the land is considered to be personal property)
- Fixtures attached to land or buildings
- Air rights

The key characteristic of real property is that is either land or things that are affixed to the land. One general test often applied to determine whether property is real property is, Would the removal of the property from the land damage or leave a scar on the land?

Of course, this test would not necessarily permit the inclusion of air rights as a type of real property. Interestingly, but not unexpectedly, the concept of air rights has changed over centuries of evolving property law. In the earliest days of property law development, it was understood that a property owner owned the property from the surface going up the sky above and to the center of the earth. Over time, environmental considerations, zoning restrictions, and air navigation needs have diluted the absolute nature of property ownership as it relates to air rights.

Ownership rights in real property

There are several types of ownership interests in real property. The most common is known in the legal world as the *fee simple absolute*. This essentially means that property ownership is exclusive and complete, subject to only government regulation or private restraints. Sometimes the fee simple can be qualified. A qualified fee simple provides for absolute ownership as long as certain conditions are met. For instance, land might be transferred to an owner on the condition that it be used only as an airport. An additional type of interest in real property is the life estate. This type of real property ownership creates an interest in real property for the duration of a person's life. Most often, the person granting the property (the grantor) grants the property for the life of the receiver (the grantee).

Multiple ownership

Whether property is real or personal, it can be owned by more than one person. This is often referred to by lawyers as *concurrent ownership*. There are four basic forms of concurrent ownership: tenancy in common, joint tenancy, tenancy by the entirety, and community property. Each of these is briefly discussed below.

In a tenancy in common, each owner gets a share of the property owned that she or he may pass on to heirs. Further, each ownership share is separate and freely transferable by its owner. Let us say that Nick and Susan (who are not married) wish to establish ownership of a maintenance hangar on the perimeter of a busy regional airport with Nick owning 50 percent and Susan owning 50 percent. In many states their concurrent ownership would automatically be treated as a tenancy in common unless otherwise stated. However, to be certain that there were no misunderstandings, Nick and Susan could create a tenancy in common by stating in the deed that the property was being transferred to "Nick and Susan, as tenants in common." In this case, if Nick dies, he can, as a general rule, leave his 50 percent share of the hangar to his heirs by will. He also has the legal right to sell his 50 percent share to any person without the need for Susan's consent. In most states, there is a presumption that property owned by more than one person or entity is owned as a tenancy in common, unless the description of the property ownership clearly states otherwise.

When two or more persons wish to set up a joint tenancy, they must expressly state that they wish to do so. Using the previous example, words creating the joint tenancy such as *Nick and Susan, as joint tenants with right of survivorship* will suffice. The unique feature of a joint tenancy is the right of survivorship vested in each owner. This means that in the joint tenancy with Nick and Susan, if Nick dies, Susan now owns 100 percent of the hangar—regardless of what Nick's will might state regarding the disposition of his property. It is also important to note that a joint tenant can transfer his or her share in a joint tenancy to another person without the consent of the other joint tenant. Let's say that Nick decides to transfer his 50 percent interest in the hangar to Amanda. He can do so without Susan's consent, but after the transfer, Amanda and Susan now own the hangar as tenants in common.

In some states, a special form of joint tenancy known as *tenancy by the entirety* occurs between husband and wife. A married couple wishing to establish a tenancy by the entirety can do so only with express language. To continue the example above, if Nick and Susan were married and wanted to create a tenancy by the entirety, they would have to state, "Nick and Susan, husband and wife, as tenants by the entireties." Everything operates the same in a tenancy by the entirety and a joint tenancy with one important exception—in a tenancy by the entirety, neither husband nor wife can transfer his or her share of the property without the consent of the other. So if Nick and Susan created a tenancy by the entirety, Nick could not transfer his share to Amanda without Susan's authorization.

Another form of concurrent ownership that applies only to married couples is community property. This form of multiple ownership is recognized only in Arizona, California, Idaho, Louisiana, Nevada, New Mexico, Texas, Washington, and Wisconsin. Community property laws require that each spouse own an equal one-half share of any assets accumulated during a marriage, regardless of who earned the income (with some exceptions for property obtained by one party through a gift or inheritance). In community property states, when one spouse dies, the spouse who survives is automatically entitled to one-half of the community property. Therefore, in the case of Nick and Susan, if they resided in a community property state and purchased the hangar in the same state, they would each be deemed to own an equal one-half share of the hangar. In a community property state, neither spouse can transfer community property to a third person without the other spouse's authorization.

Transfers of real property

There are several ways to transfer real property from one person to another. The most common methods of transfer are

- Sales
- Gifts and inheritance
- Adverse possession

In sales, gifts, and inheritance situations, the real property is transferred from one party to another by a deed. The deed is a written document that evidences the seller or donor's intention to convey the property to another person. Usually the person selling or gifting the property is called the grantor. The person receiving the property is the grantee. Once a deed is signed by the grantor and delivered to the grantee, the transfer of the property is complete. However, in virtually every state in the United States, the deed must then be recorded with a county recording office. This allows any persons who later want to purchase the property or lend money on the property to confirm that the person they are dealing with actually owns the property. A search of county land records will also confirm whether there are any liens or mortgages recorded on the property.

In certain circumstances, real property may be transferred involuntarily by adverse possession. *Adverse possession* is the unauthorized possession of another person's real property that meets certain statutory requirements. A transfer by adverse possession does not require a deed. Every state has a slightly different twist when it comes to spelling out the requirements for adverse possession. However, the common threads that run through most state statutes are as follows:

- There must be a 10- to 20-year period of unauthorized possession of the property.
- The unauthorized possession must be open and visible to the rightful owner (if the rightful owner took the time to look).
- The unauthorized possessor must actually occupy the land (living in a building or planting crops would count for this requirement).
- The unauthorized possession must be continuous (allowing for normal periods of absence for work, vacation, etc.).
- The unauthorized possession must be just that—unauthorized (the rightful owner cannot have granted rights to use the property by permission).

If all these elements are met, the adverse possessor can get clear title to the real property possessed. For instance, if a farmer planted crops for the last 25 years on a half-acre stretch adjacent to a 4000-acre airport, the farmer may be entitled to adverse possession of the half-acre parcel of land—not the entire 4000-acre airport.

To many, the adverse possession laws appear to reward the person who makes unauthorized use of another's property. That may be so. However, the adverse possession laws were designed to require landowners to diligently oversee their property and its boundaries.

Easements

Easements are interests in land that give the easement holder the legal right to make limited use of property owned by another. There are two legally recognized types of easements—easements appurtenant and easements in gross.

An easement appurtenant exists when a landowner is given the right to use the property of another for specified purposes. This type of easement is often found in situations with common driveways and rights-of-way.

An easement in gross is created when one party grants another party who does not own adjacent land the right to use the first party's land. This type of easement commonly exists with utility poles and underground cable lines.

Most easements "run with the land." That means that when an owner sells her property, the person buying will be subject to the easement.

Easements can be created in several ways. Some easements are created expressly by granting the easement to another party (usually for a price) or by reserving an easement when the property is sold. Easements can also be implied (1) when a landowner subdivides and sells property with a road, walkways, or water supply that serves all properties within the subdivision; or (2) when a property is situated between various other properties, and the only way to enter and exit the property is through the surrounding properties—this is often referred to as an *easement by necessity*. Easements can also be created by adverse possession. This is often referred to as an *easement by prescription*.

Local zoning issues

Zoning authority is the power of local government—usually at the county or municipal level—to regulate land use. To make land available for appropriate use, most local governments have adopted zoning ordinances or codes. These codes typically accomplish three things:

- Create "zones" for agricultural, residential, commercial, and industrial land use
- Establish a certain "look and feel" for buildings or other improvements made within designated zones
- Develop restrictions for building heights, size, and location in the various zones

To establish and enforce zoning requirements, many local governments create zoning commissions or planning boards that make formal recommendations to elected councils or commissioners on how land should be zoned. If landowners seek to make a use for their property that does not conform to the zoning for their land, they may seek a waiver from the zoning commission. This is often referred to as a request for a variance. Sometimes land is rezoned, and certain uses that have already been established within the designated zone do not comply—these are called nonconforming uses and are typically permitted to continue even though they do not fit the new zoning scheme.

Leasing

Leasing occurs when the owner of property permits another person to possess and use the property on a temporary basis. Many of the business relationships in aviation are based on leases and landlord-tenant law. Sometimes the leasing involves real property such as land or hangars on an airport. On other occasions, the leasing involves personal property such as aircraft or aircraft-related equipment. This discussion will focus on leasing real property.

In the leasing relationship, the property owner is called the landlord or lessor. The person leasing the property is referred to as the tenant or lessee.

A lease is a rental agreement between a landlord and a tenant. The lease may be oral, but it must be in writing (under most state laws) if it involves a lease of more than 1 year (see earlier discussion in Chap. 6 on Statute of Frauds requirements for writings). Some of the essential terms that should be included in any lease are the term of the lease, the amount of rent to be paid, and a description of the property to be used and possessed under the lease.

If a tenant wants to transfer his or her rights under a lease, the tenant may do so by either the assignment of the lease or a sublease. An assignment of a lease involves the transfer of all the tenant's rights in a lease. A transfer of anything less than all the tenant's rights under a lease is a sublease.

For instance, if Matthew's fixed base operator (FBO) leases a building and hangar from the city of Santa Anna and the lease has another 5 years until expiration, Matthew is assigning his rights in the lease if he transfers the remaining 5 years of the lease to Emily. However, if Matthew transfers only 2 years of the lease, it is a sublease.

If an assignment takes place, Emily will be responsible for paying rent to the Santa Anna, the landlord. However, that does not mean that Matthew is off the hook. He is also responsible for paying rent if Emily fails to pay.

If Matthew subleases the property to Emily, Emily has no direct obligation or relationship with the landlord, Santa Anna. Emily's only contractual obligation is with Matthew. Further, Matthew remains on the hook for all rent payments to Santa Anna.

Most leases contain clauses that prohibit a tenant from either assigning or subleasing her or his interests in a property without authorization from the landlord. Most of the time the antiassignment or sublease clause requires that the landlord not act unreasonably in deciding whether to authorize an assignment or sublease.

AIRPORTS

Many people labor with the misunderstanding that the FAA owns most of the airports in the United States. That is far from correct. In fact, the only airport the FAA owns in the entire United States is the FAA's research facility in Atlantic City, New Jersey.[1] Airport owners are typically state, county, municipal, or private owners. Many airports are publicly owned for public use. Others are privately owned for public use. And still others are privately owned for private use. This discussion focuses on public use airports.

Although the FAA does not own airports, it does play a big role in assisting airport owners in improving airports. The FAA does this through its Federal Airport Aid Program (FAAP), Airport Development Aid Program (ADAP), and Airport Improvement Program (AIP). If an airport owner is granted federal funds from the FAA to make airport improvements, the airport owner is typically required to comply with certain "Sponsor's Assurances." The assurances are promises that the airport owner makes in exchange for the grant money. They are typically enforceable by the FAA for up to 20 years from the date the grant money is accepted by the airport owner.

Some of the more substantial promises made by the airport owner are as follows:

- No exclusive rights will be granted to any person providing or intending to provide services at the airport (with some exceptions allowed for FBOs in certain circumstances).
- The airport will be made available to all types, kinds, and classes of users on reasonable terms and without unjust discrimination.
- The airport owner will properly maintain and operate airport facilities and approaches to the airport.
- Good title to the airport property will be maintained.
- Revenues generated by the airport and any local taxes on aviation fuel will be expended for capital operating costs of the airport, and books and records will be kept allowing audit of such practices.
- Assurance is given that terminal airspace required for instrument and visual operations will be adequately cleared and protected.
- The airport owner will take appropriate action, to the extent reasonable, to restrict the use of land (through zoning laws) to activities and purposes compatible with normal airport operations.

These assurances provided by the airport owner are typically found in either a deed that conveys the airport property to the airport owner or grant agreements executed under one of the federal grant programs noted earlier.

A closer look at the kinds of promises requested by the FAA reveals that the FAA is seeking to ensure that money it provides to airport owners or proprietors will be used to enhance a national transportation system and not just a local airport. That is why the FAA wants assurances that the airport will be open to all classes of users, that airport money will be used for airport projects, and that the airport will do what it can to ensure that it remains viable by judiciously guarding land restrictions and zoning rules for surrounding properties.

If an airport proprietor fails to abide by its promises, then users or other interested parties may file a complaint with the FAA. Such complaints are filed in accordance with Part 16 of the Federal Aviation Regulations (FARs). The FAA investigates these Part 16 complaints and may take action against proprietors who fail to meet their obligations under the required Sponsor's Assurances.

PERSONAL PROPERTY

Besides real property, the next big category of property law relates to personal property. As indicated earlier, personal property is all property other than real property. This leaves a lot of ground to cover. First, we will give an overview of tangible personal property which includes aircraft and equipment. Next we will touch very briefly on the second type of personal property—intangible personal property.

Tangible property

Tangible personal property includes all property that can be touched and moved. For our purposes, we will focus on aircraft as an example of tangible personal property. Specifically, we will focus on (1) transferring or acquiring an interest in an aircraft; (2) registering aircraft; (3) storage arrangements including leasing and bailments; (4) leasing aircraft for use by others; and (5) multiple-ownership arrangements for aircraft.

Transferring aircraft ownership
By far, the most common way that persons acquire an interest in tangible personal property, such as an aircraft, is through a purchase. While most items of tangible personal property do not require any special documentation of a sale (e.g., purchasing groceries at the local supermarket), aircraft require formal documentation of a purchase and sale.

To alleviate confusion and uncertainty in determining aircraft ownership questions, the federal government established a system for recording documents that transfer title or affect interests in aircraft ownership. The federal system was first established in 1938. In its current form, the law requires the FAA to maintain a system for recording documents affecting title to aircraft. The system is really designed as a one-stop federal clearinghouse for all aircraft documents.[2]

The law requires that any bills of sale or other instruments affecting title to aircraft be subject to recording with the FAA's registry. Importantly, the law states that no document or conveyance will be valid against any third person until it is recorded with the FAA.[3] An interesting discussion of how the FAA's registry plays a role in aircraft ownership issues can be found in Case 8-1 below. Commentary following the case addresses some practical issues related to aircraft sales.

As a practical matter, a person who purchases an aircraft receives a bill of sale. A sample bill of sale is illustrated in Fig. 8-1. The bill of sale is filed for recordation with the FAA along with an application for aircraft registration. A sample application for registration is illustrated in Fig. 8-2. Once the FAA has a bill of sale and an application for registration in hand, it can issue a permanent registration certificate for the aircraft.

In addition to being sold, aircraft can be transferred by gift or by will. However, in all these cases, the same basic procedures are followed as outlined above in the example of an aircraft sale. The first step is to convey the aircraft by way of a bill of sale or deed of gift. The second step is to file the bill of sale or deed of gift along with an application for registration with the FAA registry for recordation.

Registering aircraft

Under U.S. law, it is not legal to operate an aircraft that is eligible for registration unless the aircraft is registered with the FAA.[4] Under the FARs, an aircraft can be registered under the following conditions:

1. It is not registered under the laws of any foreign country.[5] and
2. The aircraft is owned by a "citizen of the United States."[6] or
3. The aircraft is owned by a citizen of a foreign country who is a permanent U.S. resident (green card holder).[7] or
4. The aircraft is owned by a corporation that is not a citizen of the United States, if the corporation is organized under the laws of the United States and the aircraft is based and primarily used in the United States.[8] or
5. The aircraft is owned by the federal government or a state, territory, or possession of the United States.[9]

The regulations provide a detailed definition of who qualifies as a "citizen of the United States." The definition includes individual citizens, partnerships in which all partners are individual citizens, and corporations where two-thirds or more of the board of directors and 75 percent or more of the shareholders are U.S. citizens.

Storing aircraft

With aircraft another important issue is storage. In most cases aircraft are either stored in a hangar or secured at a tie-down facility. If the aircraft is stored at a hangar, the arrangement usually takes the form of a lease in which the aircraft owner and the hangar owner arrange for the aircraft to be stored at a hangar in exchange for rental payments. With a leasing arrangement, the aircraft owner is responsible for any damages incurred by the aircraft unless the hangar owner (landlord) acts in a negligent manner that causes

Figure 8-1
FAA bill of sale.

---

*U.S.GPO:1999-769-500

UNITED STATES OF AMERICA	FORM APPROVED
U.S. DEPARTMENT OF TRANSPORTATION FEDERAL AVIATION ADMINISTRATION	OMB NO. 2120-0042

AIRCRAFT BILL OF SALE

FOR AND IN CONSIDERATION OF $ _____ THE UNDERSIGNED OWNER(S) OF THE FULL LEGAL AND BENEFICIAL TITLE OF THE AIRCRAFT DESCRIBED AS FOLLOWS:

UNITED STATES REGISTRATION NUMBER **N 9999B**

AIRCRAFT MANUFACTURER & MODEL **C-172**

AIRCRAFT SERIAL No. **12345**

DOES THIS _____ DAY OF _____ 19 _____ HEREBY SELL, GRANT, TRANSFER AND DELIVER ALL RIGHTS, TITLE, AND INTERESTS IN AND TO SUCH AIRCRAFT UNTO:

Do Not Write In This Block
FOR FAA USE ONLY

PURCHASER
NAME AND ADDRESS
(IF INDIVIDUAL(S), GIVE LAST NAME, FIRST NAME, AND MIDDLE INITIAL.)

Joe Buyer
2 North Avenue
Southtown, MD 99999

DEALER CERTIFICATE NUMBER

AND TO _____ EXECUTORS, ADMINISTRATORS, AND ASSIGNS TO HAVE AND TO HOLD SINGULARLY THE SAID AIRCRAFT FOREVER, AND WARRANTS THE TITLE THEREOF.

IN TESTIMONY WHEREOF _____ HAVE SET _____ HAND AND SEAL THIS _____ DAY OF _____ 19 _____

SELLER

NAME (S) OF SELLER (TYPED OR PRINTED)	SIGNATURE (S) (IN INK) (IF EXECUTED FOR CO-OWNERSHIP, ALL MUST SIGN.)	TITLE (TYPED OR PRINTED)
Seller Industries Inc.	Sam Seller	President

ACKNOWLEDGMENT (NOT REQUIRED FOR PURPOSES OF FAA RECORDING; HOWEVER, MAY BE REQUIRED BY LOCAL LAW FOR VALIDITY OF THE INSTRUMENT.)

ORIGINAL: TO FAA

AC Form 8050-2 (9/92) (NSN 0052-00-629-0003) Supersedes Previous Edition

Figure 8-2
FAA aircraft registration application.

U.S. GOVERNMENT PRINTING OFFICE: 2000-569-450

FORM APPROVED
OMB No. 2120-0042

UNITED STATES OF AMERICA DEPARTMENT OF TRANSPORTATION
FEDERAL AVIATION ADMINISTRATION-MIKE MONRONEY AERONAUTICAL CENTER
AIRCRAFT REGISTRATION APPLICATION

CERT. ISSUE DATE

UNITED STATES REGISTRATION NUMBER **N** 9999B

AIRCRAFT MANUFACTURER & MODEL
C-172

AIRCRAFT SERIAL No.
12345

FOR FAA USE ONLY

TYPE OF REGISTRATION (Check one box)

☒ 1. Individual ☐ 2. Partnership ☐ 3. Corporation ☐ 4. Co-owner ☐ 5. Gov't. ☐ 8. Non-Citizen Corporation

NAME OF APPLICANT (Person(s) shown on evidence of ownership. If individual, give last name, first name, and middle initial.)

Joe Buyer

TELEPHONE NUMBER: (243) 555-1212

ADDRESS (Permanent mailing address for first applicant listed.)

Number and street: 2 North Avenue

Rural Route: P.O. Box:

CITY	STATE	ZIP CODE
Southtown	Maryland	99999

☐ **CHECK HERE IF YOU ARE ONLY REPORTING A CHANGE OF ADDRESS**

ATTENTION! Read the following statement before signing this application. This portion MUST be completed.

A false or dishonest answer to any question in this application may be grounds for punishment by fine and / or imprisonment (U.S. Code, Title 18, Sec. 1001).

CERTIFICATION

I/WE CERTIFY:

(1) That the above aircraft is owned by the undersigned applicant, who is a citizen (including corporations) of the United States.

(For voting trust, give name of trustee: __________), or:

CHECK ONE AS APPROPRIATE:

a. ☐ A resident alien, with alien registration (Form 1-151 or Form 1-551) No. __________

b. ☐ A non-citizen corporation organized and doing business under the laws of (state) __________ and said aircraft is based and primarily used in the United States. Records or flight hours are available for inspection at __________

(2) That the aircraft is not registered under the laws of any foreign country; and

(3) That legal evidence of ownership is attached or has been filed with the Federal Aviation Administration.

NOTE: If executed for co-ownership all applicants must sign. Use reverse side if necessary.

TYPE OR PRINT NAME BELOW SIGNATURE

EACH PART OF THIS APPLICATION MUST BE SIGNED IN INK.			
	SIGNATURE Joe Buyer	TITLE	DATE
	SIGNATURE	TITLE	DATE
	SIGNATURE	TITLE	DATE

NOTE Pending receipt of the Certificate of Aircraft Registration, the aircraft may be operated for a period not in excess of 90 days, during which time the PINK copy of this application must be carried in the aircraft.

AC Form 8050-1 (12/90) (0052-00-628-9007) Supersedes Previous Edition harm to the aircraft. However, there is no specific duty that a landlord incurs to ensure the safety of the aircraft stored at the landlord's facility.

On the other hand, with tie-down arrangements, a question often arises as to whether the arrangement is a bailment or a lease. A bailment is a legal arrangement whereby a person who owns personal property (such as an aircraft) delivers the personal property to another for a specific purpose (e.g., storage). If a bailment is entered into, there is an implied or express agreement by the person being paid for the bailment that personal property will be safely stored and returned to its owner intact. The specific legal duty by the person undertaking the bailment is the duty to exercise ordinary and reasonable care for the preservation of the property. In the context of an aircraft tie-down, this can be an important distinction.

To illustrate, assume Frank's Air Charter ties down two light twin-engine aircraft at Seabreeze FBO. A strong storm sweeps through the area, and one of Frank's aircraft is damaged when garbage containers from Seabreeze's facility blow onto the ramp and dent the empennage on his twin Cessna 310. If the arrangement between Frank and Seabreeze is a lease, Frank will have to establish that Seabreeze was negligent in failing to secure the garbage containers. If the arrangement is a bailment, the burden on Frank is somewhat lessened—Frank will now have to establish that Seabreeze failed to exercise ordinary and reasonable care for the preservation of the aircraft. Most tie-down arrangements tend to be looked upon in the courts as bailments.[10]

Leasing aircraft
Personal property can be leased just as real property can be leased. Aircraft leasing arrangements are a very common way of defraying aircraft expenses and allowing flexible approaches to financing aircraft. An aircraft lease is a way for an aircraft owner to grant another person the right to possession and use of the owner's aircraft for a specified period of time in return for periodic payments (e.g., rent). As indicated earlier in the chapter, while a lease permits possession and use, it does not convey title to an aircraft.

As a general rule, a lease is a lease. However, in the world of aviation we can generally break down leasing activity into three categories with respect to timing of the lease:

- Nonexclusive leases in which the owner leases an aircraft for short-term periods such as weeks, days, or even hours
- Exclusive leases in which the party leasing the aircraft is provided continuous use and possession of an aircraft for a specified, and generally longer, stretch of time
- So-called leasebacks with flight schools and/or FBOs in which the owner purchases the aircraft from an FBO or school and then leases it back for use in flight instruction or charter services

Another set of categories for aircraft leasing arises from the type of lease involved. Some aircraft leases are referred to as "wet" and others as "dry."

In a wet lease, the aircraft owner provides the aircraft, pilots and other crewmembers, maintenance for the aircraft, and insurance. Such leases can go for any length of time.

Dry leases involve the leasing of an aircraft without crews, maintenance, insurance, etc. As with wet leases, there are no time criteria—the lease can be long- or short-term.

Often, the designation of a lease as wet or dry can have an impact on whether the aircraft's operations are deemed to be subject to Part 91 or for compensation or hire. The NTSB has ruled that obtaining both a flight crew and an airplane from the same source (as in a wet lease) is usually conclusive evidence of an operation for compensation or hire, and thus subject to regulation under Part 135 or 121 rather than Part 91.[11]

The FAA generally does not regulate the leasing of aircraft. However, when a large aircraft (aircraft weighing more than 12,500 pounds maximum certificated takeoff weight) is leased, the FAA requires that the lease be in writing and contain the following:[12]

1. Identification of the FARs controlling the operation and maintenance of the aircraft for the preceding 12 months
2. Name and address of the person maintaining operational control of the aircraft
3. A statement that questions related to the operational control of the aircraft can be answered by an FAA Flight Standards District Office.

Multiple-ownership arrangements

There are several approaches to sharing the cost of operating an aircraft through multiple-ownership arrangements. Traditionally, persons have gotten together and created flying clubs that operate as coownerships, partnerships, or limited liability companies or corporations (see Chap. 7 for more detailed discussion of these entity forms).

In recent years, the concept of fractional ownership has played a major role in corporate aircraft ownership. Fractional ownership programs involve pools of aircraft, and each aircraft is owned by several parties. All the aircraft are placed in dry lease exchange "pools," and the aircraft are made available to any program participant who owns an aircraft that is not available at the time it is needed. An important feature of a fractional ownership program is the management company that runs the aircraft pool, providing the exchange services, handling maintenance, crew selection, and hiring, and doing the recordkeeping.[13]

As the popularity of fractional ownership programs grew, the question of regulating these programs came to the forefront. Should they continue to be regulated under Part 91 standards, or should they be treated as commercial operations and regulated under Part 135 or Part 121? The FAA opted to treat fractional ownership programs under Part 91 and recently adopted Subpart K to Part 91, titled "Fractional Ownership Operations." Some of the most significant requirements of Subpart K include

- A contractual acknowledgment by the fractional owners that they are the operators of aircraft used in their program (and are therefore responsible for airworthiness and all other aspects of the aircraft's operation)
- Formal contracts with fractional program managers
- FAA-approved operating manuals
- Background checks on pilots
- Six- and twelve-month pilot check rides

In the end, much of Subpart K brings formalized requirements that much of the corporate aviation world already had in place. However, by keeping fractional ownership requirements within the purview of Part 91, fractional programs are able to avoid some of the constraints of a commercial carrier, such as additional runway length requirements, the ability to make instrument approaches at airports without weather reporting, flight-time limitations, and drug and alcohol testing programs.[14]

Another type of personal property is intangible property. Intangible property is property that does not take a physical form. The most common types of intangible personal property are copyrights, patents, trademarks, and franchise rights.

Intangible property

Intangible property can have great value. Each airline may have a name or logo in the form of a trademark that creates a "brand" for that airline. Trademark laws protect such branding. Aircraft manufacturers and equipment makers may develop systems or aircraft designs that are unique. If patents can be obtained, their intellectual property will be protected.

INSURANCE

Insurance involves a special kind of contract between the insurer and the insured. The contract in an insurance context is commonly referred to as a policy. In the insurance policy, an insurer agrees to defend and indemnify the insured upon the occurrence of a contingency or unknown occurrence as specified in the insurance agreement. The agreement to defend means that the insurer will provide legal counsel to the insured to defend the insured against any claims that might arguably be covered by the policy. The agreement to indemnify involves an agreement to make the person insured whole (within the limits of a policy) in the case of a loss.

The whole concept of insurance involves the spreading of risk among a large pool of individuals who might experience a loss. Persons seeking insurance pay regular premiums to keep their policies in force. The insurance companies use those premiums to pay on the losses that their policyholders might experience. Of course, it is the insurance company's goal to collect at least as much in premium dollars as it needs to pay off any legitimate claims, pay its operating expenses, and make a profit. Insurance companies are often large-scale investors, placing excess premiums collected in stocks and bonds in order to grow their assets and ensure their ability to pay on future claims. Sometimes, insurers spread risk even further by engaging reinsurers who will, in turn, insure the insurers. This is done by selling a portion of a policy's risk and right to receive premiums to other insurance companies.

Insurance is sold in three ways. (1) Insurance agents work exclusively for one insurance company and act as agents of that company. (2) Insurance brokers are typically independent contractors who act as agents of the insured by selling them policies from any number of qualified insurance companies. (3) Many insurance companies will sell insurance directly to the insured by direct mail or via the Internet.

Insurance and insurers play a large role in the aviation environment. The discussion in this chapter is designed to make you more aware of the way insurance works in general and to give you some insight into some of the thornier issues that have faced aviation insurers and their policyholders.

The two most common types of aviation-related insurance are hull insurance and liability insurance. The first type, hull insurance, is often referred to as *physical damage* insurance. As the name implies, this insurance covers damage to an aircraft. The second type is liability insurance, which covers the contingency that an aircraft mishap may cause damage to someone else's person or property.

Hull insurance and liability insurance

Hull insurance

Hull insurance comes in different versions. One version is the *in-flight* or *in-motion* policy. The other version is the *not-in-flight* or *not-in-motion* version. When hull insurance is purchased on an in-flight basis, it covers damage to the aircraft from the time it starts its takeoff run to the time it has landed and stopped or exited the runway. In-motion insurance is a bit more comprehensive, covering the aircraft any time an aircraft is moving under its own power. Obviously, neither of these policies will provide any coverage if an aircraft is damaged while parked or tied down. For this reason, many private and commercial operators choose all-risk ground and flight insurance to cover any damages incurred to the aircraft hull. Often banks or financing companies that financed the purchase of an aircraft will require the owner to provide all-risk coverage and name them as insureds—this will allow them to be the first in line to recover if the aircraft is damaged while the loan is not fully paid.

Liability insurance

As indicated above, liability insurance is designed to cover the possibility that an aircraft could do harm to someone else's person or property. As a general rule, there are two types of policies for aircraft liability insurance. The first type is *sublimited* coverage. The second is *smooth limits* insurance.

Sublimited coverage sets two limits. The higher amount is often referred to as an *occurrence limit*. This amount is the maximum amount that an insurance company will be liable for any single accident. The lower limits usually refer to the maximum coverage that may be available for each person injured or killed.

Let us take a look at how this would work in the context of a typical general aviation aircraft policy. We will assume a policy covering the aircraft owner for $1,000,000 per occurrence with $100,000 sublimits per injury or death to each person involved in any single occurrence. This means that if the covered aircraft is involved in an accident, the absolute top dollar amount that the insurer will cover is $1,000,000. However, the insurer will only cover the aircraft owner for up to $100,000 for each person injured or killed in the accident.

Smooth limits provide the aircraft owner with the broadest coverage. This coverage usually combines property damage and liability limits. With smooth coverage of $1,000,000 per occurrence, the aircraft owner is covered whether the damage or injury was to passengers or persons or property on the ground. The entire $1,000,000 is available to cover the losses from the same accident. While smooth limits provide greater protection, they do command a higher insurance premium. Determining the right combination of hull and liability insurance is an important responsibility for any aircraft owner, from the light plane general aviation aircraft owner to an international airline. Often the right combination of premium to product is a decision made in consultation with lawyers and risk management professionals.

Insurable interest

For either hull or liability insurance to be effective, the person insured must have an insurable interest. When it comes to aircraft, ownership typically creates an insurable interest. Others with insurable interests in an aircraft could be lienholders and lessees. Insurance on an aircraft is ineffective if the insured does not have an insurable interest at the time of the loss.

The insurance contract

Every insurance contract or policy has certain common elements. We will take some time now to look at the common elements found in all insurance contracts. Next we will review some of the specific provisions that can be found in the majority of aircraft hull and liability insurance policies.

Basic elements of insurance policies

Insurance policies are governed generally by the common law of contracts. However, many states regulate insurers and require that policies be written on standard forms. Most insurance policies, including aircraft policies, contain the following basic elements:

1. *Definitions for special terms in the policy.* These definitions need to be read carefully because the meaning of words in an insurance policy does not always match the commonly understood meaning or the dictionary definition of the word as it is used outside the insurance policy.

2. *Events or items covered under the policy.* In an aviation policy this generally means damage to an aircraft and payments for bodily injury and property damage to others that the insured might be liable for. One other very important duty of the insurer is the duty to defend the insured against any lawsuits that arguably involve a claim within the coverage of the policy. As indicated earlier, this duty to defend includes the duty to pay for lawyers and court costs on behalf of the insured.

3. *Exclusions to coverage under the policy.* Aviation policies often contain exclusions for certain types of operations, territorial limits, pilot qualifications, and common exclusions for war, civil revolt, etc.

4. *Duties for the person insured.* Most aircraft insurance policies include requirements for prompt notification to the insurance company, cooperation, proof of loss, and protection of insured aircraft from further damage. Of course, all policies include the either express or implied agreement of the insured to pay the premiums agreed to in the policy in a timely manner.

5. *Deductible clauses.* These clauses provide for insurance proceeds to be payable only after the insured has paid a certain amount of the damage or loss.

Beyond the basic elements directly found in an insurance policy, two common law issues are a recurring theme in various insurance policies. The first is the issue of misrepresentation or concealment in the application for an insurance policy. The second is the legal concept of subrogation.

Most insurance companies will require applicants to disclose relevant information in forms and questionnaires prior to agreeing to provide coverage. If the party to be insured makes misrepresentations or conceals material information, the insurance company may be able to avoid coverage. Usually this rule applies regardless of whether the misrepresentation or concealment was intentional. For instance, before agreeing to provide hull and liability insurance to an air carrier, the insurance company sends a questionnaire to the air carrier requesting information regarding its pilots. One of the questions inquires if any of the pilots have had their certificates suspended or revoked by the FAA for regulatory infractions that occurred within the last 3 years. The air carrier responds no to this question. However, three of the air carrier's ten pilots have had their certificates suspended or revoked within the past 3 years. These false responses may invalidate the coverage provided in the insurance policy if a court deems them to be material enough that the truthful response would have altered the insurer's decision to insure the air carrier.

One last general item for discussion is right of an insurance company to subrogation. Here's how subrogation works: Suppose an insurance company pays a claim on an aircraft that was damaged when a fuel truck owned by a local FBO crashed into it while the aircraft was parked on a ramp. After the aircraft owner is paid for her or his damages, the insurance company that paid the claim can then sue the FBO to recover the amounts it paid on the claim. In essence, subrogation allows the insurance company to step into the shoes of its policyholder and pursue legal action against the wrongdoer who caused the damage leading to the claim. In the end, this allows an insurer to be made whole. Insurers carefully guard their right to subrogation. In fact, most policies indicate that coverage will be denied if the person insured did something to deny the insurer's right to subrogation (e.g., releasing the wrongdoer from liability).

Coverage issues for aviation policies

A review of case law and trade publications reveals that there are several recurring issues in the realm of aviation-related hull and liability policies. A discussion of some of the more common issues follows.

- *Has the insured provided proper notice in accordance with policy procedures?* It is vital that an insured supply proper notice of a claim (often required in writing) in a timely manner after an incident. All notices, demands, and lawsuit filings must also be forwarded to the insurer in accordance with policy terms. Failure of the insured to do so may result in loss of coverage.
- *Has the insured misrepresented or concealed any material facts?* As indicated above, any misrepresentation or concealment of a material fact is likely to nullify insurance coverage. This becomes especially true in aviation policies in which so much depends upon written logs and certifications issued by the government.
- *Were there any restrictions on the aircraft's operations?* Most aviation policies will restrict coverage to aircraft used for "business or pleasure" or commercial operations.

If the aircraft was not being appropriately utilized at the time of an accident, that will place coverage in peril.

- *Did the pilot have the right certifications at the time of the accident? Was the pilot properly rated?* These two questions are critical in many aviation insurance cases. Most aviation insurance policies require certain ratings and do require that the pilot be current at the time of a covered incident or accident.
- *Did the pilot have the required number of logged hours?* This is another issue that trips up many insureds in getting coverage. It is imperative that logged hours match the requirements of the policy. Many claims are denied because logged hours did not meet insurance policy requirements for time in type, total time, or currency.
- *Did the aircraft have a current airworthiness certificate?* This is a simple enough item to check on. However, sometimes it can become more complex if the policy in question nullifies coverage on a special flight permit or ferry permit.
- *Was the pilot operating the aircraft with a current medical certificate?* This issue bites many insureds when they go to make a claim and discover that they or the pilot operating the aircraft at the time of an accident failed to have a current medical certificate.
- *Did the failure of the insured to meet the policy requirements have a causal connection to the accident?* This issue often arises when an insurance company attempts to avoid coverage because the insured fails to meet a particular technical requirement in the policy. The question boils down to whether failure to meet the technical policy requirement actually caused the accident or loss in the claim. Some state courts find this inquiry relevant; others do not.

This is by no means a comprehensive list of all the major issues that come up with regularity in aviation insurance policies. However, it is meant to give you a feel for the types of things that aviation risk managers need to be on the lookout for. Some of these issues are addressed more fully in the "Cases and Commentary" section below.

Other types of insurance

The discussion above has focused on the typical hull and liability aviation policy. However, there are many other types of insurance that all professionals (whether or not they operate in the aviation environment) should be familiar with.

Health and disability

One of the pressing issues in today's society is the ability of individuals and families to obtain affordable health care insurance. Health insurance is designed to cover the cost of medical treatment, prescription medications, hospitalization, and surgical care. Dental insurance covers the cost of dental treatment. Employers have traditionally contributed a share of the cost of health and dental insurance. However, rapidly increasing costs have forced many employers to transfer more of the burden of health and dental insurance costs to their employees.

Disability insurance can be obtained by employers, employees, and self-employed persons. This type of insurance provides for regular income (usually on a monthly basis) during the time when a person is disabled and cannot work. With stringent medical requirements on pilots and air traffic controllers, disability insurance is a must in these professions.

Life insurance

Life insurance is designed to provide death benefits to persons designated by the insured person. There are three basic types of life insurance. A brief and very general description of each type follows.

Whole life insurance provides coverage throughout the life of the insured. Premiums for whole life policies are set to cover a death benefit. Some amounts are set aside from each premium payment to create a savings component otherwise known as a cash surrender value. The savings component grows at a fixed percentage of interest.

Term life insurance covers only a limited period of time—typically from 5 to 20 years. There is no savings feature associated with a term life policy. However, premiums on a term life policy are usually lower than those on whole life policies.

Universal or variable life insurance policies are a hybrid of whole life and term life policies. There is a savings component—but it grows at a variable rate often tied to equity or bond funds.

Business insurance

Several types of insurance products are used by businesses to protect themselves and employees from foreseeable contingencies. A few of the more significant types of business insurance policies are outlined below.

Business interruption insurance is critical if a business is severely damaged or destroyed due to a physical or nonphysical peril. This type of insurance replaces lost revenue and allows a company to continue paying employees and vendors until business can be resumed. In the aftermath of 9/11 several airlines turned to business interruption policies. In one recent case interpreting a business interruption policy, a Virginia state appellate court held that U.S. Airways was entitled to reimbursement from its $25 million business interruption policy based on claims that events subsequent to 9/11 caused losses covered under the policy. The insurer in that case argued that the policy did not cover U.S. Airways losses due to 9/11 because the airline did not suffer property damage. However, the court found that other provisions in the policy related to civilian or military intervention provisions were applicable and therefore the airlines claims under the policy should be honored.[15]

Workers' compensation insurance is also a big part of any employer's risk management plan. Workers' compensation insurance covers an employee if the employee is injured while performing duties within the scope of his or her employment. Many states require employers to buy workers' compensation insurance for their workers.

Key-person life insurance is often used by small businesses as they plan for the succession of the business from the current owners. With a key-person policy, the business pays the premiums on life insurance policies on the lives of key executives, shareholders, members, or partners. If one of these key people dies, the insurance proceeds can be used to pay the deceased's family or other beneficiaries. Often this type of insurance is purchased and funded by a company as part of a buy-sell agreement where the parties make plans ahead of time for what happens when a key person sells her or his interest in the company or dies.

CASES AND COMMENTARY

Case 8-1 is one of the rare aviation cases that made it all the way to the Supreme Court. Recall from Chap. 1 that the Supreme Court is not obliged to take on every case that comes its way. In essence, the Court selects the cases that it wants to hear. It is probably fair to infer from this that the Supreme Court believed that aircraft ownership and recordation of interests in aircraft with the FAA were significant enough issues to take on in this landmark case.

CASE 8-1
PHILKO AVIATION, INC, v. SHACKET ET UX.
462 U.S. 406 (1983)

OPINION BY: White, J.

OPINION: This case presents the question whether the Federal Aviation Act of 1958 (Act), [Citation], prohibits all transfers of title to aircraft from having validity against innocent third parties unless the transfer has been evidenced by a written instrument, and the instrument has been recorded with the Federal Aviation Administration (FAA). We conclude that the Act does have such effect.

On April 19, 1978, at an airport in Illinois, a corporation operated by Roger Smith sold a new airplane to respondents. Respondents, the Shackets, paid the sale price in full and took possession of the aircraft, and they have been in possession ever since. Smith, however, did not give respondents the original bills of sale reflecting the chain of title to the plane. He instead gave them only photocopies and his assurance that he would "take care of the paperwork," which the Shackets understood to include the recordation of the original bills of sale with the FAA. Insofar as the present record reveals, the Shackets never attempted to record their title with the FAA.

Unfortunately for all, Smith did not keep his word but instead commenced a fraudulent scheme. Shortly after the sale to the Shackets, Smith purported to sell the same airplane to petitioner, Philko Aviation. According to Philko, Smith said that the plane was in Michigan having electronic equipment installed. Nevertheless, Philko and its financing bank were satisfied that all was in order, for they had examined the original bills of sale and had checked the aircraft's title against FAA records. At closing, Smith gave Philko the title documents, but, of course, he did not and could not have given Philko possession of the aircraft. Philko's bank subsequently recorded the title documents with the FAA.

After the fraud became apparent, the Shackets filed the present declaratory judgment action to determine title to the plane. Philko argued that it had title because the Shackets had never recorded their interest in the airplane with the FAA. Philko relied on § 503(c) of the Act, [Citation], which provides that no conveyance or instrument affecting the title to any civil aircraft shall be valid against third parties not having actual notice of the sale, until such conveyance or other instrument is filed for recordation with the FAA. However, the District Court awarded summary judgment in favor of the Shackets, [Citation], and the Court of Appeals affirmed, reasoning that § 503(c) did not pre-empt substantive state law regarding title transfers, and that, under the Illinois Uniform Commercial Code, [Citation], the Shackets had title but Philko did not.[Citation]. We granted certiorari, [Citation], and we now reverse and remand for further proceedings.

Section 503(a)(1) of the Act, [Citation] directs the Secretary of Transportation to establish and maintain a system for the recording of any "conveyance which affects the title to, or any interest in, any civil aircraft of the United States." Section 503(c), [Citation], states:

No conveyance or instrument the recording of which is provided for by [§503(a)(1)] shall be valid in respect of such aircraft ... against any person other than the person by whom the conveyance or other instrument is made or given, his heir or devisee, or any person having actual notice thereof, until such conveyance or other instrument is filed for recordation in the office of the Secretary of Transportation.

The statutory definition of "conveyance" defines the term as "a bill of sale, contract of conditional sale, mortgage, assignment of mortgage, or other instrument affecting title to, or interest in, property." [Citation.] If § 503(c) were to be interpreted literally in accordance with the statutory definition, that section would not require every transfer to be documented and recorded; it would only invalidate unrecorded title instruments, rather than unrecorded title transfers. Under this interpretation, a claimant might be able to prevail against an innocent third party by establishing his title without relying on an instrument. In the present case, for example, the Shackets could not prove their title on the basis of an unrecorded bill of sale or other writing purporting to evidence a transfer of title to them, even if state law did not require recordation of such instruments, but they might still prevail, since Illinois law does not require written evidence of a sale "with respect to goods for which payment has been made and accepted or which have been received and accepted." [Citation.]

We are convinced, however, that Congress did not intend § 503(c) to be interpreted in this manner. Rather, § 503(c) means that every aircraft transfer must be evidenced by an instrument, and every such instrument must be recorded, before the rights of innocent third parties can be affected. Furthermore, because of these federal requirements, state laws permitting undocumented or unrecorded transfers are pre-empted, for there is a direct conflict between § 503(c) and such state laws, and the federal law must prevail.

These conclusions are dictated by the legislative history. The House and House Conference Committee Reports, and the section-by-section analysis of one of the bill's drafters, all expressly declare that the federal statute "requires" the recordation of "every transfer ... of any interest in a civil aircraft." The House Conference Report explains: "This section requires the recordation with the Authority of every transfer made after the effective date of the section, of any interest in a civil aircraft of the United States. The conveyance evidencing *each such transfer* is to be recorded with an index in a recording system to be established by the Authority." Thus, since Congress intended to require the recordation of a conveyance evidencing *each transfer* of an interest in aircraft, Congress must have intended to pre-empt any state law under which a transfer without a recordable conveyance would be valid against innocent transferees or lienholders who have recorded.

Any other construction would defeat the primary congressional purpose for the enactment of § 503(c), which was to create "a central clearing house for recordation of titles so that a person, wherever he may be, will know where he can find ready access to the claims against, or liens, or other legal interests in an aircraft." [Citations.] Here, state law does not require any documentation whatsoever for a valid transfer of an aircraft to be effected. An oral sale is fully valid against third parties once the buyer takes possession of the plane. If the state law allowing this result were not pre-empted by § 503(c), then any buyer in possession would have absolutely no need or incentive to record his title with the FAA, and he could refuse to do so with impunity, and thereby prevent the "central clearing house" from providing "ready access" to information about his claim. This is not what Congress intended.

In the absence of the statutory definition of conveyance, our reading of § 503(c) would be by far the most natural one, because the term "conveyance" is first defined in the dictionary as "the action of conveying," i.e., "the act by which title to property ... is transferred." Webster's Third New International Dictionary 499 (P. Gove ed. 1976). Had Congress defined "conveyance" in accordance with this definition, then § 503(c) plainly would have required the recordation of every transfer. Congress' failure to adopt this definition is not dispositive, however, since the statutory definition is expressly not applicable if "the context otherwise requires." [Citation] * * *

* * *

In view of the foregoing, we find that the courts below erred by granting the Shackets summary judgment on the basis that if an unrecorded transfer of an aircraft is valid under state law, it has validity as against innocent third parties. Of course, it is undisputed that the sale to the Shackets was valid and binding as between the parties. Hence, if Philko had actual notice of the transfer to the Shackets or if, under state law, Philko failed to acquire or perfect the interest that it purports to assert for reasons wholly unrelated to the sale to the Shackets, Philko would not have an enforceable interest, and the Shackets would retain possession of the aircraft. Furthermore, we do not think that the federal law imposes a standard with which it is impossible to comply. There may be situations in which the transferee has used reasonable diligence to file and cannot be faulted for the failure of the crucial documents to be of record. But because of the manner in which this case was disposed of on summary judgment, matters such as these were not considered, and these issues remain open on remand. The judgment of the Court of Appeals is reversed, and the case is remanded for further proceedings consistent with this opinion.

So ordered.

Notice that in this case, the district court and the circuit court ruled in favor of the Shackets. The Supreme Court's decision overturned the district court and circuit court rulings. The Court essentially states that if you want your interest in an aircraft to be valid against third parties, you must record your interest with the FAA in Oklahoma City. One of the key questions in this case related to the issue of preemption (see Chapter 2). If state law prevailed, the Shackets would win because under Illinois' adaptation of the U.C.C., no filing with a central FAA clearinghouse would be necessary. Ultimately the Supreme Court ruled that the federal law requiring a national central clearinghouse for recording interests in aircraft preempts any state or local laws on this issue.

Does the Supreme Court's ruling make sense? Why do aircraft need a central registry for recording documents related to transfers of title and liens? After all, most other items of real and personal property (title to cars, boats, and real estate) do not require filing with the federal government—they only require filing with the appropriate state or local agency. What is so special about an aircraft that it requires a single national registry for aircraft?

One other lesson that might be drawn from this case is the very practical need to ensure that paperwork related to interests in aircraft actually gets recorded at FAA's registry in Oklahoma City. What is the best way to do this? Many experienced lawyers prefer to take a page from the typical real estate closing by using an escrow agent to assist in closing the deal. This allows all the money and necessary documents to be handled by an expert and neutral third party. In most cases the escrowed transaction can be broken down into two phases, the setup phase and the closing phase.

In the setup phase the following occurs:

1. The buyer wire-transfers a deposit (usually 5 to 10 percent of the purchase price of the aircraft) to the escrow agent along with an application for registration for the aircraft being purchased.
2. The buyer's bank (the bank lending money to the buyer) wires all remaining funds needed to pay the purchase price to the escrow agent.
3. The seller sends the escrow agent a bill of sale transferring ownership of the aircraft to the buyer.
4. The seller's bank is contacted by the escrow agent regarding the aircraft sale. Subsequently the seller's bank sends a lien release to the escrow agent that releases the seller from any amounts the seller may still owe on the aircraft.

Once all the funds and documents from the setup phase are in place, the transaction is poised to be closed. At the closing the following occurs:

1. The escrow agent wires the necessary funds to the seller's bank to cancel any debt and release any liens by the seller's bank.
2. The escrow agent wires any remaining proceeds to the seller.
3. The escrow agent sends the following documents to the FAA registry:
   - Lien for the buyer's bank
   - Bill of sale transferring title from seller to buyer
   - Buyer's registration application
   - Release of lien by seller's bank

In most transactions, the escrow agent takes its fees in equal shares from the buyer and seller. A graphical illustration of the escrow transaction described above is seen in Figs. 8-3 and 8-4.

While many buyers and sellers forgo escrowed transactions, the lessons from *Philko v. Shacket* should remain clear—make sure the necessary documents are properly recorded with the FAA. It may be foolhardy to let a party to the transaction "take care of the paperwork."

As discussed earlier in this chapter, one of the fundamental issues in any insurance claim is whether the claimant has an insurable interest in the property insured. Case 8-2 deals with the question of whether an insurable interest in an aircraft had passed to another at the time of an accident.

Figure 8-3
Initiation of an aircraft escrow transaction.

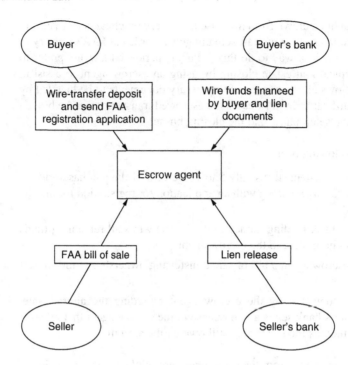

Figure 8-4
Closing an aircraft escrow transaction.

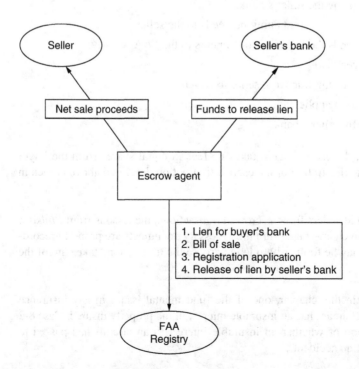

CASE 8-2
JAMES BOWMAN V. AMERICAN HOME ASSURANCE COMPANY
213 N.W. 2d 446 (1973)

OPINION BY: White, J.

OPINION: This is an action on a contract of insurance. The insured, James Bowman, seeks to have his insurer pay for damage done to an airplane covered under a policy issued by the insurer, American Home Assurance Company. The insurer argues that the insured did not have title as defined under the Uniform Commercial Code at the time of the loss and therefore asserts that the insured had no insurable interest at the time of the loss and should be denied recovery. The insured had the verdict and judgment at trial. The insurer appeals. We affirm.

James Bowman, the appellee-insured, and Keith Moeller were engaged in a partnership doing business as Bowman Hydro-Vat in Fremont, Nebraska. The partnership purchased a twin-engine Cessna in 1969. Upon purchasing the aircraft, Bowman applied for and obtained a policy of insurance issued by the appellant-insurer, American Home Assurance Company, insuring the aircraft during the period from December 23, 1969, through December 23, 1970. The named insured under the policy was James Bowman.

In December of 1970, Bowman and James Hemmer entered into negotiations for the sale of the plane to Hemmer. On December 12, 1970, Bowman and Hemmer agreed upon a price of $18,500 for the purchase of the aircraft and Hemmer paid $15,000 down. The remaining $3,500 was to be paid later in cash or through its equivalent in aircraft instrument instruction which Hemmer was to give to Bowman. Hemmer requested a bill of sale signed by both partners to protect himself, and to comply with the Federal Aviation Administration requirements for the transfer of an aircraft. Bowman testified that it was agreed that he was to remain the owner of the aircraft until "we were able to fill out the necessary paperwork." Hemmer, the buyer, testified numerous times that he was to be the owner when he received the bill of sale. This was to allow the buyer to comply with the Federal Aviation Administration requirements and make arrangements for insurance prior to the time he was to become the owner. Bowman also testified that he told Hemmer that he would leave his insurance in effect until it expired on December 23, 1970, only 11 days later. Bowman retained possession of the plane.

On December 15, 3 days later, Bowman contacted Hemmer and asked Hemmer if he wanted to go with him on a business trip to Kansas. The purpose of the flight was for Bowman to transact some business in Kansas. Upon their return to Fremont, Hemmer asked Bowman for permission to use the plane on the following Friday and Saturday. Bowman and Hemmer specifically examined Bowman's insurance policy to ascertain whether it would provide coverage while Hemmer flew the plane. Bowman then gave Hemmer permission and Hemmer took the plane to Columbus. On December 16, Hemmer flew the plane from Columbus to Fremont to obtain the bill of sale, but it had not been signed so he returned to Columbus without it.

On December 18, Hemmer flew the plane to Mitchell, South Dakota, pursuant to the permission granted by Bowman. In attempting to take off from Mitchell, the tip of a wing caught in a snow bank causing extensive damage to the aircraft.

Hemmer testified that the Federal Aviation Administration regulations require that a registration certificate be in an aircraft before title to the plane can be transferred to a new owner, and that once the bill of sale is received, it is attached to a new registration application and sent to the Federal Aviation Administration. A pink copy of the new registration is placed in the aircraft to serve as a temporary registration. This paperwork had not been completed at the time of the loss because the bill of sale had not yet been received. The registration certificate in the aircraft at the time of the accident showed James Bowman as the owner.

The signature of Bowman's partner was obtained and the bill of sale was mailed to Hemmer on December 18, the day of the accident. Hemmer received the bill of sale on December 20. The bill of sale was in blank form and had not been filled out at the time Hemmer received it. It was understood that Hemmer was to fill out the necessary information on the bill of sale. Bowman filed an accident report after the accident and indicated he was the owner.

The controversy between the parties centers around two provisions of the Uniform Commercial Code. Section 2-501, U.C.C., provides in part: "(2) The seller retains an insurable interest in goods so long as title to or any security interest in the goods remains in him * * *."

The case was tried to the jury on the theory that the seller retained an insurable interest in the goods as long as title to the goods remained with the seller. Both parties agree this was the appropriate standard for submission of the case to the jury. Section 2-401, U.C.C., details the concept of passage of title:

(2) *Unless otherwise explicitly agreed* title passes to the buyer at the time and place at which the seller completes his performance with reference to the physical delivery of the goods, despite any reservation of a security interest and even though a document of title is to be delivered at a different time or place * * *." (Emphasis supplied.)

As section 2-501, U.C.C., provides, the seller has an insurable interest until title passes to the buyer. Under section 2-401, U.C.C., title passes to the buyer (1) at the time and place where the seller completes his performance with reference to the physical delivery of the goods or (2) at any other time explicitly agreed to by the parties. As dictated by the Uniform Commercial Code, the trial court submitted two factual questions to the jury. First, whether the seller had completed physical delivery of the goods. Second, whether there was an explicit agreement between the buyer and the seller as to the time when title was to pass. * * *

There was substantial evidence from which the jury could have inferred that the seller had not completed physical delivery of the goods. The evidence shows that the buyer was only given limited use of the plane to make the trip to South Dakota. The buyer even asked the seller for permission to use the plane for this one trip. The seller had only granted a limited possession of the plane to the buyer, even though the buyer had possession of the plane for 3 days prior to the accident.

The jury could have also inferred from the evidence that the buyer and seller had an explicit agreement that title was to pass upon the completion of the "necessary paperwork."

* * *

Bowman testified that it was agreed that he was to remain the owner of the aircraft until "we were able to fill out the necessary paperwork." * * *

The evidence showed that the bill of sale sent to the buyer from the sellers on the day of the accident was not signed by both sellers until the day of the accident. The bill of sale was not received by the buyer until several days after the accident, and even at this time it remained in blank form. The parties knew that the buyer would fill in and complete the bill of sale after he received it. Thus, it is clear from the evidence that "completion of the necessary paperwork" involved more than the mere signing of the bill of sale by both sellers. It at least included receipt of the blank bill of sale by the buyer, but it could also have included the action of the buyer in filling out the necessary Federal Aviation Administration papers and completing the bill of sale. None of the above steps had been completed at the time of the accident, and therefore the time of the completion of the necessary paperwork had not occurred at the time of the accident.

In summary, the jury could reasonably have found from substantial evidence that either (1) the seller had not completed physical delivery of the goods under section 2-401, U.C.C., or (2) there was an explicit agreement for title to pass upon completion of the necessary paperwork which had not occurred at the time of the accident. Under either of these findings title had not passed to the buyer under section 2-401, U.C.C., and therefore under section 2-501, U.C.C., the sellers retained an insurable interest. For these reasons we affirm the judgment of the District Court.

Affirmed.

Whether or not you agree with the findings of the court, what would you have suggested to the parties to avoid all the confusion and subsequent litigation that followed the accident? [*Hint:* Should this agreement have been in writing? If so, what provisions would you have included?]

Cases 8-3 and 8-4 wrestle with the issue of causal connection. Each of these cases involves an insurance claim after an accident. In both cases, the pilot-in-command did not have a current medical certificate. Also in both cases, there was no apparent causal connection between the accident and the failure of the pilot-in-command to hold an effective medical certificate.

CASE 8-3
WESTERN FOOD PRODUCTS COMPANY, INC. V. UNITED STATES FIRE INSURANCE COMPANY
699 P. 2d 579 (1985)

OPINION BY: Parks, J.

OPINION: Plaintiff, Western Food Products Co., Inc., was the owner of an airplane insured by defendant, United States Fire Insurance Company. The plane was totally destroyed in a crash on January 5, 1981, and plaintiff filed a claim for its loss with the insurer. Defendant denied coverage for the loss and plaintiff filed this action. Both parties requested summary judgment based on stipulated facts and the trial court granted judgment to defendant insurer. Plaintiff appeals.

The parties stipulated that at the time of the crash, the airplane was being operated by Charles Newton Benscheidt, who had been issued a private pilot's certificate on January 14, 1976, with ratings for single engine land aircraft. Mr. Benscheidt was issued a third-class medical certificate with no limitations on October 3, 1977, which expired on

October 31, 1979. The records of the Federal Aviation Administration fail to indicate that any subsequent medical certificate was issued to Mr. Benscheidt.

The district court held that because defendant's insurance policy requires the pilot to have valid medical certification as a condition of coverage, defendant was not liable for the loss of the aircraft. Plaintiff contends that this holding is erroneous because the policy provision requiring a medical certificate did not apply to the person piloting the plane when it crashed. Arguing that this clause is ambiguous, plaintiff contends that it must be construed in its favor and held inapplicable under the circumstances. The clause at issue states as follows:

> THE PILOT FLYING THE AIRCRAFT: The aircraft must be operated in flight only by a person shown below who must have a current and proper (1) medical certificate and (2) pilot's certificate with necessary ratings, as required by the FAA for each flight. There is no coverage under the policy if the pilot does not meet these requirements.

Immediately after the above clause was typed the following language:

> STEVE BENSCHEIDT, JAMES DRISCOLL, III; OTHERWISE, PILOTS WHO HAVE A CURRENT PRIVATE OR COMMERCIAL CERTIFICATE AND A MINIMUM OF 750 LOGGED PILOT HOURS OF WHICH AT LEAST 25 HOURS HAVE BEEN IN THE SAME MAKE AND MODEL AIRCRAFT WE COVER IN ITEM 5.

It is undisputed that at the time the insured airplane crashed, it was being piloted by an individual who is not named in the above paragraph. In addition, plaintiff concedes that the pilot of the plane did not have a current medical certificate. * * *

* * *

Plaintiff contends that * * * the provision of the policy excluding coverage of a flight piloted by a person without a medical certificate * * * should not be enforced absent proof of a causal connection between the crash and the health of the pilot. Although such a holding would certainly depart from the ordinary rule mandating enforcement of an unambiguous provision according to its terms, the argument is not without some support in other jurisdictions. [Citations.] These opinions generally stress the social purpose of insurance and the unfairness to the insured if he loses his expected protection because of a technical breach of the policy which is unrelated to the risks insured. However, in each of these cases the court was able to say that the provision relied on by the insurer failed to include language of express exclusion. The opinions could fairly conclude that the provision relied upon to exclude coverage under the aviation policy was a forfeiture clause or condition subsequent which is not ordinarily deserving of enforcement. [Citation.] Thus, in these cases the court was generally concerned with a provision viewed as attempting to eliminate extended coverage rather than one which defines the intended scope of coverage in the first instance.

In this case, the provision states unequivocally, "there is no coverage under the policy if the pilot does not meet these requirements." It is not a question of the insured being forced to forego coverage for which he has already paid; no protection was ever extended to cover the circumstances of the loss. Thus, since the policy specifically excludes coverage, its application cannot be categorized as a forfeiture.

A second distinguishing point of the cases holding that a causal connection must be shown to exclude coverage in an aviation insurance policy is that in each of the cases relied upon by plaintiff, there is state law either strongly favoring coverage in the absence of causation or requiring liberal construction of exclusionary clauses even without a specific finding of ambiguity. For example, in [Citation], the court relied on a line of cases involving automobile liability and life insurance. These precedents held in varying situations that the violation of a provision limiting coverage would effectively operate to exclude liability only if a causal connection were demonstrated between the risk sought to be excluded and the actual loss.

By contrast, Kansas decisions have considered the purpose and reasonableness of an exclusionary provision in other types of insurance policies to determine its scope, but they have not demanded a relationship between the failure to satisfy the limitation and the actual loss sustained. For example, in [Citation], the Court was concerned with whether a geographical limitation in an insurance policy covering a mobile home was unambiguous and enforceable. The insured argued that there was no relevant connection between the geographical limitation and the extended coverages and that, thus, the parties did not intend this limitation to apply. The Court pointed out that there was indeed good reason for the insurer to seek to geographically limit its extended coverage and held the provision doing so to be unambiguous. The Court then refused to further consider the relationship between the limitation imposed and the actual loss suffered, stating as follows: "Further discussion of the relevance, materiality or importance of location of insured property is not warranted here, for the issue here is not whether the geographic limitation *should be* applicable, but whether it is applicable under the policy terms. We hold that it is." [Citation.]

Thus, the Court refused to discuss an argument similar to that made here—that there must be some relevant connection between the actual loss sustained and the limiting provision sought to be enforced. The connection between the limitation and the abstract risks insured may be examined to determine the scope of the limitation but there is no need for the loss to be caused by the condition named in the exclusion in order for the exclusion to be effective.

* * *

In conclusion, we remain unconvinced by plaintiff's argument that the ordinary rules of construction should be abandoned in this case. The exclusionary provision of the insurance contract is unambiguous and we refuse to alter this plain language to forestall its effect. Therefore, like the majority of jurisdictions, we must conclude that a causal connection between the accident causing the loss and the purpose of an exclusionary clause need not be proven before coverage can be denied by the aircraft insurer on the basis of the exclusion.

* * *

Affirmed.

CASE 8-4
SOUTH CAROLINA INSURANCE COMPANY V. LOIS S. COLLINS
237 S.E. 2d 358 (1977)

OPINION BY: Rhodes, J.

OPINION: This appeal presents the question of whether, in order to avoid liability under an aircraft insurance policy, the insurer is required to demonstrate a causal connection between the crash of the aircraft and the insured pilot's failure to have a valid and effective medical certificate as provided by the terms of the policy. For the reasons set forth herein, we affirm the relief granted by the special circuit judge and hold that such causal connection must be shown.

The plaintiff-appellant, South Carolina Insurance Company (hereinafter appellant), issued to Metz W. Collins, the named insured, its aircraft liability insurance policy, number AC-801297, effective for the period April 27, 1975 to April 27, 1976. The contract of insurance covered a Piper Colt aircraft, Federal Aviation Agency registration number N5723Z, owned by Collins. On May 23, 1975, Collins, while piloting the airplane described in the policy, crashed, resulting in Collins' death and injuries to one Wesley B. Nesbitt, a passenger in the airplane. Subsequently, Lois S. Collins and Evelyn C. Lee were appointed administratrices of the estate of the deceased, Metz W. Collins, and Nesbitt commenced an action against the estate seeking damages for injuries he sustained in the crash. Nesbitt's action is pending in the Court of Common Pleas of Horry County.

The appellant [insurance company] refused the demand of the administratrices to defend the insured's estate against the lawsuit instituted by Nesbitt and subsequently commenced this action for declaratory judgment pursuant to [Citation]. The appellant sought an order declaring that the policy issued to the insured was not in effect during the flight of May 23, 1975, and that it did not afford the estate of the deceased any coverage. Named as defendants in the appellant's action were both administratrices and the injured passenger, Nesbitt. (All defendants are hereafter referred to as "respondents").

The respondents answered and sought affirmative relief, demanding that the appellant's complaint be dismissed and seeking an order declaring that the aforesaid policy was in full force and effect at the time of the crash. The trial judge, after hearing arguments and considering briefs, issued an order which granted the relief prayed for by the respondents.

In the trial judge's order it is stated that the parties, through responses to requests for admissions and stipulations made before him, agreed that the following facts are not in dispute: The Federal Aviation Regulations promulgated by the Federal Aviation Agency require a medical examination of pilots under the supervision of the Federal Air Surgeon or his authorized representative. The insured held a third-class medical certificate which was valid for a twenty-four (24) month period. The last medical certificate issued to the insured was in February, 1973, and it expired on the last day of February, 1975, or nearly three (3) months before the date of the crash. The Federal Aviation Regulations also require the insured to obtain and to have in his possession a valid and effective pilot certificate. The insured had his last required flight review on November 3, 1974, or a period of six (6) months and twenty (20) days before the date of the accident. It was stipulated by the parties that at the time of the accident the insured possessed a valid and effective pilot certificate but that he did not have a valid and effective medical certificate. Moreover, for the purposes of this declaratory judgment action only, it was stipulated that the insured, to the best of the parties' knowledge, had no physical or mental defects at the time of or immediately prior to the accident and that there was no causal connection between the accident and the failure of the insured to have a valid and effective medical certificate. The above statement by the trial judge of the undisputed or stipulated facts has not been challenged on this appeal.

The appellant argues vigorously that the failure of the insured to have a valid and effective medical certificate on the date of the accident amounted to a breach of a condition subsequent or promissory warranty under the terms of the policy, thereby suspending coverage and permitting the appellant to avoid liability. The appellant contends that the trial judge's classification of the pertinent policy provisions as being merely an "exclusion" of the insured's liability was erroneous. Additionally, the appellant maintains that the case law does not support the court's holding that the insurer, in order to avoid liability on a policy such as that involved here, must show that there exists a causal connection between the resulting loss and the insured's failure to have the required effective medical certificate.

* * *

The appellant cites cases from other jurisdictions holding that under circumstances virtually identical, or closely related, to those in the instant case, the insurer need not show a causal connection between damages and injuries sustained in the crash of the aircraft and the insured's failure to comply with certain terms of the policy. These decisions hold essentially that the purpose of the exclusionary language of such policies is not that the risk is excluded if damage to the aircraft is *caused* by the failure of the pilot to be properly certificated, but that the risk is excluded absolutely if loss *occurs while* the aircraft is being flown by a pilot not properly certificated. [Citations.]

* * *

After examination of the decisions cited by the appellant, we find them unpersuasive. Only their number, not their reasoning, lends support to a reversal here. We find that the reasoning used in the [citing South Carolina cases requiring causal connection in automobile insurance policies] line of cases no less compelling when applied to an aircraft liability policy.

In view of the stipulation by the parties that there was no causal connection between the loss and injuries resulting from the crash and the failure of the insured to have a valid and effective medical certificate at the time of the accident, we hold that the lower court acted correctly in awarding the relief sought by the respondents.

Affirmed.

At last count, 19 states had case law holding that no causal connection is required for an insurer to void coverage, and 8 states had case law indicating causal connection was required for an insurer to avoid coverage on a claim.[16] Which approach do you believe to be the most appropriate? Why?

DISCUSSION CASES

1. On October 5, 1997, several organizations conducted a fly-in at Auburn airport near Auburn, Nebraska. As part of the event, persons attending the fly-in could pay $10 for a short flight piloted by Sarah Farrington in a plane owned by Auburn Flying Services ("AFS"). Money for the flights was collected at a table near the runway, which had a sign near it advertising the rides. On the ninth ride of the day, Farrington was landing with three passengers on board when her aircraft collided with a passing semi-tractor truck. All on board, including Farrington, were killed in the crash. AFS had a noncommercial liability insurance policy on the aircraft issued by Avemco Insurance Company. Farrington was a covered pilot under that policy. The policy contained the following exclusion: "This policy does not cover bodily injury, property damage, or loss ... when your insured aircraft is ... used for a commercial purpose." The term *commercial purpose* is defined in the policy as "any use of your insured aircraft for which an insured person receives or intends to receive, money or other benefits." Based on this exclusion,

Avemco denied any coverage or defense to AFS. Both Avemco and AFS agreed that while money was collected for the flights, the amounts collected did not cover the operating expenses of the flights. AFS goes to court to challenge Avemco's denial of coverage. Who will prevail? Explain.

2. In September 2001, Kevin Jensen was piloting an aircraft on approach at the Carson City Airport in Nevada. His engine quit and he crashed into the backyard of Robert Griffin. Griffin was pinned down by the wreckage and suffered extensive injuries exceeding $200,000. Jensen owned the aircraft for a few months prior to the crash and purchased liability insurance through Old Republic Insurance Co. ("Old Republic"). Griffin filed suit against Jensen for the damages he suffered as a result of the crash. Jensen turned to Old Republic for defense and coverage. However, Old Republic refused to defend or cover Jensen for this crash. Old Republic argued that Jensen was not entitled to coverage because he failed to complete an annual inspection on the aircraft as required by the insurance policy's airworthiness requirements. In the NTSB's accident report, it determined that the probable cause for the accident was Jensen's incorrect setting of a fuel valve during the aircraft's prelanding checklist. The NTSB opined that the incorrect setting resulted in fuel starvation. Jensen sues Old Republic to force Old Republic to provide him with insurance coverage in this case. Will Jensen be successful? Why or why not? What arguments are likely to surface from both sides?

3. McCarthy operated her aircraft on a ferry flight permit in order to get the aircraft to the manufacturer's facility in Georgia for a required FAA annual inspection. The aircraft's previous annual inspection expired on January 31, 2004. McCarthy made several attempts to fly her aircraft to Georgia prior to that date, but snow and other weather conditions forced her to delay past the annual expiration date. Therefore, she applied for and received an FAA authorized ferry permit to move the aircraft from Ice Breaker, Maine, to the manufacturer's factory in Georgia. Upon arrival at the Georgia airport, she lost directional control of the aircraft. The resulting ground loop caused her right main landing gear to collapse and the right wing to strike the ground. All parties agreed that McCarthy's accident and the ensuing damage were caused by pilot error. McCarthy obtained her current insurance policy through Aviation Underwriter Enterprises ("AUE"). McCarthy's original AUE insurance policy and policies up until May 2001 included the following language under Part Two (Exclusions): "This insurance does not apply: ... under any coverage ... when the aircraft is in flight: ... when a special permit or waiver is required by the FAA; *except in order to obtain an Annual Inspection or maintenance required by the Federal Aviation Regulations*" (emphasis added). Unlike the language in McCarthy's renewal policy, this language provided coverage if the aircraft was operated under a ferry permit if the ferry permit was required to fly the aircraft to obtain an FAA mandated annual inspection or other FAA required maintenance. To the best of McCarthy's knowledge and recollection, neither AUE nor her insurance broker/agent, Aviation Insurance Resources, ever notified McCarthy of the modification to her policy which after his May 2001 renewal excluded from coverage a ferry permit flight for the purpose of obtaining an FAA-required annual inspection. If McCarthy is denied coverage by AUE, will she be able to force AUE to provide coverage in the courts? Discuss.

4. Robins owned and operated a Piper Malibu Meridian airplane. The Meridian is powered by a Pratt & Whitney PT6A-42A turbine engine. Robins insured the aircraft with Associated Aviation Underwriters, Inc. ("AAU"). While attempting to start the aircraft's engine on August 10, 2001, Robins observed flames coming out of both exhaust stacks. He properly employed emergency engine shutdown procedures, and the fire was eventually extinguished. Inspections after the incident revealed that the engine had been operated at an unsafe temperature range for several seconds, and the aircraft required extensive repairs. Robins paid approximately $240,000 for repairs, engine removal and installation, and alternative transportation. Robins submitted his claim for damages to AAU. AAU denied the claim on the basis that the policy had exclusions for physical damage due to wear and tear including damage caused by heat from the operation, attempted operation, or shutdown of the engine. Based on this exclusion, AAU argued that the damage was caused by heat and not the fire. AAU further argued that the circumstances faced by Robins were not in the nature of an emergency, but instead part of the normal operation of the aircraft. Robins decides to bring this matter to court to have a judge determine whether AAU should honor his claim. Will Robins or AAU prevail? Explain.

5. Courtney and her law firm procured aircraft liability insurance from U.S. Specialty Insurance Co. ("USSI"). Courtney and one of the employees of her law office were killed when her aircraft crashed en route to a business-related function. The estate of the deceased employee filed suit against Courtney's estate, claiming gross negligence. Claims were filed with USSI. USSI denied the claims and sought a ruling from the court that it had no duty to defend or indemnify Courtney's estate or her law firm because the liability policy in question had language that specifically excluded employees acting in the scope of their employment. USSI also argued that Minnesota's Worker's Compensation Act permitted it to exclude coverage for employees injured in the course of their employment. However, the claimants (Courtney's estate, the law firm, and the estate of the deceased employee) countered by citing a Minnesota law that stated, "No policy of insurance issued or delivered in this state covering an aircraft equipped with passenger seats and covering liability hazards shall be issued excluding coverage for injury to or death of passengers or non-passengers." Will the exclusions in the USSI policy withstand this challenge? Discuss.

6. Westchester County in New York State owns and operates the Westchester County Airport, a busy regional facility for private and commercial aircraft. The airport borders a small lake in the west and the State of Connecticut to the east. The airport was originally built in the early 1940s for military use and later turned over to Westchester for civil aviation. There are two runways at the airport. Runway 16/34 is 6550 feet long and is the primary runway used by larger aircraft, both commercial and private. Runway 11/29 is the shorter runway at 4450 feet. Its western boundary ends at the lake. Its eastern boundary is only 300 feet from the Connecticut state line. As a result, when aircraft use runway 11/29, they typically fly low over land within the Town of Greenwich in Connecticut. As years passed, trees growing in Connecticut under the flight path of departing and landing air traffic for runway 11/29 grew to heights that affected safe flight paths. In a 1984 NOTAM, the FAA urged pilots landing on 11/29 to use a descent angle of 9 degrees rather than the standard 3 degree angle for landing approaches. Eventually, in 1988 or 1989, the FAA ordered Westchester County to shorten the usable

length of runway 11/29 by 1350 feet because of the tree growth on the Connecticut side of the border. Fearing that the runway will become totally unusable, Westchester County brings an action in federal district court to force the Town of Greenwich and residents owning trees that impinge on the flight path to runway 11/29 to trim the trees so they would no longer adversely affect flight safety. The district court hears the case and rules that with so many years of use over the Connecticut properties by the aircraft landing and taking off on runway 11/29, Westchester County acquired a prescriptive aviation and clearance easement to the airspace above the Connecticut property owners, and it ordered that Westchester County be permitted to cut back certain trees affecting the use of runway 11/29. The Town of Greenwich and other affected landowners in Connecticut appeal the case to the U.S. Court of Appeals. Who will win this case on appeal? Explain.

7. Durango Air Services, Inc., owned an aircraft and employed Borcher as a pilot. During a flight in 1995, Borcher was piloting Durango's Cessna aircraft when it crashed, killing Borcher and two passengers, Colt and John Ross. Both Ross families filed a wrongful death and mental anguish claims against Durango and won a judgment of $4 million. At the time of the crash, Durango held a policy with Old Republic Insurance Company. The policy provided for a maximum event limit of $1 million and a per passenger sublimit of $100,000. After Durango files a claim, Old Republic agrees that it is liable on the aviation policy. However, Old Republic claims that the claims made by the passengers' families are included within the $100,000 sublimit because these claims by the passengers' families are derived from the actions for bodily injury (which included *mental anguish* under the terms and definitions in the policy). Should the mental anguish claims by the nonpassenger families be limited by the $100,000 sublimit per passenger included in the policy? Discuss.

8. Winter spends several years constructing a home-built aircraft. After the aircraft was completed and ready for use, Winter made arrangements to keep his aircraft tied down at Aerowake FBO located at North Huntsville Airport in Alabama. There was no written agreement between Winter and Aerowake. Aerowake was a full-service FBO with employee attendance from 8:00 a.m. to 8:00 p.m. every day. Winter's plane was tied down under a covered plane port. The port was right next to the trailer where Aerowake's mechanic, Trulson, lived with his family. From time to time if a storm approached, Trulson would clear debris and ensure the aircraft at the tie-down ports were secure. One night Sunholm drove his dune buggy onto airport premises and crashed into Winter's airplane, causing substantial damage to Winter's plane. Sunholm was related by marriage to Trulson, and from time to time he and Trulson's son would ride their dune buggies in fields around the airport. Winter sues Aerowake and argues that his arrangement with Aerowake was a bailment. Aerowake argued that it was a lease. Who will prevail? If the arrangement is recognized as a bailment, what does that mean to Winter?

9. Davis is a commercial pilot. From time to time he has flown for Mathis Aviation and received compensation for his services. Mathis owns a Cessna 421 that it seeks to lease in order to defray costs. Davis sends a letter to Mike Cole, president of Chem-Nuclear, offering the use of an aircraft that Chem-Nuclear had previously chartered from a Part 135 operator (Eagle Aviation) when Davis was employed by Eagle Aviation. The letter from Davis to Mike Cole of Chem-Nuclear states as follows:

I have the Cessna 421 that you used when I was with Eagle Aviation. I am offering it on a rental basis to a few of my old customers at a greatly reduced rate. The way this works is you rent the plane from Mathis Aviation and pay the pilot separately.

The letter goes on to detail a cost work-up for three destinations and concludes with the following: "Mike, I hope we can accommodate some of your travel needs. If you have any questions, you can contact me at the above address or call me." Davis is subsequently contacted by Chem-Nuclear and flies two flights, using the Cessna 421. Mathis Aviation billed Chem-Nuclear for the rental of the aircraft, and Davis billed Chem-Nuclear separately for his pilot services. The FAA investigates the flights and concludes that Davis violated the FARs requiring these types of flights to operate under Part 135 rules. Davis argues that the arrangement with Chem-Nuclear was a dry lease and therefore subject only to Part 91. Who is correct? Explain.

ENDNOTES

1. See AOPA's "Airport Compatible Land Use," www.aopa.org.
2. 49 U.S.C. § 44107.
3. 49 U.S.C. § 44018.
4. 49 U.S.C. § 44101.
5. 14 C.F.R. § 47.3(b).
6. 14 C.F.R. § 47.3(b)(1)(A)(i).
7. 14 C.F.R. § 47.3(b)(1)(A)(i).
8. 14 C.F.R. § 47.9.
9. 14 C.F.R. § 47.3(b)(1)(B)(2).
10. See *Aerowake Aviation Inc. v. Winter and Avemco Insurance Company,* 423 So. 2d 165 (1982).
11. See *Administrator v. Poirier,* 5 NTSB 1928 (1987).
12. See 14 C.F.R. § 91.23.
13. See Eileen M. Gleimer, "When Less Can Be More: Fractional Ownership of Aircraft-The Wings of the Future," *SMU Journal of Air Law and Commerce,* Fall 1999, 64 J. Air L. & Com. 979.
14. See 14 C.F.R. Subparts D, E, and F.
15. See *U.S. Airways Inc. et al. v. Commonwealth Insurance Co.,* No. 03-587, 2004 WL 1637139 (VA. Cir. Ct. Arlington County, July 23, 2004).
16. See Paul A. Lange, "Is Causal Connection Required to Avoid Coverage on an Aviation Liability Policy?" Presentation before the Lawyer-Pilots Bar Association, Summer Meeting, Sun Valley, ID, July 20, 2000.

9 Employment Law and the Aviation Industry

AGENCY LAW   254
Defining agency   254
Employee versus independent contractor   254
Creating and terminating agency relationships   255
Duties of parties   257
Contract liability   258
Tort liability   259

LABOR UNIONS AND EMPLOYMENT   261
Major federal labor union laws   261
Basics of collective bargaining   261
What if collective bargaining fails?   262
Strikes and related issues   263

EMPLOYEE PROTECTION   263
Employment discrimination   263
Employee safety and security   265

CASES AND COMMENTARY   269

DISCUSSION CASES   276

ENDNOTES   278

In order to function, the aviation industry utilizes many highly skilled professionals. Air traffic controllers, pilots, engineers, flight attendants, mechanics, and business professionals are all part of the aviation world as we know it today. These professionals are typically employees or employers. Through the years, case law and statutory law have developed to create a body of law known as employment law.

For the sake of presentation and clarity, this chapter is broken down into three parts, with each part tackling a significant aspect of employment law. First, in a discussion of agency law we will introduce the basics of the relationship between employers and their employees and independent contractors. Second, we will review the impact of labor unions on employment relationships. Third, we will address selected laws that were put in place to protect employees and applicants for jobs from discrimination, loss of income, and unsafe working conditions.

AGENCY LAW

To operate just about any business in the modern world requires the use of third parties to act on behalf of the business owners. This applies to megacorporations such as Microsoft and small businesses such as the FBO at your local airport. Persons who are authorized to act for a business are agents, and their relationship with the business employing them is considered to be an agency relationship. Agents and agency relationships are governed by agency law, which is largely a blend of contract and tort law. In this section we will take a closer look at the fundamentals of agency law and how this body of law impacts the aviation industry.

Defining agency

Agency is defined as a fiduciary relationship in which one person is authorized to act on behalf of another person. The word *fiduciary* means that the relationship is a relationship of trust.[1] The relationship must be consensual; that is, both parties agree to the arrangement. In the final analysis, much of agency law deals with the legal proposition that persons who act through agents act themselves. Lawyers often refer to the agency relationship as a master-servant relationship. This is a somewhat antiquated way of referring to a relationship in which one party—the servant—acts as an agent on behalf of another—the master. In modern times, the relationship is often referred to as principal-agent.

Employee versus independent contractor

The agency relationship can take two forms—employer-employee or employer-independent contractor. As a general matter, the employer-employee relationship is one in which the employer controls the methods and means by which the employee gets his or her job accomplished. Most persons who work for others are engaged in an employer-employee relationship. Airline pilots, air traffic controllers, flight attendants, and anyone who works during hours prescribed by someone else and who uses someone else's equipment to perform the work is considered an employee.[2]

On the other hand, independent contractors are required to do the job they've contracted to perform. Unlike the employee, the independent contractor controls the methods and the means used to do the job.[3] In the world of aviation, aviation medical examiners, owners of small charter operations, and certain flight instructors who do freelance work may be considered to be independent contractors.

Sometimes it is not easy to determine whether someone is an independent contractor or an employee. Over the years, the courts have determined a list of tests that can be used in analyzing this question. This list includes (but is not limited to) the following tests:

- How much control does the employer exercise over the details of the work?
- Is the person employed engaged in a distinct business or profession?
- Is the work performed part of the regular work of the employer?
- What is the skill and training level required of the person performing the work?
- Does the employer provide the equipment or tools for getting the work completed?

Why is it important to properly distinguish between an independent contractor and an employee? There are several reasons. First, for tax purposes, an employer dealing with employees must withhold income taxes and Social Security and Medicaid taxes while paying a share of the tax to the government. On the other hand, someone who employs an independent contractor has no such duty. Independent contractors have the duty to pay their own taxes directly to the government, including all Social Security and Medicaid taxes. Second, as we will see later in this chapter, there are many federal and state protections built into the law for employees. Often, these protections such as workers' compensation insurance, unemployment insurance, and unemployment compensation do not apply to independent contractors. Third (as discussed in greater detail later), as a general rule, employers may be held liable for the torts of their employees, but not for those of their independent contractors.

Creating and terminating agency relationships

Agency relationships are consensual, which means that the parties must be competent to enter an agreement and both must consent to being a part of an agency relationship.[4] As discussed in greater detail below, the agency relationship can be expressed or implied by the parties. In most cases, there are no special formalities to satisfy to create an agency relationship. However, in some cases, an agency relationship must be in writing if required by the Statute of Frauds. One particular case in which a written agency agreement is required is the real estate agent contract. The reason for the writing requirement is the need to have any agreement dealing with the sale of real property reduced to writing.

Creating the agency relationship

In express agency relationships, the agent and principal have explicitly agreed that they will enter into a fiduciary relationship wherein the agent will act on behalf of the principal. The agreement can be in writing or strictly oral. In some cases, the relationship can also be exclusive. That means the principal cannot employ any other agent to serve a particular purpose. In a nonexclusive agency relationship, the principal can hire as many agents as desired to complete a task. Once one of the agents gets the job accomplished, all other agency agreements are terminated.

Most employees and independent contractors enter into an express agency relationship. If an A&P mechanic is hired by Larrivee's Air Repair, it is expressly understood that

that she will work as an A&P mechanic for Larrivee's Air Repair and Larrivee will pay her for her services (as negotiated on a per hour or salaried basis).

Implied agencies are not nearly so common in the aviation industry. These agency relationships occur when an agency relationship is implied or inferred by the conduct of the parties. The inference of an agency relationship can come from past dealings with clients or customers, industry custom, or other relevant factors. For instance, if Tara Air brings its Part 135 Pilatus to Golden's AvRepair Shop for an annual inspection and each year at the time an annual is due, Golden tows the aircraft to its shop for the annual without any express authorization from Tara Air, and Tara Air pays for the services, then an implied agency relationship was formed whereby Tara Air, the principal, impliedly directs Golden to perform the annual on its Pilatus.

Terminating the agency relationship

The law recognizes two ways that an agency relationship may be terminated. The first is by acts of the parties. The second is by operation of law.

Termination by acts of the parties Because an agency relationship requires the consent of both parties, the relationship can be terminated by the withdrawal of either party. These are some of the typical ways in which an agency relationship might be terminated by acts of the principal or agent:

- The principal and agent mutually agree to termination of the relationship.
- A specified or reasonable amount of time has lapsed. Many agencies are established for a specified period of time. Once that time is up, or a reasonable time has passed, the agency is terminated.[5]
- The purpose of the agency is fulfilled.[6]
- The principal revokes the agent's authority.[7]
- The agent gives up his or her authority.

Termination by operation of law Sometimes an agency relationship can be terminated automatically by the law. Usually these events make it impossible for either the agent or the principal to perform the duties required by an agency relationship. Some of the events that will terminate an agency relationship by operation of law are listed below.

- Death of either the principal or agent.
- Incapacity of the principal or agent. If either the principal or the agent is incapacitated mentally or physically so as to be unable to perform her or his duties, the agency relationship is terminated.
- Bankruptcy of principal. In most cases the agent's bankruptcy does not terminate the agency relationship.
- Disloyal actions of an agent.
- Destruction of the subject matter of the agency (e.g., the aircraft that Bob was hired to sell as an agent of John is destroyed in a crash).

Ordinarily, the principal-agent relationship is created by a contract, whether oral or written. Each contract may spell out specific duties of the parties. However, regardless of what any individual contract might provide for, the agent and the principal are responsible to each other for various duties imposed by law. A brief review of the legal duties implied in an agency relationship follows.

Duties of parties

Duties of agent

Agents have a fundamental legal duty of obedience to their principals.[8] This duty simply requires that an agent follow all reasonable directions presented by the principal. For instance, George's Flight School has a standing rule that no flights are to depart when wind gusts exceed 20 knots. Emilio is a veteran flight instructor at George's. He has always complained about the "20-knot rule" because it limits student exposure to gusty wind conditions. Emilio disregards the rule and takes a commercial student on an instructional flight when wind gusts exceed 25 knots. During one takeoff after a touch-and-go, the wind suddenly shifts, and Emilio's aircraft departs the runway and is substantially damaged. Emilio can be held liable for the damages done to George's aircraft and business as a result of his failure to obey George's directive on departures during windy conditions.

Another primary duty of agents is the performance of their duties with reasonable care and skill. If special skills are required for the agent's duties, those skills must also be exercised by the agent.[9] For example, if George's Flight School hires Kristen as a flight instructor, it can expect that she will demonstrate the skill and expertise required by a certificated flight instructor. If Kristen damages an aircraft while demonstrating a relatively routine soft-field landing, she may be liable to George's for her breach of duty.

Agents also have a duty to inform their principals regarding matters relevant to their services.[10] To illustrate, suppose Mike, one of George's flight instructors, learns that one of his regular students has recently filed for bankruptcy. Mike would have a duty to inform George about the situation because George might decide that he can no longer provide services to this student unless they are paid for in cash.

Other general and fiduciary duties owed by an agent to a principal are as follows:

- Avoidance of conflicts of interest
- No self-dealing
- Duty not to compete
- Duty to keep information confidential

Any dereliction of these fiduciary duties by an agent could allow the principal to recover damages.[11]

Duties of principal

The duties of a principal in an agency relationship are fairly limited, but nonetheless important. There are essentially three duties owed by a principal to an agent:

- Compensation
- Reimbursement
- Indemnification

The duty of compensation requires that the principal pay the agent for his or her services in accordance with the agreement between the parties. Compensation means payment for services provided. The payment can be cast as an annual salary, an hourly wage, or commissions on sales. If the parties have not agreed on a specific amount of compensation, the principal will have a duty to pay the agent for the reasonable value of services provided.[12]

A principal also has a duty to reimburse an agent for any expenses the agent has incurred on behalf of the principal.[13] For instance, Trudy is a flight instructor for George's Flight School. She takes a student on a cross-country flight and lands at the second destination of the day. She and her student need additional fuel to get back to their home base. Trudy pays for the fuel, using her personal credit card. Trudy will be entitled to reimbursement from George for her credit card payment.

Indemnification is the duty of the principal to make the agent whole for any losses the agent may suffer while acting at the direction of the principal. For example, suppose that Cole is employed as an agent to sell Ashley's twin-engine Seneca aircraft. Ashley does not inform Cole that she does not possess title to the aircraft. Cole sells the aircraft, and shortly thereafter he is sued by the rightful owner of the aircraft, Sandra. Cole pays damages to Sandra. Cole will be entitled to indemnification from Ashley for the damages he incurred due to his actions at Ashley's direction.

Contract liability

One of the most common issues in the agency relationship is the sorting out of the contract liability of the principal and agent to third parties that they deal with in the course of doing business. If the agent is acting on behalf of a principal, is the agent personally liable on contracts made for the principal? Is the principal liable? Are both liable? In this portion of our discussion, first we will look at the liability of the principal. Next, we will look at whether an agent can become personally liable on a contract entered into on behalf of a principal.

Principal's contract liability

As a general rule, a principal who authorizes an agent to enter into contracts on his or her behalf will assume full liability for the contract. That means that whether a principal wants to be or not, the principal will be responsible for performing on a contract if the contract was otherwise valid. The theory behind this rule is that the third party entering into the contract believes—reasonably—that the agent has the authority to enter into the contract. The third party should be secure in knowing that her or his deal is with the principal and that the principal can be held liable.

For example, Kreke Air Cargo hires Chappy to manage its corporate aviation department, and Chappy enters into a contract with Hook Avionics for four new moving map displays. Even if Kreke is unhappy with the contract Chappy entered with Hook because another supplier is less expensive and faster, Kreke is still obligated under the contract with Hook.

Agent's contract liability

An agent's liability to third persons under a contract depends on whether the agent's relationship with the principal is (1) fully disclosed, (2) partially disclosed, or (3) undisclosed.

A fully disclosed relationship is one in which the third party entering into the contract is fully aware that (1) the agent is working for a principal and (2) the third party also knows the identity of the principal. In the vast majority of cases, the agent's relationship with the principal is fully disclosed. In the example above, Hook was aware that Chappy was entering into a deal on behalf of Kreke. In cases like this, the principal, Kreke, remains liable on the contract. However, Chappy has no liability on the contract. He was simply acting in his capacity as Kreke's agent. Of course, this rule makes practical sense for most agency relationships. There would be very few employees willing to work in today's environment if they believed they could be held liable for every contract they entered into behalf of their boss.

On some rare occasions, an agency relationship may be partially disclosed or undisclosed. A partially disclosed agency exists when a third party dealing with an agent knows that the agent is acting for a principal. However, the third party does not know the identity of the principal. A partially disclosed agency is most likely to exist because the principal does not want to be identified. In some unusual cases, it may exist because the agent neglects to inform the third party of the principal's identity. An undisclosed agency exists when the principal does not even want his or her existence to be known to the third party. The third party is led to believe that the agent is the principal. Partially disclosed and undisclosed agencies often result when a principal believes that knowledge of their existence will make it more difficult to fairly negotiate a contract. For instance, if Sollenberger were a rich and famous entertainer, he might prefer to have an agency that is undisclosed or partially disclosed as he goes out searching for a certain rare war bird. If sellers knew his identity, it might be hard to bargain or get a good price.

In the case of both undisclosed and partially disclosed agencies, both the principal and the agent are liable on any contracts. If the agent is forced to pay anything on a contract, the agent can recover, calling on indemnification from the principal.

Tort liability

Another significant issue for employers, employees, and independent contractors is tort liability. Can an employer be liable for a tort committed by employees? Can an employer be liable for a tort committed by an independent contractor? These are significant questions, especially in the high-dollar, high-exposure world of aviation.

Principal's liability for torts of employees

Of course, an employer and an employee are personally liable for their own torts. The bigger question for employers is whether they can be held liable for the torts of their employees. In companies with hundreds and even thousands of employees (such as many aviation enterprises) this question looms even larger. Often, the answer to these questions lies in the type of tort committed—was it a negligent or intentional tort?

Employer's liability for employee negligence The general rule is that employers will be held liable for the negligence of their employees when the employees are acting within the scope of their employment. The legal philosophy behind this rule is that an employer gains the benefit from having others acting on the employer's behalf. Therefore, it is not unreasonable to require the employer to bear the responsibility for the employees' negligence if it occurs while they are operating within the scope of their employment.

Lawyers often refer to this general rule as *respondeat superior*. This means "let the master answer." The concept of *respondeat superior* arose from the notion of vicarious liability—making a person liable without fault because that person is responsible for the conduct of another.

The sometimes tricky part in applying this general rule lies in determining whether an employee was operating within the scope of her or his employment. For instance, Tim is a corporate pilot. He has finished the first leg of his flight and safely delivered his passengers in Tulsa, Oklahoma, with the next leg of the trip scheduled for the following morning. Tim's company makes arrangements for him to get lodging and food at the nearest airport hotel. Tim stays at the hotel and decides to go out for a late dinner at a nearby restaurant. Tim drives his rental car from the hotel. On the way to the restaurant, he negligently misses a stop sign and injures a pedestrian. There is no question that Tim is liable for his negligence. However, is his employer liable as well?

There is no one test to determine whether an employee is acting within the scope of his or her employment. However, these are some factors that should be considered:[14]

- Was the act specifically authorized by the employer?
- Did the negligence occur while the employee was within the time period of employment authorized by the employer?
- Did the negligence occur near or in close proximity to the employer's place of business?
- Was the employee working to advance the employer's business at the time of the negligent act?

Employer's liability for employee's intentional torts   As a general rule, employers will also be held liable for the intentional torts (e.g., assault, battery, false imprisonment) of employees that are committed within the scope of their employment. The usual test applied is the "work-related test." This test requires that for an employer to be held liable for the intentional torts of employees, the employee must commit the intentional tort while within a work related place or time. Again, this all gets back to the basic question of whether the employee was acting within the scope of the employee's work.

Principal's liability for torts of independent contractors

Recall from the discussion above that independent contractors are different from employees. They are usually "outsiders" with special expertise who are engaged to work on special projects on an as-needed basis. Independent contractors supply their own tools and equipment and have the right to control the methods and means they will employ to get a job done.

Because of the "separateness" between a principal and an independent contractor, a principal is generally not held liable for the torts of an independent contractor. The independent contractor is solely liable for any injuries caused by its negligent or intentional torts.

For example, suppose Fox Flyers hires Bornarth Builders to put an addition on its hangar and office buildings at the municipal airport. If one of Bornarth's trucks negligently strikes a parked aircraft on the ramp, Bornarth will be solely liable for the damages done to the aircraft.

There are some exceptions to this general rule. A principal may be held liable for the torts of an independent contractor when (1) the principal assigns the independent contractor "inherently dangerous activities" or (2) the principal is negligent in the selection of an independent contractor.

These exceptions have some implications for aviation businesses. It has been held that crop dusting is an "inherently dangerous activity."[15] Could this be extended to make liable the aviation employers who hire independent contractors for rescue operations, air ambulance services, or firefighting activities?

LABOR UNIONS AND EMPLOYMENT

Until the late 1800s, workers were largely at the mercy of their employers when it came to pay and working conditions. Employees eventually turned to unions to represent their interests through bargaining leverage, often referred to as *collective bargaining*. In this section of the chapter, we will give an overview of some fundamentals of labor union law.

There are several major federal laws that codify the right of workers to organize and bargain with employers. Three of the major pieces of union legislation passed in the 1900s are as follows:

Major federal labor union laws

- *Railway Labor Act (RLA)*. The RLA was passed in 1926 and amended in 1934. This act relates to railroad and airline workers.[16]
- *Norris-LaGuardia Act*. This law was enacted in 1932. It protects the rights of employees to organize.[17]
- *Wagner Act*. This law is also known as the *National Labor Relations Act* (NLRA). The NLRA was passed as law in 1935. It protects the rights of employees to create, join, and collectively bargain with employers through labor unions.[18]

Once a labor union is established by election, its most potent tool is collective bargaining—the ability of the union to represent all workers within a plant, business, or other employer unit. Laws protecting the right to collective bargaining require that employers and unions bargain in good faith.

Basics of collective bargaining

Collective bargaining agreements often contain certain provisions that are so fundamentally important to the employer-employee relationship that they are considered compulsory to negotiations. Compulsory subjects include wages, hours, safety rules, pension and retirement plans, and similar issues.

Subjects that are not compulsory, but that may be a part of employer-union negotiations include management/supervisory structures, plant locations, and reorganization of the employer's business. These noncompulsory subjects are permitted to become part of a collective bargaining agreement.

What if collective bargaining fails?

As this chapter is being drafted, mechanics at Northwest Airlines walked off their jobs after union officials and management failed to reach an agreement on Northwest's demands for $178 million in wage and benefit cuts. Unions and employers may attempt to act in good faith during collective bargaining agreements. However, attempts to come to agreement may not always be successful. That leaves the question, What next?

To respond to this question, we will focus on the process as it affects most employees in the aviation industry. As indicated above, most pilots, dispatchers, mechanics, and other air carrier employees are covered under the Railway Labor Act.

The RLA was enacted in 1926 to allow railway laborers to organize and collectively bargain while protecting railroads and the general public from "wildcat" or unauthorized strikes that had crippling effects on commerce in the early days of the labor movement. In 1936 the RLA was expanded and amended to include employees from the growing airline industry.

The RLA's provisions regarding breakdowns in collective bargaining negotiations require that workers continue to perform their duties, even while collective bargaining efforts may falter. This is sometimes referred to as the "work now, grieve later" rule. The philosophy behind the rule is to avoid wholesale disruptions of commerce while unions and employees try to sort out their differences.

The first step in the collective bargaining process is usually direct negotiation, that is, negotiations without the presence of any third parties.

If the parties reach an impasse during direct negotiations, either party can prepare an application for mediation before the National Mediation Board (NMB). The NMB will assign a mediator to the case. Most mediators have labor or management backgrounds with extensive experience in the airline industry. The role of the mediator is to move the parties to productive discussions of the issues in the hopes of reaching compromise. The NMB mediators will continue to work with the parties until an agreement can be reached or until the mediator concludes that all reasonable attempts at a voluntary agreement have failed. At that point, the NMB will offer the opportunity for the parties to submit any remaining disagreements to arbitration.

If either party rejects arbitration, the NMB will then release the parties from mediation. This begins a 30-day "cooling off" period. Near the end of the cooling off period, the parties are joined again by the NMB for another round of intense mediation in an attempt to reach an agreement.

If this fails, the parties are free to proceed with "self-help" efforts. For the employer, this may mean an air carrier can forcibly impose its last offer, shut down temporarily, or hire replacement workers. For the union, self-help means a strike or other legal self-help activity.

In certain instances, the NMB may recommend a Presidential Emergency Board during the 30-day cooling off period. This may be done if the dispute "threaten(s) substantially to interrupt interstate commerce to a degree such as to deprive any section of the country of essential transportation service."[19] If a Presidential Emergency Board is convened, it has 30 days to present a proposed agreement to the parties, after which another 30-day cooling off period begins. If this proposal is rejected by either party, self-help can begin after the second cooling off period. In serious enough cases, Congress can intervene and impose a settlement, usually by taking the recommendations of the Presidential Emergency Board and enacting them as law.

As discussed above, at times the collective bargaining process reaches a point where the parties are unable to reach an agreement and one or both parties choose to engage in self-help. The most potent (and legal) self-help tool available to union employees is the strike.

Strikes and related issues

To be legal, a strike must be authorized by a majority vote of union members. Strikers usually set up picket lines and demonstrations outside the employer's place of business. Union members and representatives carry signs announcing and supporting the strike. The technique of striking is designed to place pressure on an employer to settle any differences in a manner more favorable to the union.

Strikes are a legal form of self-help. However, they become illegal if they are (1) accompanied by violent actions or (2) preventing nonstriking employees, customers, and vendors from entering and doing business with the employer.

Sometimes strikes can last for extended periods of time. Individual employees may decide from the start that they do not want to strike. At other times they may decide they do not want to strike after the strike has been underway for some time. In either case, if the employees return to work, they are referred to as crossover workers.

In some instances, employers who reasonably anticipate a strike can lock out union workers and hire replacements. The replacements can be either permanent or temporary. There is no requirement that a replacement employee be terminated once a strike is over.

EMPLOYEE PROTECTION

Under common law, and continuing in today's world, most employees are employees at will. This means that they can be dismissed by an employer with or without cause for just about any reason. While the fundamental principle of employment at will continues, it has been eroded by federal and state statutes and judicial doctrine that give employees, and prospective employees, greater rights and protections under the law.

In reviewing the changing landscape of employee protection law, first we will review laws that protect employees and applicants for employment from discrimination. Then we will review laws designed to provide employees with greater safety and security.

Over the course of the last 40 years, both federal and state lawmakers have tackled the issue of discrimination against employees and applicants for employment. In this segment of the chapter, we will focus on the most prominent federal statutes that seek to protect employees from discrimination in the workplace.

Employment discrimination

Fair employment practices act

The Fair Employment Practices Act is part of the Civil Rights Act of 1964. This act is more commonly known as *Title VII* because it is Title VII to the Civil Rights Act of 1964. Title VII prohibits job discrimination for current employees and applicants based on the following protected classifications:

- Race
- Color
- Nationality
- Gender
- Religion

Title VII does not apply to all employers. However, it does apply to many, especially larger entities including the following:

- Private employers with more than 15 employees
- Federal, state, and local governments
- Employment agencies
- Labor unions with more than 15 employees

Title VII specifically prohibits employers within its scope from discrimination in hiring decisions; payment of salary, wages, and benefits; determinations regarding dismissal; and all other terms, conditions, or privileges of employment.

When workers want to bring Title VII complaints, they must do so with the Equal Employment Opportunity Commission (EEOC). Some states require that the claim be filed first with their state or local office that serves in the same capacity at the EEOC. The EEOC or local agency will investigate the complaint and decide if the claim has merit. If it is decided that the case has merit, the EEOC or local agency can sue on behalf of the claimant. If the EEOC or local agency decides that it does not wish to file suit, it will issue a "right to sue" letter to the person who filed the complaint, thus giving that person the right to sue the employer privately.

An employee or applicant for employment who is successful in a Title VII claim might recover back pay and reasonable attorney fees. Beyond monetary awards, the claimant could get reinstatement or other remedies whereby the court forces changes in the employer's behavior and/or hiring practices.

Equal pay act

The Equal Pay Act seeks to eliminate gender discrimination in matters of compensation. As implied in the title of the law, it seeks to ensure that if a man and a woman are performing essentially equal work functions, they are similarly paid. Someone who feels that he or she has been paid less than someone of the opposite gender will be able to establish a basic (or prima facie) case of discrimination if evidence can be presented that someone of the opposite gender who works a similar job at the same employer is paid more. Once the basic case is established, it is up to the employer to prove that the differential in pay is justifiable. Under the Equal Pay Act, an employer has justification for unequal pay if the inequality is based on

- Seniority
- Merit or performance
- A verifiable measure of production by quantity or quality
- Any reasonable factor other than gender

If an employer loses an Equal Pay Act case, the employer can be liable for back pay, court orders to cease the unequal treatment of workers, and payment of the claimant's reasonable attorney fees to bring the claim.[20]

Age discrimination in employment act (ADEA)

The ADEA was passed in 1967. Its objective is to protect employees and job applicants over age 40 from discrimination in the workplace. The ADEA applies to all government employers and private employers with more than 20 employees.[21] The ADEA also prohibits (with some exceptions for highly placed executives) a mandatory retirement age.

Employers have several defenses against an ADEA charge. The first defense is that the age requirement imposed is a bona fide occupational qualification (or BFOQ). This means that the employer is able to establish that there are reasonable and appropriate grounds to require age limits on certain jobs due to physical and mental demands of the job. A second defense for an employer is a bona fide seniority system. Such a system would likely have the effect of compensating junior employees at a lesser rate than more experienced (and likely older) employees. A final defense will be any other reasonable action taken by the employer.

If an employer violates the ADEA, it may be liable for back pay, injunctions against the employer, and reinstatement. As indicated in Case 9-2 below, the issue of employment of pilots over the age of 60 in the airline industry has been litigated several times.

Americans with disabilities act (ADA)

The ADA became law in 1990. By 1994 it was applicable to all employers with more than 15 employees. The ADA prohibits employers from discriminating against current employees or applicants for employment because the employee or applicant has a disability.[22]

The law further requires that employers provide for accommodation for employees, customers, and clients who may be disabled. The accommodations typically take the form of wheelchair ramps, interpreters, larger-screen computers, etc. These accommodations must be made unless the employer can establish that to do so would create an "undue burden."

A chart outlining the provisions of the federal employment discrimination laws is found in Fig. 9-1.

Over time, the federal government and state and local governments provided greater protections for the safety and security of employees. Some of the major enactments and provisions related to worker safety and security are outlined below.

Employee safety and security

Figure 9-1 Summary of employee protection provisions.

	Equal Pay Act	Title VII	ADEA	ADA
Type of Discrimination Targeted	• Sex	• Nationality • Religion • Color • Race • Sex	• Age	• Disability
Conduct Prohibited	• Wages	• Employment terms or conditions	• Employment terms or conditions	• Employment terms or conditions
Allowable Defenses	• Merit • Seniority • Considerations other than sex	• Demonstrated Ability • BFOQ • Seniority	• Seniority • BFOQ	• Job-required criteria • Safety considerations • Undue hardship to employer
Typical Remedies for Successful Claimant	• Back pay • Lawyers' fees • Back pay	• Back pay • Injunction • Compensation for damages • Reinstatement • Lawyers' fees	• Back pay • Injunction • Reinstatement • Lawyers' fees	• Back pay • Injunction • Reinstatement • Lawyers' fees

Occupational safety and health act

The Occupational Safety and Health Act became law in 1970. Its purpose is to create and preserve safe working environments for employees. The act created the Occupational Safety and Health Administration (OSHA), a government agency charged with enforcing worker safety standards.

OSHA typically performs its functions through inspections of work sites, reviews of employer records, and (if necessary) citations against violators. Penalties for violations of the act's requirements can be civil or criminal depending on the nature and severity of the violations.

If a state wants to regulate its own workers' safety, it may do so by filing a plan that is approved by OSHA. The state plan must meet the minimum standards applied by OSHA. However, the state plan does not have to conform line by line to OSHA standards. Many states now regulate worker safety through their own plans that have been approved by OSHA.

Employee privacy

Employee privacy has become an important issue in recent years. Employers have the ability to read employee emails, monitor phone calls, initiate electronic surveillance, and (in certain cases, especially relevant to aviation employees) perform drug testing.

As a general rule, it is a federal crime to intentionally intercept another person's wire or electronic communications.[23] However, employers are armed with substantial exceptions to this general rule. As a matter of law, employers are permitted to monitor employee electronic communications upon the implied or express consent of employees (which is typically not very difficult to obtain). These exceptions routinely permit employers to monitor employees' emails and business-related telephone conversations.

For employees who serve in safety-sensitive functions in the aviation environment, drug testing is a part of the job. Appendix I to FAR Part 121 requires preemployment testing, periodic testing, random testing, testing based on reasonable cause, return-to-duty testing, and postaccident testing. The testing is applicable to employees working for Part 135 and Part 121 operations. The specific employees subject to testing include

- Flight crewmembers
- Flight attendants
- Flight instructors
- Aircraft dispatchers
- Aircraft maintenance personnel
- Ground security personnel
- Air traffic controllers

Employees who fail or refuse to take required drug tests are subject to certificate suspensions, revocations, and/or mandatory drug treatment.[24]

Workers' compensation insurance

On occasion, employees are injured while performing their duties. Under traditional common law, an employee could sue an employer for negligence or any other theory appropriate to the case. This approach to compensating injured workers had obvious limits. First, it pitted employee against employer. Second, it did not guarantee that the injured employee would be compensated for injuries.

Because of the drawbacks to the traditional common law system, virtually every state has a workers' compensation law. Workers' compensation statutes typically require employers to obtain workers' compensation insurance for the benefit of their employees. The insurance is often purchased from third-party providers, or employers may be permitted to self-insure if they have adequate resources.

If an employee is injured on the job, the employee files a claim with the state's workers' compensation board or commission. That agency is tasked with determining the validity of the claim. Appeals are often made a part of the system. Often the biggest hurdle the employee faces in a claim is being able to establish that an injury was work-related. Injuries incurred while a worker is performing work duties are clearly covered. If the employee were injured at a company gymnasium or involved in an accident while taking a client to lunch, it would be within the scope of employment. However, accidents that took place off the employer's premises while the employee was at lunch or after work hours would not be covered.

It is important to note that workers' compensation is considered to be an exclusive remedy. That means that once an employee selects workers' compensation as her remedy, she must forgo any other remedy, including taking legal action against her employer. One significant exception to this rule occurs if an employee was intentionally injured by her employer. An employee can also sue a responsible third party for injuries without being barred by a workers' compensation claim.

Social Security and unemployment insurance

Congress enacted Social Security in 1935. Initially the law was put in place to provide modest retirement benefits to eligible employees. Over time, the scope of Social Security grew. Today, the system covers the following benefits:

- Old-Age and Survivors Insurance (monthly payments of retirement benefits based on working years and compensation earned)
- Disability insurance
- Hospitalization insurance (Medicare)
- Supplemental Security Income

The Social Security system is financed through withholding taxes paid for by both employers and employees. The current rate is approximately 7.65 percent with both employers and employees each paying 7.65 percent of the employees' salaries up to a certain cap. The rate and caps are subject to change by Congress.

In recent years there has been substantial talk regarding reformation of Social Security designed to allow the program to remain solvent over the long term. Consideration is being given to raised rates and caps, allowing employees the option to create private accounts where their tax contributions will be invested, and other combinations of these approaches.

The federal unemployment system was also put into place in 1935. This system was a part of the Social Security legislation. Employers pay into this system a percentage of workers' earned income. The system provides for several months of payments to a worker who is out of work not due to his or her own fault. Most states also have an unemployment program in place that corresponds to the federal unemployment system.

Family and Medical Leave Act

In 1993 Congress passed the Family and Medical Leave Act (FMLA).[25] This law applies to employers (including federal and state governments) with 50 or more employees. The purpose of FMLA is to allow employees unpaid leave for medical emergencies. Not all employees are covered by this act. All covered employees must have served their employers for at least 1 year and performed a minimum of 1250 hours of work in the preceding 12 months. Any covered employees are allowed up to 12 weeks of unpaid leave for the following situations:

- Birth and early care of children
- Placement of a child for adoption or foster care
- Serious health problems
- Care for a spouse, child, or parent with serious health problems

Generally, an employee who takes leave under FMLA must be restored to her or his previous position at the same pay as when the employee left. One exception to this rule applies to salaried employees who are in the highest-paid 10 percent of the employer's workforce.

CASES AND COMMENTARY

A survey of Title VII case law reveals that a substantial number of the cases involve some form of harassment in the workplace. Case 9-1 illustrates how the courts dealt with a claim involving racial discrimination.

CASE 9-1
DAVID E. HOLLINS V. DELTA AIRLINES
238 F. 3d 1255 (2001)

OPINION BY: SEYMOUR

David E. Hollins appeals from the order of the district court granting summary judgment to his former employer, Delta Airlines, on his racial harassment claim under [Citation to Title VII]. For the reasons stated below, we affirm.

I

David E. Hollins began his employment as an associate customer service agent with Delta Airlines in December 1995. On February 4, 1996, a white co-worker, Rex Fidler told Mr. Hollins the following joke: "How can you tell when a person is well-hung?" Answer: "When you can't get two fingers between his neck and the rope." [Citation.] Mr. Hollins immediately reported the joke to two Delta supervisors, Dennis Jacobson and Carla Sutera. Mr. Jacobson and Ms. Sutera spoke with Mr. Fidler about the inappropriateness of the joke and requested written statements from Mr. Fidler, Mr. Hollins, and another Delta employee who had witnessed the incident. Delta thereafter gave Mr. Fidler a warning letter and placed it in his employment file.

At some point prior to his telling of the "well-hung" joke, Mr. Fidler had told a group of employees, including Mr. Hollins, the following joke: "If you have a Black, a Mexican, and a Tongan in a car, who is the driver?" Answer: "The Sheriff." [Citation to Record]. No employee ever reported this joke to a Delta supervisor.

Sometime after the "well-hung" joke incident, Mr. Hollins noticed several hangman's nooses dangling from the ceiling above his work area. He also noticed nooses hanging in two other areas. One was hung in such a way that it swung down when a door was opened. The other was hung in an area where an African-American employee worked. Mr. Hollins did not complain about these nooses to anyone. However, Charles Wilson, an African-American co-worker of Mr. Hollins, complained about the nooses to Tom Brothers, the immediate supervisor of both Mr. Hollins and Mr. Wilson. Mr. Brothers immediately removed all the ropes and then held a meeting with the employees in which he indicated the ropes were offensive and would not be tolerated. Shortly thereafter, an employee named Stan White told Mr. Brothers that it was he who had tied the ropes. He stated he tied ropes to pass the time and had not intended to offend anyone. Mr. White was given a warning letter and a copy was placed in his employment file.

According to Mr. Hollins, Mr. Brothers' treatment of him dramatically changed after he complained about the "well-hung" joke. Mr. Brothers began to follow him during his meal breaks, warning him and other African-American employees that "You'd better be back to work before I'm through eating." [Citation to Record.] He followed and "intently" watched African-American employees while they ate. [Citation to Record.] Mr. Brothers also began to stand near Mr. Hollins while he was at work and to scrutinize his work closely. He followed Mr. Hollins and used the restroom at the same time. Mr. Hollins also contends Mr. Brothers began to write him up for minor infractions in the workplace that, while violations of Delta policy, were often ignored by supervisors. However, Mr. Hollins never complained of Mr. Brother's conduct to a supervisor or to Delta's Equal Employment Opportunity (EEO) officer.

In granting summary judgment for Delta, the district court focused on whether the treatment Mr. Hollins received from his co-workers and supervisors amounted to a racially hostile work environment, relying on this Court's analysis in *Bolden v. PRC, Inc.*, 43 F. 3d 545 (10th Cir. 1994). The district court held that Mr. Hollins failed to meet the *Bolden* requirement that the harassment be "pervasive or severe enough to alter the terms, conditions, or privileges of employment" and that it be "racial or stem from racial animus." [Citation.] The court concluded "there is no viable evidence that either the hanging joke or rope incidents were racial or stemmed from racial animus," and that while the "'sheriff' joke appears to have racial overtones ... it clearly was an isolated incident." [Citation to Record]. The district court held it "uncontroverted that one of Delta's employees had a habit of tying knots of various kinds in ropes found in the area and that he sometimes would throw the ropes over pipes near the ceiling. The employee did not view the ropes as racial symbols." *Id.* Alternatively, the court held that Delta was entitled to summary judgment in any event because it took prompt remedial action whenever it learned about the offensive conduct.

II

* * *

The issue presented for our resolution is whether the district court erred in concluding that Mr. Hollins was not subjected to a racially hostile work environment and that even if he were, Delta was not liable for the harassment. For the reasons discussed below, we affirm the district court's grant of summary judgment. In so doing, we reject the district court's conclusion that the "well-hung" joke and the presence of hangmen's nooses were uncontrovertedly innocuous. The joke appears to be a facially racist remark, and the nooses may also have been racially motivated, regardless of the rope-tying employee's representations to Delta. These are genuine issues of material fact that preclude summary judgment on the hostile work environment issue. However, we agree with the district court that Delta can not be held liable for the asserted harassment on this record.

An employer may be liable for the racially harassing conduct of its employee under three theories, all of which are derived from the common law of agency: the negligence theory, under which the employer fails to remedy a hostile work environment it "knew or should have known about;" the actual authority theory, under which an employee harasses another employee within the scope of his employment; or the apparent authority theory, under which the harassing employee acts with apparent authority from the employer. [Citations.]

Mr. Hollins contends the situation here implicates all three theories of liability. Pursuant to the negligence theory, he argues that Delta knew or should have known about the hostile work environment he suffered because Delta was notified about the "well-hung" joke and the nooses. Alternatively, he asserts that Delta is vicariously liable for the acts of Mr. Brothers because, as a supervisor, Mr. Brothers acted with either actual or apparent authority. We examine each of these contentions.

A. Negligence Theory

Employers are not automatically liable for harassment perpetrated by their employees. [Citation.] In order to prevail on a negligence-based hostile work environment claim, Mr. Hollins "bears the burden of establishing that the employer's conduct was unreasonable." [Citation.] He must prove that Delta was itself negligent because "it knew or should have known about the conduct and failed to stop it." [Citation.]. Thus, the focus is not on whether the employer is liable for the bad acts of others, but whether the employer itself is responsible for failing to intervene.

At the outset, we note that Delta has a written harassment policy. Although this fact is more directly relevant to evaluating Mr. Hollins' vicarious liability claims, it is also relevant to his actual liability argument. Delta's harassment policy encourages employees to go to their supervisors or, if they "cannot resolve the matter" within their own department, to go directly to Delta's EEO director. [Citation.] This policy satisfies the requirements placed on harassment policies by the Supreme Court in its recent hostile work environment decisions. [Citation.]

The "sheriff" joke which Rex Fidler told Mr. Hollins was never reported by Mr. Hollins or any other Delta employee to any Delta supervisor, as Mr. Hollins concedes. Delta cannot be held liable for this incident because there is no evidence it could have reasonably known of it. The "well-hung" joke was timely reported by Mr. Hollins to two supervisors, who acted immediately by taking written statements, reprimanding Mr. Fidler, and sending Mr. Fidler a warning letter which went on his employment record at Delta. Mr. Hollins does not claim Mr. Fidler told him any other racially offensive jokes after this intervention by Delta. Finally, as soon as the nooses were brought to the attention of Mr. Brothers, he immediately removed the ropes, warned his employees against such offensive and intolerable conduct, and reprimanded the employee involved. Mr. Hollins does not contend that any other ropes were found in the workplace after this action was taken by Delta.

Although Mr. Hollins asserts that Mr. Brothers' conduct was racially harassing, Mr. Hollins never complained about Mr. Brothers to a supervisor or to Delta's EEO officer. Consequently, Delta did not have actual knowledge of the situation. Moreover, knowledge cannot reasonably be imputed to Delta given that it had no prior notice of any kind that Mr. Hollins believed he was being harassed by a supervisor.

In each of the above incidents, when Delta was presented with a potentially harassing situation, it immediately investigated, took corrective action, and disciplined any offending employees. Delta conducted itself as a reasonable employer. [Citation.]

B. Vicarious Liability

Mr. Hollins' alternative theory of liability is that Mr. Brothers acted with either the actual or apparent authority of Delta, making Delta vicariously liable for his acts. We cannot properly evaluate this argument because Mr. Hollins offered no evidence whatsoever that would show Mr. Brothers was acting at the behest of Delta, with actual authority. He also offered no evidence that Mr. Brothers was a "management level employee" who could be said to be acting under the apparent authority of his employer. [Citations.] Delta, conversely, provided evidence on the two issues it must prove in order to mount an affirmative defense to vicarious liability: "(a) that the employer exercised reasonable care to prevent and correct promptly any ... harassing behavior, and (b) that the plaintiff employee unreasonably

> failed to take advantage of any preventive or corrective opportunities provided by the employer …" [Citation.] Under these circumstances, Delta's responses were reasonable, and it may not be held liable for any racial harassment that may have occurred in the workplace.
>
> We AFFIRM the judgment of the district court.

Note that in the *Hollins* case the circuit court had to wrestle with two issues. First, the court had to deal with the question of whether Hollins was the victim of a "hostile or abusive" work environment. While the district court rejected this claim, the court of appeals determined that jokes and rope tying were enough to at least arguably assert a hostile work environment.

However, the second issue was whether Delta should be held liable for any Title VII violations. What two theories did Hollins assert in making the claim that Delta was liable? How did the court address each of these theories?

One of the most often claimed Title VII violations is sexual harassment that creates a hostile or abusive work environment. While there is no bright line test available for determining whether a hostile or abusive work environment exists, the Supreme Court has provided the following guidance:

> We can say that whether an environment is "hostile" or "abusive can be determined only by looking at all the circumstances. These may include the frequency of the discriminatory conduct; its severity; whether it is physically threatening or humiliating, or a mere offensive utterance; and whether it unreasonably interferes with an employee's work performance.[26]

Against this backdrop, one question that many employers have to address is how to avoid liability for sexual harassment claims. In two relatively recent cases, the Supreme Court has ruled that employers cannot be held strictly liable for sexual harassment. From these cases we can glean that if an employer exercises reasonable care to prevent and remediate conduct that might be considered harassment, then that employer has a reasonable defense in a sexual harassment case. The Supreme Court has also indicated that a plaintiff's claims may fail if he or she has not taken advantage of any reasonable opportunities presented by the employer to prevent or correct the inappropriate behavior.[27]

If an employer wants to be able to effectively defend itself in a sexual harassment case, it will most likely have to be able to establish at least some of the following preventive measures:

- A clearly enunciated antiharassment policy
- A fair and easy-to-access complaint process
- Regular communications to sensitize employees to antiharassment policies

Case 9-2 deals with the very thorny issue of the so-called Age 60 rule that requires airline captains to retire upon reaching 60 years of age. Does the Age 60 rule discriminate against older pilots?

CASE 9-2
PROFESSIONAL PILOTS FEDERATION, ET AL. V. FAA
118 F. 3d 758 (1997)

OPINION: GINSBURG, *Circuit Judge*: The Professional Pilots Federation and two individual pilots petition for review of two decisions of the Federal Aviation Administration: not to institute a rulemaking to relax the FAA Rule that requires commercial airline pilots to retire at age 60, and to extend application of the Rule to commuter airline operations. The Pilots contend, first, that the Rule unlawfully requires airlines to violate the Age Discrimination in Employment Act, [Citation], and, second, that the FAA acted arbitrarily and capriciously, in violation of the Administrative Procedure Act, when it decided to retain and expand the scope of the Rule. Finding merit in neither contention, we deny the petitions for review.

I. BACKGROUND

The FAA first promulgated the Age 60 Rule in 1959 pursuant to its mandate under the Federal Aviation Act of 1958 to ensure air safety. [Citation] (authorizing Administrator to promulgate "regulations in the interest of safety for the ... periods of service of airmen"); [Citation] (requiring Administrator to regulate "in a way that best tends to reduce or eliminate the possibility or recurrence of accidents in air transportation"); [Citation] (requiring Administrator to consider "the duty of an air carrier to provide service with the highest possible degree of safety" when issuing an airman, air carrier, or other certificate); [Citation]. The agency concluded that the Rule would promote air safety after finding "that available medical studies show that sudden incapacitation due to heart attacks or strokes becomes more frequent as men approach age sixty and present medical knowledge is such that it is impossible to predict with accuracy those individuals most likely to suffer attacks." [Citation]. The Second Circuit, reasoning that it was not for a court to substitute its own "untutored judgment for the expert knowledge" of the agency, accepted this conclusion and dismissed an early challenge to the Rule. [Citation.]

The FAA has reconsidered the Rule on several occasions. In the early 1960s, the agency began, but never completed, a study to determine the feasibility of testing individual pilots over the age of 60 in order to determine whether they remained fit to fly. [Citation.] In 1970 the Air Line Pilots Association called upon the FAA to replace the blanket prohibition of the Age 60 Rule with a regime of individualized performance tests and medical evaluations, but the agency decided to retain the Rule because "an increase in the number of medical examinations administered to a given pilot ... would not be an effective deterrent to incapacitation inasmuch as the indices of such incapacitation are not now sufficiently developed." [Citation.]

In 1979 the Congress directed the National Institutes of Health to determine whether the Rule was still medically warranted. [Citation.] In its final report, the NIH concluded that there was "no special medical significance to age 60 as a mandatory age for retirement of airline pilots" but recommended that the age 60 limit be retained nonetheless because there was still no "medical or performance appraisal system that can single out those pilots who would pose the greatest hazard because of early, or impending, deterioration in health or performance." [Citation.]

In 1982 the FAA considered relaxing the Rule in order to allow a small group of pilots to continue flying until age 62 in order to generate data on their performance under actual operating conditions. [Citation.] The FAA ultimately determined, however, that "no medical or performance appraisal system can be identified that would single out pilots who would pose a hazard to safety."

[Citation.] Unable "to distinguish those pilots who, as a consequence of aging, present a threat to air safety from those who do not," the agency decided not to experiment with changing the Rule. [Citation.]

The present litigation was stimulated, at least in part, by a 1993 study of the Age 60 Rule that was performed by Hilton Systems, Inc. for the FAA's Civil Aeromedical Institute. The Hilton Study correlated accident data for the period from 1976 to 1988 with pilot age and flying time. This analysis revealed "no support for the hypothesis that pilots of scheduled air carriers had increased accident rates as they neared the age of 60." Hilton Study at 6-2. On the contrary, the study found a "slight downward trend" in accident rates as pilots neared the age of 60. The authors cautioned, however, that this decrease might have resulted from "the FAA's rigorous medical and operational performance standards screening out, over time, pilots more likely to be in accidents."

Shortly after publication of the Hilton Study the FAA announced that it was again considering whether to institute a rulemaking concerning the Age 60 Rule and invited comments from the public on various aspects of the Hilton Study. (April 20, 1993). The agency held a public hearing in September 1993 at which 46 members of the public made presentations. The agency also received more than a thousand written comments.

In July 1993 the Professional Pilots Federation filed with the FAA a rulemaking petition to repeal the Rule. The Pilots maintained that "time and empirical evidence have shown that the blanket elimination of the country's most experienced pilots is not justified in the interests of safety and, therefore, is arbitrary and capricious, and violates this country's policy of prohibiting employment discrimination on the basis of age."

In early 1995 after a series of accidents involving commuter airlines, the FAA proposed in a separate rulemaking to bring certain commuter operations, previously conducted under Part 135, under Part 121. [Citation.] These operations would then become subject to the more stringent safety standards of Part 121, including the Age 60 Rule, relaxation of which the agency was still considering in the wake of the Hilton Study.

In December 1995 the FAA denied the Pilots' petitions to repeal the Age 60 Rule and decided not to institute a rulemaking in response to the Hilton Study. [Citation.] The agency determined that the "concerns regarding aging pilots and underlying the original rule have not been shown to be invalid or misplaced," and concluded that the Rule was still warranted as a safety measure. [Citation.] The FAA therefore retained the Rule, which provides that: *No certificate holder may use the services of any person as a pilot on an airplane engaged in operations under [Part 121] if that person has reached his 60th birthday. No person may serve as a pilot on an airplane engaged in operations under [Part 121] if that person has reached his 60th birthday.* [italics added]

[Citation.] In addition the FAA adopted its proposed rule bringing under Part 121 certain commuter operations previously conducted under Part 135. [Citation.] As a result, these commuter operations became newly subject to the Age 60 Rule. The Pilots petitioned this court for review of both rulemaking decisions.

II. ANALYSIS

The Pilots challenge the FAA's decision not to institute a rulemaking to repeal the Age 60 Rule and its decision to apply the Rule to commuter airlines as violations of both the ADEA and the APA. First, the Pilots assert that by requiring the airlines to discriminate on the basis of age the Rule is in "direct conflict" with the ADEA. Second, they claim that the agency violated the APA by: (1) not affording adequate consideration to the reasonable alternatives proposed by various commenters; (2) reaching a decision that is against the weight of the evidence; and (3) failing to provide any reasoned basis for treating older pilots differently than other groups of pilots who create as great or greater a safety risk.

A. The ADEA

The Pilots argue that the Age 60 Rule violates the ADEA because it requires the airlines to discriminate against older pilots and because the FAA need not have relied upon an age-based Rule in order to achieve its objective of air safety. The agency, we are told, could instead have implemented a scheme of medical evaluations and individualized testing in order to determine whether each pilot remains fit to fly. In any event, in the ADEA the Congress spoke directly to the role that age may play in employment decisions and the FAA cannot—as a matter of logic if not of statutory interpretation—countermand that clear statutory command through an exercise of its rulemaking authority.

The FAA responds that the ADEA speaks only to employers—including federal agencies acting in their role as employers—and therefore places no substantive limitation upon the agency's power to regulate airline safety pursuant to the mandate of the Federal Aviation Act. In the alternative, the FAA contends that if the ADEA does apply to the air safety rules it promulgates, then the Age 60 Rule comes within the exception in [Citation] of that statute for a bona fide occupational qualification. [Citation].

The FAA bases its first point upon the central provision of the ADEA itself, which states that "it shall be unlawful for an employer" to discriminate in employment upon the basis of age. [Citation.] The FAA argues that it promulgated the Age 60 Rule in its capacity not as an employer but as a regulator; in that capacity the agency is specifically authorized, *inter alia,* to prescribe "regulations in the interest of safety for the maximum hours or period of service of airmen." [Citation.] Absent a provision in the ADEA comparably specific or otherwise capable of over-riding the authorization of [Citation]—"not withstanding any other provision of law" comes to mind—the ADEA places no limitation upon the rulemaking authority of the FAA.

The FAA also contrasts the ADEA with the Rehabilitation Act, in which the Congress expressly subjected the programmatic activities of the Government to the stricture against discrimination. [Citation.] Absent a similarly plain and unequivocal expression of intent, the FAA urges that we ought not lightly infer that the Congress intended to compromise its single-minded pursuit of safety in the air.

We agree with the FAA that the ADEA places no substantive limitation upon the agency's authority to act as a regulator of the airline industry. The statute prohibits both an employer in the private sector and an agency of the federal government from discriminating upon the basis of age in making employment decisions. [Citation.] Nothing in the Act can plausibly be read to restrict the FAA from making age a criterion for employment when its acts in its capacity as the guarantor of public safety in the air. The general prohibition of the ADEA, addressed as it is to employers, should not be read by mere implication to override the specific grants of authority to the FAA in [Citation]. If the Congress intends to limit the means available to the FAA in its pursuit of air safety, we trust it will say so rather than leave the matter to the courts to infer. Therefore, we conclude that the ADEA does not limit the authority of the FAA to prescribe a mandatory retirement age for pilots; as a result, we need not reach the question whether the Age 60 Rule constitutes a bona fide occupational qualification within the meaning of [Citation] of that Act.

* * *

[In this portion of the opinion the Court of Appeals addresses the Pilots' arguments that the APA is violated by FAA's Age-60 Rule.]

III. CONCLUSION

We hold that the ADEA does not limit the authority of the FAA to regulate air carriers in the interest of safety. Because we also conclude that the FAA was not arbitrary and capricious, in violation of the APA, in deciding not to conduct a rulemaking for the purpose of amending the Age 60 Rule, the petitions for review are *Denied.*

This case provides a good overview of the background and history to the Age 60 rule. Regardless of whether you think the Age 60 rule reflects good policy, what do you think of the court's decision? Can the FAA be held responsible for ADEA violations? Is age an appropriate consideration for an airline pilot job? These questions are becoming even more pressing in modern times as airline pilots have suffered pay cuts and less job-security than in previous times. The problem is compounded by the fact that most Americans can—and may need to—work well past 60 years of age. Could the Age 60 rule become a deterrent to encouraging young people entering the profession?

DISCUSSION CASES

1. Butler County Airport contracted with Kauffman Excavating Company to prepare site work and necessary drainage for a 1000-foot runway addition to its airport. According to the terms of the contract, Butler retained the right to inspect the property and Kauffman's work to ensure compliance with contract terms. However, Butler did not have any authority to guide Kauffman with respect to the methods and means used to do its work. Shannon Pugh, an employee of Kauffman, was killed on the job when the sides of an excavation she was working on caved in over her. Investigations later revealed that the excavation site was not properly shored up and violated minimum safety standards. Pugh's parents and estate seek to recover from Butler County Airport. Will Pugh's family and estate prevail? Why or why not?

2. Chipman was a flight instructor and charter pilot for Barrickman's Aero Services, Inc. Barrickman's chief pilot requested that Chipman fly an airplane to its maintenance base at another airport and fly back with an aircraft that had just completed its annual. Chipman invited his girlfriend, Estell, along for the ride without Barrickman's knowledge. On the return flight, the aircraft was involved in an accident due to Chipman's negligence and Estell was injured. Estell sued Chipman and Barrickman's. At a trial court, Estell's case against Barrickman's was dismissed after the trial court noted Barrickman's had a "no rider" rule. Estell appeals the trial court decision. You are the appellate judge ruling on this case. How will you rule? Explain.

3. Rodriquez was the sole flight instructor for SafeFlight, Inc., an FBO at a small municipal airport. She was usually the only person at the FBO, and the company's owner visited the FBO on an occasional basis only. Although she had no hiring authority,

Rodriguez hired Amy on a salary basis to act as a receptionist for the FBO. After Amy worked a bit over a month and did not get paid, she sued SafeFlight for her unpaid compensation. Will she be successful? Explain.

4. Chauvinista Airlines is a highly profitable small airline running passengers and freight service in the Caribbean. The senior management of the company recently passed an employment rule requiring that any new applicants for first officer be at least 5 feet, 7 inches tall. They have come to you for legal advice on whether the requirement will pass muster when measured against the requirements of Title VII. Records from the meeting when the measure was discussed indicate that management felt the rule should be defensible because it is neutral and applies to anyone who applies for a first officer position with the airline. What advice would you give the airline's management? Explain.

5. Sandra is a highly accomplished business executive with a CPA and MBA degrees from Harvard, and on top of all that she's a commercial pilot with instrument and flight instructor ratings. She applies for a job with Forgang Air Safety Group, a world-renowned flight training company. She interviews with Bill, a senior vice president. Bill indicates that he's very impressed with all Sandra's credentials and lets her know that she is the last candidate to be interviewed and clearly the best of the bunch. However, he concludes the interview by confiding in Sandra that he will nonetheless be unable to offer her a position with the company. Sandra is surprised and disappointed, so she asks Bill why she is not being offered the position. Bill then tells Sandra that the problem is a bit awkward for him to bring up, but he feels that she is simply too attractive. He explains that her physical attractiveness would likely cause distractions and loss of productivity in the mostly male environment of the company. Sandra is angered by all this and files a complaint with the EEOC, claiming that Forgang violated Title VII. Will Sandra succeed in her claim? Why or why not?

6. Victor is a tug operator employed by Albatross Airlines. While known to be very careful and professional while towing airliners, Victor has been known to hot-rod around carelessly on tugs that are not being used. He's been cited by management on a prior occasion for horsing around on the tugs. While on the job, he decides to hot-rod around the ramp during a slow period. During his antics, he crashes the tug into a parked supply truck and sustains serious injuries. Will he be able to recover for his injuries? If so, how?

7. Wilson was employed by the City of Emmittown as a police officer and pilot for one of the city's helicopter rescue and patrol units. He was a member of one of the city's special emergency reaction teams (SERTs) that is trained to handle highly dangerous rescue missions. SERT membership was strictly voluntary, and no additional pay or benefits were provided by the city. To be a member of SERT, each officer was required to pass rigorous physical tests involving running, pull-ups, push-ups, sit-ups, etc., four times a year. Officers not a part of SERT were not required to take these physical tests. One evening, after completing his patrol flights, Wilson closed out his paperwork, changed clothes, and drove to a nearby college running track to do some running and physical training. While running, he injured his ankle. Wilson filed for workers' compensation benefits. The City of Emmittown contested the claim. Who will win? Explain.

8. Ned Naïve is a newly minted flight instructor with CFI-A and CFII ratings. He starts a new position as a flight instructor with Albatross Air ("Albatross"), an FBO and flight school. The owner of Albatross, Sammy Slick, provides Ned with a weekly schedule that clearly indicates the start and end hours of each workday for Ned. Sammy pays Ned only for the hours that he is flying with a student or providing ground instruction for a student. Ned gets his first paycheck after two weeks. He notices that he has no taxes withheld from his paycheck. He asks Sammy why no withholding had ever taken place. Sammy replies that Ned is an independent contractor; therefore, Ned is responsible for paying quarterly self-employment and income taxes. Ned is a bit confused and comes to you for some guidance. He wants to know if you think he's an employee or an independent contractor. What do you think? Explain.

ENDNOTES

1. *Restatement (Second) Agency* § 1.
2. Id. at § 2.
3. Id. at § 2(3).
4. Id. at § 15.
5. Id. at § 105.
6. Id. at § 106.
7. Id. at § 119.
8. Id. at §§ 383 and 385.
9. Id. at § 379.
10. Id. at § 381.
11. Id. at § 469.
12. Id. at § 443.
13. Id. at § 438.
14. Id. at § 228.
15. *A. R. Boroughs et al. v. Leo Joiner,* 337 So. 2d 340 (1976).
16. 45 U.S.C. §§ 151–162 and 181–188.
17. 29 U.S.C §§ 101–110 and 113–115.
18. 29 U.S.C. §§ 151–169.
19. Railway Labor Act § 160.
20. 29 U.S.C. § 206.
21. 29 U.S.C. §§ 621–634.
22. 42 U.S.C. § 1201 et seq.
23. 18 U.S.C. § 2701 et seq.
24. 14 C.F.R. § 61.14.
25. 29 U.S.C. §§ 2601, 2611–2619, and 2651–2654.
26. *Harris v. Forklift Systems,* 510 U.S. 17 (1993).
27. See *Farragher v. City of Boca Raton,* 524 U.S. 775 (1998) and *Burlington Industries, Inc. v. Ellerth,* 524 U.S. 742 (1998).

10 International Aviation Law

FUNDAMENTALS OF INTERNATIONAL LAW 280
Origins of international law 280

PUBLIC INTERNATIONAL AVIATION LAW 281
The beginnings of public international aviation law 281
The Chicago Convention 282
Open skies agreements 283

PRIVATE INTERNATIONAL AVIATION LAW 283
The Warsaw Convention 284
The Cape Town Convention 286

INTERNATIONAL CIVIL AVIATION ORGANIZATION 287

CASES AND COMMENTARY 287

DISCUSSION CASES 295

ENDNOTES 297

Modern aircraft routinely traverse international boundaries for private and commercial purposes. While the safety and ease of air transportation make international travel simpler, it sometimes complicates things when it comes to legal questions. If an aircraft that belongs to a U.S. carrier is involved in an accident in France, do U.S. or French laws apply? How can an aircraft with South African registry gain access to airports in Mexico?

As air travel grew in speed and sophistication, it became more and more obvious that nations would need to cooperate to establish a body of law applicable to international aviation activities. Over time, two distinct bodies of international aviation law have developed—private and public international aviation law. In this chapter we will provide a brief overview of these two subject areas along with a summary and update of activities related to the International Civil Aviation Organization (ICAO). However, before getting to the specific aviation applications of international law, we will discuss the basics of international law.

FUNDAMENTALS OF INTERNATIONAL LAW

Some legal scholars have questioned whether there is really such a thing as international law. Those questioning the existence of a true body of international law typically cite the fact that there is (1) no single legislative source, (2) no single court or judicial body assigned to interpreting such law, and (3) no single executive organization or branch with authority to enforce international law.

Despite these questions, there is widespread acceptance that a body of international law exists and continues to become increasingly relevant in today's more connected world. A review of the varied sources of international law and of the impact of international law on the U.S. legal system follows.

Origins of international law

There are five widely recognized sources of international law:[1]

- Treaties
- Conventions
- Custom
- Court decisions and treatises
- General law

Treaties are agreements between sovereign nations that have been formally adopted under the laws of each nation involved. Sometimes treaties are referred to as *bilateral* when they involve two nations and *multilateral* when they involve multiple nations.

Conventions are a form of treaties. The distinction is that a convention is sponsored by an international organization. For that reason, conventions typically have many nations involved as signatories.

Custom is a course of dealing that develops over time between nations. For custom to rise to the level of international law, it must be established that (1) the custom has been consistently applied in the nations' dealings for a substantial length of time and (2) the nations involved have come to treat the custom as law between them.

Court decisions and treatises by noted legal scholars also create a source of international law. Although international courts are not bound by their previous decisions (there is no doctrine of *stare decisis*), they often turn to past decisions and the writings of experts in the law to help shape opinions in particular cases.

General law is a body of law that emanates from the general principles of common and statutory law developed by civilized nations. It is hard to get a precise fix on where these general principles of law might be derived in any given case. However, in the absence of treaties, conventions, or customs directly addressing an issue, international courts may turn to this nebulous source of international law for guidance.

PUBLIC INTERNATIONAL AVIATION LAW

Public international law, in an aviation context, refers to agreements and treaties among various nations related to issues such as

- Landing rights
- Overflight authorizations
- Security and registration
- Communications

The earliest origins of public international aviation law are rooted in the period following the end of World War I. Before and during the war, aircraft were largely viewed as military weapons. In the aftermath of the war, lawyers, judges, and politicians from all over the world recognized the profound impact that air travel would have in challenging traditional notions of borders and "ownership" of airspace. Some of the more significant aviation treaties, conventions (international agreements), and compacts related to public aviation law are outlined below.

After World War I, the Paris Convention of 1919 was drafted. The Paris Convention marked the first formal efforts at establishing a rule of law related to sovereignty over airspace, registration of aircraft, standards for pilots, and movement of military aircraft. The Paris Convention also created the first formal organization for the oversight of international aviation activities, the Commission Internationale de Navigation Aerienne.

The beginnings of public international aviation law

Although the Paris Convention was a start in the right direction, it became apparent that more extensive cooperation and legal infrastructure might be necessary to support a growing aviation industry. The Havana Convention of 1928 built on much of what was started in the Paris Convention and established several new legal principles upon which international aviation would be governed.

The Chicago Convention

Perhaps one of the most significant agreements in public international aviation law is the Chicago Convention of 1944. One of the most noteworthy achievements of the Chicago Convention was the establishment of the International Civil Aviation Organization, which continues to operate today. When the Chicago Convention came into effect in 1947, it resulted in the termination of the Paris Convention of 1919 and the Havana Convention of 1928.

It is probably fair to state that the Chicago Convention created the foundation for our current system of international transportation by air. The Chicago Convention states that each "state has complete and exclusive sovereignty over the airspace above its territory."[2] Care was taken during the drafting of the Chicago Convention to specifically exclude all military, police, customs, and other state-operated aircraft from the operation of the convention.[3]

The key provisions for civil aircraft permit aircraft that are not engaged in scheduled air service to[4]

- Make flights into or in transit nonstop across its territory
- Make stops for nontraffic purposes (fuel and maintenance) without the necessity of obtaining prior permission.

However, the convention clearly states that aircraft operating in scheduled international air service are prohibited from operating "over or into the territory of a contracting State, except with the special permission or other authorization of that State."[5]

Other noteworthy provisions of the Chicago Convention include

- The rights of each state to establish restricted and prohibited areas as long as the restrictions and prohibitions apply equally to domestic and international aircraft[6]
- Establishment of responsibility for each contracting state to maintain radio and air navigation services and facilities[7]
- The adoption of a standard system of communication procedures[8]

Some commentators have noted that while the Chicago Convention did establish the two fundamental freedoms of overflight and stops for fuel and maintenance, it failed to establish three additional freedoms sought in an amendment or annex to the Chicago Convention.[9] The three freedoms sought in the unsuccessful annex were

1. The right of a country's airlines to transport passengers and/or cargo from its home nation to a second nation without special authorization
2. The right of a country's airlines to freely transport cargo and passengers from a second nation back to its home nation
3. The right of a country's airlines to freely transport passengers and cargo between a second and third nation

In the end, the Chicago Convention left these issues up to the individual nations involved to negotiate directly. If all these freedoms were codified in the Chicago Convention, the result would have been a sort of "open skies" agreement among the more developed nations of the world.

Although the multilateral Chicago Convention failed to create an open skies environment for international air transportation, it did spawn several bilateral agreements (agreements between individual nations) that have effectively created a more open skies type of approach over time. The first of these bilateral agreements was known as the Bermuda I agreement. Bermuda I was entered into in 1946, and the sole parties were the United States and Great Britain. The Bermuda I agreement permitted the airlines of United States and Great Britain to operate to and from each country—but only to designated "gateway" airports. Each airline would be allowed as many flights as it desired. Agreements similar to Bermuda I were entered into between the United States and many other nations. Similar agreements were also entered into by various nations throughout the world. Later, the Bermuda I agreement was superseded in 1967 by Bermuda II.

In 1992, the Department of Transportation initiated an "open skies" initiative that would allow for a more liberal framework for air route selection, capacity determinations, fare setting, and frequency of flights. The first open skies agreement was entered into in October 1992 between the United States and the Netherlands. Subsequently, the United States entered into open skies agreements with 13 European nations. Open skies agreements were subsequently entered into with Canada, South America, Peru, Malaysia, Taiwan, New Zealand, and Singapore, among others.

Open skies agreements

Some of the newer bilateral open skies agreements have even opened up what are referred to as *beyond rights*. These beyond rights permit air carriers to fly cargo to a partner country and then fly directly from the partner country to a third nation with no requirement that the flights first return to the United States.

Other open skies initiatives target the logistics of hundreds of bilateral agreements between individual countries. The trend appears to be moving in the direction of more multilateral open skies agreements such as the agreement entered into in 2002 between the United States, Brunei, Chile, New Zealand, and Singapore. In one recent effort, the United States and the European Union (EU) have entered into an agreement that will allow any U.S. or EU airline to operate flights to any U.S. or EU destination.[10]

PRIVATE INTERNATIONAL AVIATION LAW

Private international aviation law is the body of law relating to agreements and treaties between different countries in which the liability of a party in one country to an injured party in another country can be established. In many ways, the development of private international aviation law is an effort to sort out the uncertainties of jurisdiction and liability when persons from various sovereign nations are involved.

Fundamentals of Aviation Law

The formation of a body of private international aviation law began in 1925 at the First Conference on Private Air Law in Paris. The conference established an International Technical Committee of Aerial Legal Experts. This committee was charged with providing an ongoing study of the issues involved with private liability stemming from international air transportation. The committee studied the legal landscape of international air transportation, and as a result the Warsaw Convention was convened in Warsaw, Poland, in 1929. The Warsaw Convention is the centerpiece of private international aviation law. It is discussed below along with some relatively new developments from the Montreal Convention of 1999 and the Cape Town Convention of 2001.

The Warsaw Convention

The Warsaw Convention as we know it today includes the work of the 1929 Warsaw Convention along with subsequent protocols that amended the original convention. Currently, more than 135 nations are parties to the Warsaw Convention. The delegates to the Warsaw Convention had two major objectives. The first was the creation of a uniform system of regulation for issues such as ticketing, baggage transport, movement of cargo, and claims by passengers or customers concerning lost or damaged luggage or cargo. The second primary goal of the convention was to cap the amount of damages an air carrier could incur in an accident with the offsetting limitation to the defenses that air carriers could invoke to avoid liability. In the end, the overarching concern of the Warsaw Convention drafters was protection of the fledgling international air transportation industry.

In the discussion below, we will provide an overview of some of the key provisions, agreements, and protocols of the Warsaw Convention.

Air carrier liability

Articles 17, 18, and 19 of the Warsaw Convention are of key importance. Article 17 states that an international air carrier will be liable for a passenger's death or injury resulting from an "accident" that takes place when a passenger is (1) on an airplane, (2) boarding an airplane, or (3) disembarking an airplane. Article 18 imposes liability on an air carrier for baggage that is checked and goods that are damaged while in the care and custody of the air carrier. Exposure to liability for baggage was expanded in the Montreal Convention of 1999. The Montreal Convention modified the Warsaw Convention by defining baggage as both checked and unchecked (carry-on) baggage. Article 19 states that an air carrier is liable for any damages resulting from delays of passengers, cargo, or baggage.

While Articles 18 and 19 are fairly straightforward in application, the use of the word *accident* to trigger liability under Article 17 has sometimes spawned conflicting views. In a 1985 case, the U.S. Supreme Court defined the term *accident* as "an unexpected or unusual event or happening that is external to the passenger."[11]

Defenses available to airlines

Airlines have two primary defenses available to liability claims under the Warsaw Convention. The first defense is based on Article 20. Article 20 permits an airline to completely avoid liability if it took "all necessary measures" to avoid an accident, or if it was impossible for the carrier to avoid the accident. Warsaw Convention Article 21 permits an airline to mitigate its damages if the injuries to a passenger were caused in part or in their entirety by the contributory negligence of the injured passenger. As discussed below, the defenses available to air carriers under the original provisions of the Warsaw Convention were significantly modified by the Montreal Convention of 1999.

Dollar limits to liability

One of the linchpins of the Warsaw Convention was a limit on an air carrier's liability for injury or death of a passenger. The limit is found in Article 22. Article 22 limits an airline's liability to 125,000 francs (or approximately $8300). Naturally, this limitation has engendered controversy—especially in view of the fact that a passenger injured in a domestic accident could recover damages in the millions of dollars.

The $8300 limitation on liability was doubled by the Hague Protocol of 1955. Nonetheless, the United States was still displeased with the relatively low value placed on a human life while traveling overseas on an air carrier. After threatening to denounce the Warsaw Convention, air carriers agreed to a $75,000 limitation of liability in the Montreal Convention of 1966. However, this increased limit applied to only flights going to and from the United States.

The latest revisions to liability limitations came with the Montreal Convention of 1999. The Montreal Convention created a two-tiered liability structure and modified Articles 21 and 22 of the Warsaw Convention.

First, an air carrier is held strictly liable up to approximately $140,000[12] for injuries and/or death of a passenger due to an accident. For claims up to $140,000 the only defense for the carrier is contributory negligence of the passenger.

In claims that exceed $140,000 the air carrier can now be held liable for unlimited amounts. However, the carrier can defend itself by using the defense that the harm caused to a passenger was not due to the carrier's negligence or wrongful act. In claims larger than $140,000 the air carrier can also claim that the accident occurred due to circumstances out of the carrier's control.

Forum for trial

Another thorny question that arises in the context of international air transportation is the place for a trial. Article 28 of the Warsaw Convention provides the four possible places where a plaintiff may bring an action against an air carrier:

- The place where the air carrier is domiciled
- The primary place of business for the air carrier
- The country where the contract of travel was made (as long as the air carrier does business in that country)
- The destination country

Interestingly, the Warsaw Convention did not permit a plaintiff to bring suit in his or her country. This rule tended to act as a bar to plaintiff's filing suit due to the inconvenience and cost of having to file a lawsuit in a foreign country.

Ultimately, dissatisfaction with the inability of a passenger to bring a lawsuit in his or her home country led to modification of these provisions in the 1999 Montreal Convention. Under Article 33 of the 1999 Montreal Convention, a plaintiff may file a lawsuit in the county of the "principal and permanent residence" of the passenger. However, to make use of this newly available forum, the carrier must lease or own property in the passenger's home country and fly to and from that country.

Statute of limitations

Article 29 of the Warsaw Convention provides for a 2-year statute of limitations for bringing an action against an air carrier. Some courts have calculated the 2-year limitation strictly, and others have allowed for suspended periods to the time limits.

The Cape Town Convention

Because aircraft are easily moved from one country to another, there are inherent difficulties in financing arrangements and securing the interests of lenders. A lender in France may wish to use a newly purchased aircraft as security for a loan on that aircraft. However, the value of that aircraft as collateral for the loan is diminished by the fact that the aircraft can be easily moved to another country where the lender's interests may not be protected. Because of this type of issue, several nations participated in the UN International Institute for the Unification of Private Law (the "Cape Town Convention").

At the Cape Town Convention, the participants sought to draft an agreement that would create a more stable and enforceable legal rubric for aircraft title, security interests in aircraft, and aircraft leasing. Fifty-three nations adopted the Cape Town Convention in 2001. The Cape Town Convention addresses airframes, aircraft engines, and helicopters that exceed certain weight thresholds.[13]

The Cape Town Convention was signed by the United States on May 9, 2003. Some of the more significant features of the Cape Town Convention include the following:

1. The right of creditors to repossess or sell aircraft in case of default on a loan
2. A high-tech international aircraft registry that gives first-in-time priority to creditors who file security interests in an aircraft (in the United States the FAA will act as the entry point for the International Registry)
3. Creation of a system of protections for creditors and debtors that closely mirrors the current U.S. system
4. Permitting of creditors to deregister an aircraft when a debtor defaults and to procure the export of the aircraft
5. Giving to creditors the ability to take possession or control of an aircraft upon a default

The hope is that the Cape Town Convention will result in lower financing charges and provide easier funding of aircraft transactions. The convention became effective on April 1, 2004, and was signed into U.S. law on August 9, 2004.

INTERNATIONAL CIVIL AVIATION ORGANIZATION

As indicated earlier, the ICAO was created during the Chicago Convention in 1944. The aims and purposes of the ICAO are outlined in a U.S. government bulletin as follows:[14]

> Under the terms of the convention there will now come into being a permanent International Civil Aviation Organization, which will be brought into relationship with the United Nations into accordance with the Charter of the [United Nations] and will form a part of the general pattern of international cooperation. This organization has as its aim the establishment of principles and techniques of international air navigation and the fostering of the development of international air transport to the end that it will—
> a. Insure the safe and orderly growth of international civil aviation throughout the world;
> b. Encourage the arts of aircraft design and operation for peaceful purposes;
> c. Encourage the development of airways, airports, and air-navigation facilities for international civil aviation;
> d. Meet the needs of the peoples of the world for safe, regular, efficient, and economical air transport;
> e. Prevent economic waste caused by unreasonable competition;
> f. Insure the that rights of contracting states are fully respected and that every contracting state has a fair opportunity to operate international airlines;
> g. Avoid discrimination between contracting states;
> h. Promote safety of flight in international air navigation;
> i. Promote generally the development of all aspects of international civil aeronautics.

ICAO is a special agency of the United Nations. It is governed by an Assembly and a Council. The Assembly meets at least once every 3 years at meetings convened by the Council. Each member state of the ICAO is entitled to one vote on the Assembly where majority rule prevails. The Council has 36 member states that are elected by the Assembly for 3-year terms. Most of the day-to-day work of the ICAO is accomplished by the Council.

A major duty of the Council lies in adopting "International Standards and Practices" for the benefit of all member states. These standards can include safety measures, airport infrastructure issues, airworthiness issues, aircraft operation, search and rescue, environmental concerns, new technology, and security.

Currently the ICAO is working on a number of safety and security issues. One of the more recent ICAO issues is the creation of standard passports by using biometric identification features. The ICAO is also reviewing ways to communicate accident investigation databases in a more efficient and timely manner to all member states to enhance air safety.

CASES AND COMMENTARY

As indicated earlier, one of the most litigated issues created by the Warsaw Conference is the question of whether a passenger has been involved in an accident. In Case 10-1, the U.S. Supreme Court tackles the issue of just what constitutes an accident.

CASE 10-1
AIR FRANCE v. SAKS
470 U.S. 392 (1985)

OPINION BY: O'CONNOR

Article 17 of the Warsaw Convention makes air carriers liable for injuries sustained by a passenger "if the accident which caused the damage so sustained took place on board the aircraft or in the course of any of the operations of embarking or disembarking." We granted certiorari, [Citation], to resolve a conflict among the Courts of Appeals as to the proper definition of the word "accident" as used in this international air carriage treaty.

I

On November 16, 1980, respondent Valerie Saks boarded an Air France jetliner in Paris for a 12-hour flight to Los Angeles. The flight went smoothly in all respects until, as the aircraft descended to Los Angeles, Saks felt severe pressure and pain in her left ear. The pain continued after the plane landed, but Saks disembarked without informing any Air France crew member or employee of her ailment. Five days later, Saks consulted a doctor who concluded that she had become permanently deaf in her left ear.

Saks filed suit against Air France in California state court, alleging that her hearing loss was caused by negligent maintenance and operation of the jetliner's pressurization system. [Citation to record.] The case was removed to the United States District Court for the Central District of California. After extensive discovery, Air France moved for summary judgment on the ground that respondent could not prove that her injury was caused by an "accident" within the meaning of the Warsaw Convention. The term "accident," according to Air France, means an "abnormal, unusual or unexpected occurrence aboard the aircraft." [Citation.] All the available evidence, including the postflight reports, pilot's affidavit, and passenger testimony, indicated that the aircraft's pressurization system had operated in the usual manner. Accordingly, the airline contended that the suit should be dismissed because the only alleged cause of respondent's injury—normal operation of a pressurization system—could not qualify as an "accident." In her opposition to the summary judgment motion, Saks acknowledged that "[the] sole question of law presented ... by the parties is whether a loss of hearing proximately caused by normal operation of the aircraft's pressurization system is an 'accident' within the meaning of Article 17 of the Warsaw Convention...." [Citation.] She argued that "accident" should be defined as a "hazard of air travel," and that her injury had indeed been caused by such a hazard.

Relying on precedent which defines the term "accident" in Article 17 as an "unusual or unexpected" happening, [Citation], the District Court granted summary judgment to Air France. [Citation to case holding that normal cabin pressure changes are not "accidents" within the meaning of Article 17]. A divided panel of the Court of Appeals for the Ninth Circuit reversed. [Citation.] The appellate court reviewed the history of the Warsaw Convention and its modification by the 1966 Montreal Agreement, a private agreement among airlines that has been approved by the United States Government. [Citation.] The court concluded that the language, history, and policy of the Warsaw Convention and the Montreal Agreement impose absolute liability on airlines for injuries proximately caused by the risks inherent in air travel. The court found a definition of "accident" consistent with this history and policy in Annex 13 to the Convention on International Civil Aviation, [Citation] conformed to in 49 CFR § 830.2 (1984): "an occurrence associated with the operation of an aircraft which takes place between the time any person boards the aircraft with the intention of flight and all such persons have disembarked...." [Citation]. Normal cabin pressure changes qualify as an "accident" under this definition. A dissent agreed with the District Court that "accident" should be defined as an unusual or unexpected occurrence. [Citation.] We disagree with the definition of "accident" adopted by the Court of Appeals, and we reverse.

* * *

III

We conclude that liability under Article 17 of the Warsaw Convention arises only if a passenger's injury is caused by an unexpected or unusual event or happening that is external to the passenger. This definition should be flexibly applied after assessment of all the circumstances surrounding a passenger's injuries. [Citation.] For example, lower courts in this country have interpreted Article 17 broadly enough to encompass torts committed by terrorists or fellow passengers. [Citations.] In cases where there is contradictory evidence, it is for the trier of fact to decide whether an "accident" as here defined caused the passenger's injury. [Citations.] But when the injury indisputably results from the passenger's own internal reaction to the usual, normal, and expected operation of the aircraft, it has not been caused by an accident, and Article 17 of the Warsaw Convention cannot apply. The judgment of the Court of Appeals in this case must accordingly be reversed.

The judgment of the Court of Appeals is reversed, and the case is remanded for further proceedings consistent with this opinion.

It is so ordered.

After reading the case, do you agree with the Court's assessment of how an accident should be defined?

After the *Saks* case, the Supreme Court was faced with another intriguing Warsaw Convention issue. If the plaintiff is not permitted to recover damages under the terms of the Warsaw Convention, does the plaintiff still have the ability to reach outside the Warsaw Convention and sue in an action based on local law? See the Court's response to this question in Case 10-2.

CASE 10-2
EL AL ISRAEL AIRLINES, LTD. V. TSUI YUAN TSENG
525 U.S. 155 (1999)

OPINION BY: GINSBURG

Plaintiff-respondent Tsui Yuan Tseng was subjected to an intrusive security search at John F. Kennedy International Airport in New York before she boarded an El Al Israel Airlines May 22, 1993 flight to Tel Aviv. Tseng seeks tort damages from El Al for this occurrence. The episode-in-suit, both parties now submit, does not qualify as an "accident" within the meaning of the treaty popularly known as the Warsaw Convention, which governs air carrier liability for "all international transportation." Tseng alleges psychic or psychosomatic injuries, but no "bodily injury," as that term is used in the Convention. Her case presents a question of the Convention's exclusivity: When the Convention allows no recovery for the episode-in-suit, does it correspondingly preclude the passenger from maintaining an action for damages under another source of law, in this case, New York tort law?

The exclusivity question before us has been settled prospectively in a Warsaw Convention protocol (Montreal Protocol No. 4) recently ratified by the Senate. In accord with the protocol, Tseng concedes, a passenger whose injury is not compensable under the Convention (because it entails no "bodily injury" or was not the result of an "accident") will have no recourse to an alternate remedy. We conclude that the protocol, to which the United States has now subscribed, clarifies, but does not change, the Convention's exclusivity domain. We therefore hold that recovery for a personal injury suffered "on board [an] aircraft or in the course of any of the operations of embarking or disembarking," Art. 17, 49 Stat. 3018, if not allowed under the Convention, is not available at all.

The Court of Appeals for the Second Circuit ruled otherwise. In that court's view, a plaintiff who did not qualify for relief under the Convention could seek relief under local law for an injury sustained in the course of international air travel. [Citation.] We granted certiorari, [Citation.]), and now reverse the Second Circuit's judgment. Recourse to local law, we are persuaded, would undermine the uniform regulation of international air carrier liability that the Warsaw Convention was designed to foster.

I

We have twice reserved decision on the Convention's exclusivity. In *Air France* v. *Saks,* [Citation.], we concluded that a passenger's injury was not caused by an "accident" for which the airline could be held accountable under the Convention, but expressed no view whether that passenger could maintain "a state cause of action for negligence." [Citation.] In *Eastern Airlines, Inc.* v. *Floyd,* [Citation], we held that mental or psychic injuries unaccompanied by physical injuries are not compensable under Article 17 of the Convention, but declined to reach the question whether the Convention "provides the exclusive cause of action for injuries sustained during international air transportation." [Citation.] We resolve in this case the question on which we earlier reserved judgment.

At the outset, we highlight key provisions of the treaty we are interpreting. Chapter I of the Warsaw Convention, entitled "SCOPE — DEFINITIONS," declares in Article 1(1) that the "Convention shall apply to all international transportation of persons, baggage, or goods performed by aircraft for hire." [Citation.] Chapter III, entitled "LIABILITY OF THE CARRIER," defines in Articles 17, 18, and 19 the three kinds of liability for which the Convention provides. Article 17 establishes the conditions of liability for personal injury to passengers: "The carrier shall be liable for damage sustained in the event of the death or wounding of a passenger or any other bodily injury suffered by a passenger, if the accident which caused the damage so sustained took place on board the aircraft or in the course of any of the operations of embarking or disembarking." [Citation.]

Article 18 establishes the conditions of liability for damage to baggage or goods. [Citation.] Article 19 establishes the conditions of liability for damage caused by delay. [Citation.] Article 24, referring back to Articles 17, 18, and 19, instructs:

> (1) In the cases covered by articles 18 and 19 any action for damages, however founded, can only be brought subject to the conditions and limits set out in this convention.
>
> (2) In the cases covered by article 17 the provisions of the preceding paragraph shall also apply, without prejudice to the questions as to who are the persons who have the right to bring suit and what are their respective rights. [Citation.]

II

With the key treaty provisions as the backdrop, we next describe the episode-in-suit. On May 22, 1993, Tsui Yuan Tseng arrived at John F. Kennedy International Airport (hereinafter JFK) to board an El Al Israel Airlines flight to Tel Aviv. In conformity with standard El Al preboarding procedures, a security guard questioned Tseng about her destination and travel plans. The guard considered Tseng's responses "illogical," and ranked her as a "high risk" passenger. Tseng was taken to a private security room where her baggage and person were searched for explosives and detonating devices. She was told to remove her shoes, jacket, and sweater, and to lower her blue jeans to midhip. A female security guard then searched Tseng's body outside her clothes by hand and with an electronic security wand.

After the search, which lasted 15 minutes, El Al personnel decided that Tseng did not pose a security threat and allowed her to board the flight. Tseng later testified that she "was really sick and very upset" during the flight, that she was "emotionally traumatized and disturbed" during her month-long trip in Israel, and that, upon her return, she underwent medical and psychiatric treatment for the lingering effects of the body search. [Citation.]

Tseng filed suit against El Al in 1994 in a New York state court of first instance. Her complaint alleged a state law personal injury claim based on the May 22, 1993 episode at JFK. Tseng's pleading charged, *inter alia*, assault and false imprisonment, but alleged no bodily injury. El Al removed the case to federal court.

The District Court, after a bench trial, dismissed Tseng's personal injury claim. [Citation.] That claim, the court concluded, was governed by Article 17 of the Warsaw Convention, which creates a cause of action for personal injuries suffered as a result of an "accident ... in the course of any of the operations of embarking or disembarking." [Citation.] Tseng's claim was not compensable under Article 17, the District Court stated, because Tseng "sustained no bodily injury" as a result of the search, [Citation], and the Convention does not permit "recovery for psychic or psychosomatic injury unaccompanied by bodily injury." [Citation]. The District Court further concluded that Tseng could not pursue her claim, alternately, under New York tort law; as that court read the Convention, Article 24 shields the carrier from liability for personal injuries not compensable under Article 17. [Citation.]

The Court of Appeals reversed in relevant part. [Citation] The Second Circuit concluded first that no "accident" within Article 17's compass had occurred; in the Court of Appeals' view, the Convention drafters did not "aim to impose close to absolute liability" for an individual's "personal reaction" to "routine operating procedures," measures that, although "inconvenient and embarrassing," are the "price passengers pay for ... airline safety." [Citation.] In some tension with that reasoning, the Second Circuit next concluded that the Convention does not shield the very same "routine operating procedures" from assessment under the diverse laws of signatory nations (and, in the case of the United States, States within one Nation) governing assault and false imprisonment. [Citation.]

Article 24 of the Convention, the Court of Appeals said, "clearly states that resort to local law is precluded only where the incident is 'covered' by Article 17, meaning where there has been an accident, either on the plane or in the course of embarking or disembarking, which led to death, wounding or other bodily injury." [Citation.] The court found support in the drafting history of the Convention, which it construed to "indicate that national law was intended to provide the passenger's remedy where the Convention did not expressly apply." [Citation.] The Second Circuit also rejected the argument that allowance of state-law claims when the Convention does not permit recovery would contravene the treaty's goal of uniformity. The court read our decision in *Zicherman* v. *Korean Air Lines Co.,* [Citation.], to "instruct specifically that the Convention expresses no compelling interest in uniformity that would warrant ... supplanting an otherwise applicable body of law." [Citation.]

III

* * *

The Court of Appeals looked to our precedent for guidance on this point, but it misperceived our meaning. It misread our decision in *Zicherman* to say that the Warsaw Convention expresses no compelling interest in uniformity that would warrant preempting an otherwise applicable body of law, here New York tort law. [Citation.] *Zicherman* acknowledges that the Convention centrally endeavors "to foster uniformity in the law of international air travel." [Citation.] It further recognizes that the Convention addresses the question whether there is airline liability *vel non*. [Citation.] The *Zicherman* case itself involved auxiliary issues: who may seek recovery in lieu of passengers, and for what harms they may be compensated. [Citation.]. Looking to the Convention's text, negotiating and drafting history, contracting states' postratification understanding of the Convention, and scholarly commentary, the Court in *Zicherman* determined that Warsaw drafters intended to resolve *whether there is liability*, but to leave to domestic law (the local law identified by the forum under its choice of law rules or approaches) determination of the compensatory damages available to the suitor. See id., at 231.

A complementary purpose of the Convention is to accommodate or balance the interests of passengers seeking recovery for personal injuries, and the interests of air carriers seeking to limit potential liability. Before the Warsaw accord, injured passengers could file suits for damages, subject only to the limitations of the forum's laws, including the forum's choice of law regime. This exposure inhibited the growth of the then-fledgling international airline industry. [Citation.] Many international air carriers at that time endeavored to require passengers, as a condition of air travel, to relieve or reduce the carrier's liability in case of injury. [Citation.] The Convention drafters designed Articles 17, 22, and 24 of the Convention as a compromise between the interests of air carriers and their customers worldwide. In Article 17 of the Convention, carriers are denied the contractual prerogative to exclude or limit their liability for personal injury. In Articles 22 and 24, passengers are limited in the amount of damages they may recover, and are restricted in the claims they may pursue by the conditions and limits set out in the Convention.

Construing the Convention, as did the Court of Appeals, to allow passengers to pursue claims under local law when the Convention does not permit recovery could produce several anomalies. Carriers might be exposed to unlimited liability under diverse legal regimes, but would be prevented, under the treaty, from contracting out of such liability. Passengers injured physically in an emergency landing might be subject to the liability caps of the Convention, while those merely traumatized in the same mishap would be free to sue outside of the Convention for potentially unlimited damages. The Court of Appeals' construction of the Convention would encourage artful pleading by plaintiffs seeking to opt out of the Convention's liability scheme when local law promised recovery in excess of that prescribed by the treaty. [Citation.] Such a reading would scarcely advance the predictability that adherence to the treaty has achieved worldwide.

* * *

For the reasons stated, we hold that the Warsaw Convention precludes a passenger from maintaining an action for personal injury damages under local law when her claim does not satisfy the conditions for liability under the Convention. Accordingly, we reverse the judgment of the Second Circuit.

It is so ordered.

What do you think about the Court's ruling in this case? What problems could ensue if the Court ruled that the Warsaw Convention did not provide the exclusive remedy?

This final case involves another difficult issue facing international travelers who want to file a claim against an airline: Where should the claim be filed? See Case 10-3 for a discussion of how the proper forum for a case is determined under the Warsaw Convention.

CASE 10-3
Resham Jeet Singh, Gursharan Jeet Kaur, Individually and as Guardians of Gurpreet Kaur v. Tarom Romanian Air Transport
88 F. Supp. 2d 62 (2000)

OPINION BY: David G. Trager

Defendant foreign airline moves to dismiss plaintiff passengers' personal injury claims for lack of subject matter jurisdiction under the Warsaw Convention.

Background

Plaintiffs (the "Singhs") are citizens of India admitted to the United States for permanent residence, who reside in Queens, New York. In or about May 1998, Resham Jeet Singh's father purchased three tickets for plaintiffs from Globe Travel, a travel agency in Jackson Heights, New York. The tickets were for round trip travel on defendant Tarom Romanian Air Transport ("Tarom") from Delhi, India to New York and back. Plaintiffs explain that round trip tickets were purchased in order to save money on a short trip back to India that plaintiffs intended to take later in the year. Although the tickets were paid for in New York, the tickets were issued by Bajaj Travels in Delhi, India, and were picked up by plaintiffs in Delhi.

Plaintiffs' flight from Delhi to New York included a stop-over in Bucharest, Romania. Although plaintiffs had the proper documentation for entry into the United States, when the flight arrived in Bucharest, agents of Tarom refused to permit the Singhs to continue travel to New York. Plaintiffs allege that Tarom employees then confined plaintiffs in the customs area of the airport for six days and deprived them of sufficient food and bathing facilities before allowing them to continue their travel to New York.

Plaintiffs subsequently brought this action, complaining that Tarom's conduct during the six-day detention constituted a violation of Articles 17 and 19 of the Warsaw Convention. In addition, plaintiffs claim that the detention constituted "malicious prosecution" under New York law.

Defendant moves to dismiss: (1) plaintiffs' state law claim on the ground that it is preempted by the Convention, and (2) plaintiffs' Warsaw Convention claim for lack of subject matter jurisdiction.

Discussion

(1)

As a treaty of the United States, the Warsaw Convention is supreme law of the land. [Citations.] It is well-established that where the provisions of the Warsaw Convention apply to a claim, the Convention exclusively governs the rights and liabilities of the parties, and, thus, preempts state law. [Citations.] In this case, plaintiffs concede—perhaps unwisely—that the Convention governs their claims. [Citation to record.] Accordingly, plaintiffs' state law malicious prosecution claim is preempted by the Convention and must be dismissed.

(2)

Defendant moves to dismiss plaintiffs' remaining Warsaw Convention action on the ground that this court has no subject matter jurisdiction over the action under the provisions of the Warsaw Convention. Article 28(1) of the Convention specifies that actions arising out of international transportation governed by the Convention must be brought in one of four clearly identified fora: (1) the domicile of the carrier, (2) its principal place of business of the carrier, (3) the forum in which the carrier has a place of business through which the contract was made, or (4) the place of destination. It is well-established that unless one of the specified fora is in the United States, a federal district court lacks jurisdiction over the claim under the terms of the Convention and, hence, lacks federal subject matter jurisdiction over the controversy. [Citation.] Because plaintiffs have conceded that the Convention does apply, [Citation to record] the dispositive question for the present motion is whether the Eastern District of New York is a proper forum for plaintiffs' action under the terms of the Convention. Each of the possible bases for subject matter jurisdiction under Article 28(1) will be considered in turn.

First, the "domicile" of a carrier within the meaning of Article 28(1) is the carrier's place of incorporation. [Citation.] There is no dispute that Tarom is organized under the laws of the Republic of Romania. Romania is, therefore, Tarom's domicile for the purposes of the Convention. Accordingly, the domicile of the carrier clause does not provide a basis for this court to exercise jurisdiction over the Singhs' claims.

Second, for the purposes of Article 28(1), a foreign corporation has only one "principal place of business." [Citation.] Tarom's corporate headquarters are located at the Otopeni Airport in Bucharest, Romania. Moreover, Tarom operates the majority of its flights out of Bucharest. Tarom's principal place of business is, therefore, Bucharest, Romania. Accordingly, the carrier's principal place of business clause does not provide a basis for this court to exercise jurisdiction under the Convention.

Third, the "destination" of a round trip international airline ticket within the meaning of Article 28(1) is the starting point of the journey. [Citation.] In determining the "destination" of a journey covered by the Convention, the Second Circuit has made clear that the unexpressed intentions of the passenger are not relevant if the instrument evidencing the contract is unambiguous. [Citation including quote—"When a person purchases a roundtrip ticket, there can be but one destination, where the ticket originated."]. Here, plaintiffs' tickets clearly indicate that Delhi was the starting point of plaintiffs' travel. Thus, notwithstanding plaintiffs' claim that they intended to remain in New York for an extended period of time—a fact they did not communicate to Tarom or its agents, Article 28(1)'s destination clause does not give a basis for this court to exercise jurisdiction under the Convention.

The only remaining possible basis for this court's jurisdiction, then, is the clause of Article 28(1) relating to the carrier's place of business through which the contract was made. Defendant argues that the place of business through which it made the contract was the travel agency that issued and delivered the tickets, viz., Bajaj Travels in Delhi. [Citations—with quotation (holding that the "'place of business through which the contract has been made' is the place where the passenger ticket was issued"]. In this regard, it should be noted that the tickets themselves indicate that they were issued in Delhi. Plaintiffs, however, argue that the place of business through which the contract was made is Globe Travel, the New York travel agent to whom the purchase price of the tickets was paid.

* * *

In absence of any evidence of a principal-agent relationship between Globe Travel and Bajaj Travel or Tarom, it can only be found that the place of business through which the contract was made was the agency that actually issued the tickets, Bajaj Travel. Since Bajaj Travel is located in Delhi, India, the place-of-business clause of Article 28(1) does not provide a basis for this court to exercise jurisdiction over plaintiffs' Warsaw Convention claim.

Because the Eastern District of New York is not an appropriate forum for this action under any of the bases specified in Article 28(1), defendant's motion to dismiss plaintiffs' Warsaw Convention claim for lack of subject matter jurisdiction must be granted.

Conclusion

For the reasons set forth above, plaintiffs' state law malicious prosecution claim is preempted by the Warsaw Convention, and this court has no subject matter jurisdiction over plaintiffs' remaining claim under the Convention. Accordingly, defendant's motion to dismiss is granted. The Clerk of Court is directed to enter judgment and close the case.

Do you think the decision in this case is a fair result? Should the intent of the passengers be taken into consideration when determining what their destination may have been? Should the fact that the plaintiffs purchased a round trip ticket back to India mean that the U.S. was not the real destination (and therefore a possible site for the lawsuit)? Would the results of this case have been any different if the Montreal Convention of 1999 been in force?

DISCUSSION CASES

1. In February 2001, Eric Searcy was en route from New Orleans, Louisiana, to Ecuador. He was on board an American Airlines passenger jet. Searcy is a paraplegic. On the date of the flight, he was accompanied by his mother, Rhonda Searcy. The first leg of their trip was a flight from New Orleans to Miami. In Miami, both Rhonda and Eric Searcy disembarked their first aircraft so they could transfer to another plane for the final leg of their trip. Eric was physically injured while being assisted by American Airlines employees onto the aircraft that would depart Miami for Ecuador. Both Eric and Rhonda Searcy file claims against American. Eric claims that he was physically injured due to negligence by the American employees sent to assist him. Rhonda claims that she saw the whole episode, including Eric's injury. She files a suit against American for mental damages she experienced when she witnessed her son's injury.

2. The Coyles, a married couple in Portland, Oregon, purchase airline tickets through a travel agent in Portland. The intended travel route was United States to Taipai, Taipai to Jakarta, Jakarta to Singapore, and from Singapore back to the United States. Sometime after the couple arrived in Jakarta, they decided on a side trip and purchased a round-trip ticket from Garuda Airlines from Jakarta to Medan (on the Indonesian island of Sumatra) and back to Jakarta. The couple was killed when their flight from Jakarta to Medan crashed in Medan. The surviving family sued the airline. The question raised was whether the family could sue in a U.S. court. How would this case be decided under the Warsaw Convention Article 28? Would it be decided any differently under the Montreal Convention of 1999? Explain.

3. Neil Scala was an American Airlines passenger on a flight from Aruba to Puerto Rico. During the flight he was offered a drink by the flight attendant. He requested cranberry juice. However, as a result of a mixup, he was accidentally served a mixture of cranberry juice and alcohol. Scala alleged that the alcohol caused him to suffer damage to his heart due to a preexisting heart condition. There is no question that the incident took place on an international flight. There is also no question regarding Scala's preexisting heart condition. The question faced by the U.S. District Court is whether the incident qualifies as an "accident" under the Warsaw Convention. You have been assigned to review this case. What is your opinion? Explain.

4. Caroline Neischer suffered from asthma. She booked a trip from Los Angeles to Georgetown, Guyana. The domestic leg of the trip from Los Angeles to New York was flown by American Airlines. British West India Airline (BWIA) flew the leg of the trip from New York to Guyana. During the period of time when Neischer had to connect to the BWIA flight, she was forced to give up her carry-on bag that contained her asthma medicine to an employee of either American Airlines or BWIA. The employee told her the bag would be promptly returned to her once the flight reached Guyana. However, her baggage did not reach her until approximately 48 hours had elapsed from the time she arrived in Guyana. Neischer became very ill, was admitted to a Guyana hospital, and died a few days later. Would Neischer's family have a cause of action against the airline(s) responsible? Discuss.

5. Clariza Luna lived with her husband and two children in Houston, Texas. She purchased a ticket for a round trip between Houston and Cali, Columbia, on June 3, 1992. The ticket was purchased through a travel agency, Excelsior Travel. The ticket provided for the following route of flight: Houston to Panama City (via Continental Airlines), Panama to Cali, Columbia [with the leg from Panama City to Cali on Compania Panamena De Aviacion (COPA)]. The return flight followed the same route with the same carriers. On June 6, 1992, the COPA flight crashed in the vicinity of Tuciti, Panama, and killed all passengers on board, including Luna. Luna's family sought to have the case heard in the U.S. District Court for the Southern District of Texas. The family argued that the company that issued the COPA ticket, Excelsior, was a U.S. company and the ticket was purchased as part of a standard airline interline agreement. Under the interline agreement, other carriers, such as Continental, could make reservations and issue tickets for COPA flights. The family also cited the fact that COPA had aircraft engines overhauled by a Texas company and had previously sent employees to Texas in connection with the overhauls. COPA argued that it should not be subject to jurisdiction in the United States. It argued that it was a Panamanian corporation with no place of business in the United States and no flights into the United States. Does the U.S. District Court have jurisdiction to hear this case? Discuss.

6. Dr. Hanson and his wife, Rubina Husain, boarded an international flight returning to the United States in December 1997. Because Dr. Hanson suffered from severe asthma and was sensitive to second-hand smoke, he specifically requested to be seated in nonsmoking sections during the flight. Upon boarding, Dr. Hanson learned that although he was seated in a nonsmoking section of the airplane, he was only three rows in front of the economy-class smoking section. Ms. Husain advised a flight attendant that her husband would have to be moved. The flight attendant told Ms. Husain to "have a seat." Prior to takeoff, Ms. Husain again requested that her husband be moved and specifically noted his allergies to second-hand smoke. The flight attendant replied that Dr. Hanson could not be moved because the aircraft was fully loaded and she was "too busy." After takeoff, passengers in the smoking section began to smoke and Dr. Hanson found himself surrounded by ambient smoke. Ms. Husain spoke to the same flight attendant for a third time, stating, "You have to move my husband from here." Again the flight attendant refused to assist, but indicated that Dr. Hanson could switch seats if another passenger was willing. After 2 hours, the smoking increased behind Dr. Hanson. He ultimately moved himself to the front of the airplane to get some fresher air and collapsed while leaning against a chair in the galley. Despite immediate medical intervention by an allergist onboard the aircraft, Dr. Hanson died. His wife sues the airline for wrongful death in a U.S. district court. Who will prevail? Explain.

ENDNOTES

1. Statute of the International Court of Justice (I.C.J. Acts & Docs.), art. 38(1).
2. Convention on International Civil Aviation, December 7, 1944, art. 3(d), 61 Stat. 1180, 1181, 15 U.N.T.S. at 298.
3. Id.
4. Id. at 1181, 15 U.N.T.S. at 298.
5. Id. at 1182, 15 U.N.T.S. at 300.
6. Id. at 1182, 15 U.N.T.S. at 302.
7. Id. at 1188, 15 U.N.T.S. at 314.
8. Id.
9. Gabriel S. Meyer, *U.S.–China Aviation Relations: Flight Path to Open Skies,* 35 Cornell Int'l L. J. 427, 432–433 (Fall 2002).
10. Laura Meckler, *EU and U.S. Approve an Open Skies Agreement,* Wall St. J. November 19–20, 2005 at A2.
11. See *United States v. Saks,* 470 U.S. 292, 405 (1985).
12. This is an approximate dollar value based on the legal limit of 100,000 SDRs (Special Drawing Rights) as this standard is used in the Montreal Convention. The SDR is an international measurement of a reserve asset. The measurement was created by the International Monetary Fund (IMF) in 1969. SDRs are allocated to various countries based on their proportion of IMF quotas. Its value is based on a basket of key international currencies. The current SDR rate for U.S. dollars indicates that one SDR equals approximately $1.40.
13. Sean D. Murphy, *Contemporary Practice of the United States Relating to International Law: Private International Law: Cape Town Convention on Financing of High-Value, Mobile Equipment,* The American Society of International Law American Journal of International Law, 98 A.J.I.L. 852 (October 2004).
14. See *Department of State Bulletin,* vol. XVI, no. 403, March 23, 1947, at 530.

Appendix A — Case Briefs

From time to time your instructor may request that you prepare written case briefs. The case briefs, if prepared properly, should help you to develop a better understanding of the legal issues involved in each case you are assigned.

In essence, a case brief is a summary of the case. The typical structure of a case briefing will include the following elements:

1. Case citation (name of the case with references)
2. Summary of relevant facts
3. Issues presented
4. Case holding
5. Summary of court's analysis

As you prepare to brief a case, keep in mind the following legal terms and definitions:

- The *plaintiff* is the party who initially brings a lawsuit.
- The *defendant* is the party who has been sued.
- A *petitioner* or *appellant* is the party appealing that decision of a lower court or trial court. An appellant could be either a plaintiff or a defendant, depending on who prevailed or lost at the lower court or trial court.
- A *respondent* or *appellee* is the party responding to the appeal. Again, an appellee could have been either a plaintiff or a defendant in the lower court or trial court. Typically the appellee was the prevailing party at the lower court or trial court level. However, there may be cases in which both the plaintiff and the defendant are in disagreement with aspects of the lower court or trial court decision.

To provide you with some guidance on how to brief a case, read the following edited case and a sample brief following the case.

WACKENHUT CORPORATION and DELTA AIRLINES, INC. v. LIPPERT
SUPREME COURT OF FLORIDA

609 So. 2d 1304

December 3, 1992, Decided

OPINION: GRIMES, J.

Felice Lippert and her husband bought tickets on a Delta Airlines' flight from West Palm Beach to New York. On her way to board the flight, Ms. Lippert took a handbag containing valuable jewelry through a security checkpoint at Palm Beach International Airport. Because this portion of the airport was designated exclusively for Delta flights, Delta was responsible for maintaining the checkpoint. Delta contracted with the Wackenhut Corporation to act as its agent in the operation of the checkpoint. The checkpoint included a magnetometer scan of baggage and other carry-on items as well as a scan of the person that occurs as the person walks through a specially designed archway. Ms. Lippert placed her bag on the conveyor belt as required and walked through the archway. When the archway magnetometer alarm sounded, Ms. Lippert was briefly inspected by Wackenhut personnel. She then walked forward to collect her handbag containing the jewelry at the end of the conveyor, but it was missing. A search of the area for the missing handbag was unsuccessful.

Ms. Lippert sued Delta and Wackenhut for the value of the lost jewelry. As an affirmative defense, Delta and Wackenhut asserted a $1,250 limitation on liability which is set forth on her ticket and on Delta's published tariff. The trial judge initially entered a partial summary judgment for Delta and Wackenhut, upholding the limitation on liability to the maximum amount of $1,250. The judge reasoned that Ms. Lippert had delivered her property into Delta's custody through its agent, Wackenhut, and had thereby invoked the liability limitation. At the time of trial, a new judge was assigned to the case. The second judge permitted the jury to consider the total amount of damages claimed by Ms. Lippert, though throughout the trial the judge remarked that he would be bound to enter a final judgment for no more than $1,250 despite the jury's verdict. The jury returned a verdict in favor of Ms. Lippert for $431,000. At this point, the judge changed his mind regarding the applicability of the liability limitation and entered a final judgment against both defendants for $431,000.

The district court of appeal held that the limitation on liability contained in the ticket and the tariff did not apply under the facts of the case. The court also found that a bailment for the mutual benefit of both the passenger and the airline had been created when Ms. Lippert relinquished possession of her valuables to go through the x-ray machine. Therefore, the trial court was correct in applying the ordinary negligence standard. However, the court felt that the defendants had been unduly prejudiced by the judge's assurances throughout the pretrial proceedings and the trial that the potential judgment could not exceed $1,250. Thus, the case was remanded for a new trial with the proviso that the limitation of liability would not apply.

On petition for review in this Court, Delta and Wackenhut argue for the $1,250 limitation. In addition, they contend that, because the airport security check was mandated by law, they were gratuitous bailees, who could only be held liable if grossly negligent. Ms. Lippert cross-petitions to review the granting of a new trial.

The airline ticket purchased by Ms. Lippert provided in pertinent part:

> [Defendants] shall be liable for the loss of, damage to, or delay in the delivery of a fare-paying passenger's baggage, or other property (including carry-on baggage, if tendered to

[Defendant's] in flight personnel for storage during flight or otherwise delivered into the custody of [Defendants].) Such liability, if any, for the loss, damage or delay in the delivery of a fare-paying passenger's baggage or other property (whether checked or otherwise delivered into the custody of [Defendants]), shall be limited to an amount equal to the value of the property, plus consequential damages, if any, and shall not exceed the maximum limitation of USD 1,250.00 for all liability for each fare-paying passenger (unless the passenger elects to pay for higher liability as provided for in paragraph 3) below). The passenger shall not be automatically entitled to USD 1,250.00 but must prove the value of losses or damages.

2) Exclusions From Liability

[Defendants][are] not responsible for jewelry, cash, camera equipment, or other similar valuable items contained in checked or unchecked baggage, unless excess valuation has been purchased. These items should be carried by the passenger.

Ms. Lippert seems to argue that under the emphasized portion of section 1 of her ticket, quoted above, an article only becomes baggage, and therefore triggers the limitation on liability, when it reaches the cargo compartment or the cabin of the aircraft. However, this interpretation would lead to the dubious conclusion that passenger's property in transit to the airplane after being delivered to the airline at the check-in point where tickets are purchased should not be considered baggage. The phrase in the ticket's definition of baggage— "whether checked in the cargo compartment or carried in the cabin" …"—s more realistically construed as emphasizing that, for purposes of Delta's contract with its passengers, there is no difference between "carry-on" and "checked" baggage. Thus, the ticket's references to the cargo compartment and the cabin are merely descriptive of the words "checked" or "carried," and there can be no doubt that Ms. Lippert's handbag was a passenger's "article or other property … acceptable for transportation … whether checked or carried …." We believe that a ticketed passenger's property, destined for an airplane and in transit between the airport's security checkpoint and the actual airplane, constitutes "baggage" as defined by the ticket.

We cannot accept Ms. Lippert's contention that our interpretation will mean that an article carried by a nonticketed person which had gone through the checkpoint screening process would be subject to the $1,250 limitation. Nonticketed persons have not contracted to a limitation of liability as have those persons who purchase tickets. Therefore, the liability for articles of nonticketed passengers would be determined by the ordinary laws of bailment.

We hold that the $1,250 baggage limitation of liability was applicable to the loss of Ms. Lippert's handbag while it was in the possession of Delta's agent at the airport security checkpoint.

It is so ordered.

Here's what a case briefing for this case might look like:

Brief of the Case

1. *Case citation*: The Wackenhut Corporation and Delta Airlines, Inc. v. Lippert, 609 SO. 1304 (1992); Supreme Court of Florida

2. *Summary of facts*:

   - Felice Lippert deposits her handbag (carrying approximately $431,000 in jewelry) with airport security screeners (employed by Wackenhut, a Delta contractor) as she passes through a Delta airport security checkpoint.
   - The handbag is lost.
   - Ms. Lippert and her husband sued Delta and Wackenhut for the full value of the jewelry contained in the handbag.
   - Delta and Wackenhut claim that they are only liable for $1250, the limitation of liability for luggage contained in writing in Ms. Lippert's ticket and airline tariff.

3. *Issue presented*: Is Ms. Lippert's handbag "baggage" and therefore subjecting the Lipperts to a $1250 limitation in damages?

4. *Holding*: Yes.

5. *Summary of court's analysis*: The court rejects Ms. Lippert's argument in that her handbag does not become "baggage" (subject to the liability limitation in her ticket and tariff) until it reaches the aircraft cargo compartment or the aircraft cabin. The court reasons that this line of argument would also mean that check-in luggage is not considered "baggage" until it reaches the aircraft. Accordingly, the court held that the $1250 limitation of liability for baggage was applicable to this case.

Author's Note: This illustration provides just one example of how you can put together a case briefing. Your instructor may provide you with more detailed instructions. This illustration is merely meant to provide you with some basic guidelines.

Appendix B

The Constitution of the United States of America

We the People of the United States, in Order to form a more perfect Union, establish Justice, insure domestic Tranquility, provide for the common defense, promote the general Welfare, and secure the Blessings of Liberty to ourselves and our Posterity, do ordain and establish this Constitution for the United States of America.

Article I

Section 1.

All legislative Powers herein granted shall be vested in a Congress of the United States, which shall consist of a Senate and House of Representatives.

Section 2.

Clause 1: The House of Representatives shall be composed of Members chosen every second Year by the People of the several States, and the Electors in each State shall have the Qualifications requisite for Electors of the most numerous Branch of the State Legislature.

Clause 2: No Person shall be a Representative who shall not have attained to the Age of twenty five Years, and been seven Years a Citizen of the United States, and who shall not, when elected, be an Inhabitant of that State in which he shall be chosen.

Clause 3: Representatives and direct Taxes shall be apportioned among the several States which may be included within this Union, according to their respective Numbers, which shall be determined by adding to the whole Number of free Persons, including those bound to Service for a Term of Years, and excluding Indians not taxed, three fifths of all other Persons. The actual Enumeration shall be made within three Years after the first Meeting of the Congress of the United States, and within every subsequent Term of ten Years, in such Manner as they shall by Law direct. The Number of Representatives shall not exceed one for every thirty Thousand, but each State shall have at Least one Representative; and until such enumeration shall be made, the State of New Hampshire shall be entitled to choose [sic] three, Massachusetts eight, Rhode Island and Providence Plantations one, Connecticut five, New York six, New Jersey four, Pennsylvania eight, Delaware one, Maryland six, Virginia ten, North Carolina five, South Carolina five, and Georgia three.

Clause 4: When vacancies happen in the Representation from any State, the Executive Authority thereof shall issue Writs of Election to fill such Vacancies.

Clause 5: The House of Representatives shall choose their Speaker and other Officers; and shall have the sole Power of Impeachment.

Section 3.

Clause 1: The Senate of the United States shall be composed of two Senators from each State, chosen by the Legislature thereof, for six Years; and each Senator shall have one Vote.

Clause 2: Immediately after they shall be assembled in Consequence of the first Election, they shall be divided as equally as may be into three Classes. The Seats of the Senators of the first Class shall be vacated at the Expiration of the second Year, of the second Class at the Expiration of the fourth Year, and of the third Class at the Expiration of the sixth Year, so that one third may be chosen every second Year; and if Vacancies happen by Resignation, or otherwise, during the Recess of the Legislature of any State, the Executive thereof may make temporary Appointments until the next Meeting of the Legislature, which shall then fill such Vacancies.

Clause 3: No Person shall be a Senator who shall not have attained to the Age of thirty Years, and been nine Years a Citizen of the United States, and who shall not, when elected, be an Inhabitant of that State for which he shall be chosen.

Clause 4: The Vice President of the United States shall be President of the Senate, but shall have no Vote, unless they be equally divided.

Clause 5: The Senate shall choose their other Officers, and also a President pro tempore, in the Absence of the Vice President, or when he shall exercise the Office of President of the United States.

Clause 6: The Senate shall have the sole Power to try all Impeachments. When sitting for that Purpose, they shall be on Oath or Affirmation. When the President of the United States is tried, the Chief Justice shall preside: And no Person shall be convicted without the Concurrence of two thirds of the Members present.

Clause 7: Judgment in Cases of Impeachment shall not extend further than to removal from Office, and disqualification to hold and enjoy any Office of honor, Trust or Profit under the United States: but the Party convicted shall nevertheless be liable and subject to Indictment, Trial, Judgment and Punishment, according to Law.

Section 4.

Clause 1: The Times, Places and Manner of holding Elections for Senators and Representatives, shall be prescribed in each State by the Legislature thereof; but the

Congress may at any time by Law make or alter such Regulations, except as to the Places of chusing [sic] Senators.

Clause 2: The Congress shall assemble at least once in every Year, and such Meeting shall be on the first Monday in December, unless they shall by Law appoint a different Day.

Section 5.

Clause 1: Each House shall be the Judge of the Elections, Returns and Qualifications of its own Members, and a Majority of each shall constitute a Quorum to do Business; but a smaller Number may adjourn from day to day, and may be authorized to compel the Attendance of absent Members, in such Manner, and under such Penalties as each House may provide.

Clause 2: Each House may determine the Rules of its Proceedings, punish its Members for disorderly Behavior, and, with the Concurrence of two thirds, expel a Member.

Clause 3: Each House shall keep a Journal of its Proceedings, and from time to time publish the same, excepting such Parts as may in their Judgment require Secrecy; and the Yeas and Nays of the Members of either House on any question shall, at the Desire of one fifth of those Present, be entered on the Journal.

Clause 4: Neither House, during the Session of Congress, shall, without the Consent of the other, adjourn for more than three days, nor to any other Place than that in which the two Houses shall be sitting.

Section 6.

Clause 1: The Senators and Representatives shall receive a Compensation for their Services, to be ascertained by Law, and paid out of the Treasury of the United States. They shall in all Cases, except Treason, Felony and Breach of the Peace, be privileged from Arrest during their Attendance at the Session of their respective Houses, and in going to and returning from the same; and for any Speech or Debate in either House, they shall not be questioned in any other Place.

Clause 2: No Senator or Representative shall, during the Time for which he was elected, be appointed to any civil Office under the Authority of the United States, which shall have been created, or the Emoluments whereof shall have been increased during such time; and no Person holding any Office under the United States, shall be a Member of either House during his Continuance in Office.

Section 7.

Clause 1: All Bills for raising Revenue shall originate in the House of Representatives; but the Senate may propose or concur with Amendments as on other Bills.

Clause 2: Every Bill which shall have passed the House of Representatives and the Senate, shall, before it become a Law, be presented to the President of the United States; If he approve he shall sign it, but if not he shall return it, with his Objections to that House in which it shall have originated, who shall enter the Objections at large on their Journal, and proceed to reconsider it. If after such Reconsideration two thirds of that House shall agree to pass the Bill, it shall be sent, together with the Objections, to the other House, by which it shall likewise be reconsidered, and if approved by two thirds of that House, it shall become a Law. But in all such Cases the Votes of both Houses shall be determined by yeas and Nays, and the Names of the Persons voting for and against the Bill shall be entered on the Journal of each House respectively. If any Bill shall not be returned by the President within ten Days (Sundays excepted) after it shall have been presented to him, the Same shall be a Law, in like Manner as if he had signed it, unless the Congress by their Adjournment prevent its Return, in which Case it shall not be a Law.

Clause 3: Every Order, Resolution, or Vote to which the Concurrence of the Senate and House of Representatives may be necessary (except on a question of Adjournment) shall be presented to the President of the United States; and before the Same shall take Effect, shall be approved by him, or being disapproved by him, shall be repassed by two thirds of the Senate and House of Representatives, according to the Rules and Limitations prescribed in the Case of a Bill.

Section 8.

Clause 1: The Congress shall have Power To lay and collect Taxes, Duties, Imposts and Excises, to pay the Debts and provide for the common Defence and general Welfare of the United States; but all Duties, Imposts and Excises shall be uniform throughout the United States;

Clause 2: To borrow Money on the credit of the United States;

Clause 3: To regulate Commerce with foreign Nations, and among the several States, and with the Indian Tribes;

Clause 4: To establish an uniform Rule of Naturalization, and uniform Laws on the subject of Bankruptcies throughout the United States;

Clause 5: To coin Money, regulate the Value thereof, and of foreign Coin, and fix the Standard of Weights and Measures;

Clause 6: To provide for the Punishment of counterfeiting the Securities and current Coin of the United States;

Clause 7: To establish Post Offices and Post Roads;

Clause 8: To promote the Progress of Science and useful Arts, by securing for limited Times to Authors and Inventors the exclusive Right to their respective Writings and Discoveries;

Clause 9: To constitute Tribunals inferior to the supreme Court;

Clause 10: To define and punish Piracies and Felonies committed on the high Seas, and Offenses against the Law of Nations;

Clause 11: To declare War, grant Letters of Marque and Reprisal, and make Rules concerning Captures on Land and Water;

Clause 12: To raise and support Armies, but no Appropriation of Money to that Use shall be for a longer Term than two Years;

Clause 13: To provide and maintain a Navy;

Clause 14: To make Rules for the Government and Regulation of the land and naval Forces;

Clause 15: To provide for calling forth the Militia to execute the Laws of the Union, suppress Insurrections and repel Invasions;

Clause 16: To provide for organizing, arming, and disciplining, the Militia, and for governing such Part of them as may be employed in the Service of the United States, reserving to the States respectively, the Appointment of the Officers, and the Authority of training the Militia according to the discipline prescribed by Congress;

Clause 17: To exercise exclusive Legislation in all Cases whatsoever, over such District (not exceeding ten Miles square) as may, by Cession of particular States, and the Acceptance of Congress, become the Seat of the Government of the United States, and to exercise like Authority over all Places purchased by the Consent of the Legislature of the State in which the Same shall be, for the Erection of Forts, Magazines, Arsenals, dock-Yards, and other needful Buildings;—And

Clause 18: To make all Laws which shall be necessary and proper for carrying into Execution the foregoing Powers, and all other Powers vested by this Constitution in the Government of the United States, or in any Department or Officer thereof.

Section 9.

Clause 1: The Migration or Importation of such Persons as any of the States now existing shall think proper to admit, shall not be prohibited by the Congress prior to the Year one thousand eight hundred and eight, but a Tax or duty may be imposed on such Importation, not exceeding ten dollars for each Person.

Clause 2: The Privilege of the Writ of Habeas Corpus shall not be suspended, unless when in Cases of Rebellion or Invasion the public Safety may require it.

Clause 3: No Bill of Attainder or ex post facto Law shall be passed.

Clause 4: No Capitation, or other direct, Tax shall be laid, unless in Proportion to the Census or Enumeration herein before directed to be taken.

Clause 5: No Tax or Duty shall be laid on Articles exported from any State.

Clause 6: No Preference shall be given by any Regulation of Commerce or Revenue to the Ports of one State over those of another: nor shall Vessels bound to, or from, one State, be obliged to enter, clear, or pay Duties in another.

Clause 7: No Money shall be drawn from the Treasury, but in Consequence of Appropriations made by Law; and a regular Statement and Account of the Receipts and Expenditures of all public Money shall be published from time to time.

Clause 8: No Title of Nobility shall be granted by the United States: And no Person holding any Office of Profit or Trust under them, shall, without the Consent of the Congress, accept of any present, Emolument, Office, or Title, of any kind whatever, from any King, Prince, or foreign State.

Section 10.

Clause 1: No State shall enter into any Treaty, Alliance, or Confederation; grant Letters of Marque and Reprisal; coin Money; emit Bills of Credit; make any Thing but gold and silver Coin a Tender in Payment of Debts; pass any Bill of Attainder, ex post facto Law, or Law impairing the Obligation of Contracts, or grant any Title of Nobility.

Clause 2: No State shall, without the Consent of the Congress, lay any Imposts or Duties on Imports or Exports, except what may be absolutely necessary for executing it's [sic] inspection Laws: and the net Produce of all Duties and Imposts, laid by any State on Imports or Exports, shall be for the Use of the Treasury of the United States; and all such Laws shall be subject to the Revision and Controul [sic] of the Congress.

Clause 3: No State shall, without the Consent of Congress, lay any Duty of Tonnage, keep Troops, or Ships of War in time of Peace, enter into any Agreement or Compact with another State, or with a foreign Power, or engage in War, unless actually invaded, or in such imminent Danger as will not admit of delay.

Article II

Section 1.

Clause 1: The executive Power shall be vested in a President of the United States of America. He shall hold his Office during the Term of four Years, and, together with the Vice-President, chosen for the same Term, be elected, as follows

Clause 2: Each State shall appoint, in such Manner as the Legislature thereof may direct, a Number of Electors, equal to the whole Number of Senators and Representatives to which the State may be entitled in the Congress: but no Senator or

Representative, or Person holding an Office of Trust or Profit under the United States, shall be appointed an Elector.

Clause 3: The Electors shall meet in their respective States, and vote by Ballot for two Persons, of whom one at least shall not be an Inhabitant of the same State with themselves. And they shall make a List of all the Persons voted for, and of the Number of Votes for each; which List they shall sign and certify, and transmit sealed to the Seat of the Government of the United States, directed to the President of the Senate. The President of the Senate shall, in the Presence of the Senate and House of Representatives, open all the Certificates, and the Votes shall then be counted. The Person having the greatest Number of Votes shall be the President, if such Number be a Majority of the whole Number of Electors appointed; and if there be more than one who have such Majority, and have an equal Number of Votes, then the House of Representatives shall immediately choose [sic] by Ballot one of them for President; and if no Person have a Majority, then from the five highest on the List the said House shall in like Manner chuse the President. But in choosing [sic] the President, the Votes shall be taken by States, the Representation from each State having one Vote; A quorum for this Purpose shall consist of a Member or Members from two thirds of the States, and a Majority of all the States shall be necessary to a Choice. In every Case, after the Choice of the President, the Person having the greatest Number of Votes of the Electors shall be the Vice President. But if there should remain two or more who have equal Votes, the Senate shall chuse from them by Ballot the Vice President. *(See Note 8)*

Clause 4: The Congress may determine the Time of chusing the Electors, and the Day on which they shall give their Votes; which Day shall be the same throughout the United States.

Clause 5: No Person except a natural born Citizen, or a Citizen of the United States, at the time of the Adoption of this Constitution, shall be eligible to the Office of President; neither shall any Person be eligible to that Office who shall not have attained to the Age of thirty five Years, and been fourteen Years a Resident within the United States.

Clause 6: In Case of the Removal of the President from Office, or of his Death, Resignation, or Inability to discharge the Powers and Duties of the said Office, the Same shall devolve on the Vice President, and the Congress may by Law provide for the Case of Removal, Death, Resignation or Inability, both of the President and Vice President, declaring what Officer shall then act as President, and such Officer shall act accordingly, until the Disability be removed, or a President shall be elected.

Clause 7: The President shall, at stated Times, receive for his Services, a Compensation, which shall neither be encreased nor diminished during the Period for which he shall have been elected, and he shall not receive within that Period any other Emolument from the United States, or any of them.

Clause 8: Before he enter on the Execution of his Office, he shall take the following Oath or Affirmation:—"I do solemnly swear (or affirm) that I will faithfully execute the Office of President of the United States, and will to the best of my Ability, preserve, protect and defend the Constitution of the United States."

Section 2.

Clause 1: The President shall be Commander in Chief of the Army and Navy of the United States, and of the Militia of the several States, when called into the actual Service of the United States; he may require the Opinion, in writing, of the principal Officer in each of the executive Departments, upon any Subject relating to the Duties of their respective Offices, and he shall have Power to grant Reprieves and Pardons for Offences against the United States, except in Cases of Impeachment.

Clause 2: He shall have Power, by and with the Advice and Consent of the Senate, to make Treaties, provided two thirds of the Senators present concur; and he shall nominate, and by and with the Advice and Consent of the Senate, shall appoint Ambassadors, other public Ministers and Consuls, Judges of the supreme Court, and all other Officers of the United States, whose Appointments are not herein otherwise provided for, and which shall be established by Law: but the Congress may by Law vest the Appointment of such inferior Officers, as they think proper, in the President alone, in the Courts of Law, or in the Heads of Departments.

Clause 3: The President shall have Power to fill up all Vacancies that may happen during the Recess of the Senate, by granting Commissions which shall expire at the End of their next Session.

Section 3.

He shall from time to time give to the Congress Information of the State of the Union, and recommend to their Consideration such Measures as he shall judge necessary and expedient; he may, on extraordinary Occasions, convene both Houses, or either of them, and in Case of Disagreement between them, with Respect to the Time of Adjournment, he may adjourn them to such Time as he shall think proper; he shall receive Ambassadors and other public Ministers; he shall take Care that the Laws be faithfully executed, and shall Commission all the Officers of the United States.

Section 4.

The President, Vice President and all civil Officers of the United States, shall be removed from Office on Impeachment for, and Conviction of, Treason, Bribery, or other high Crimes and Misdemeanors.

Article III

Section 1.

The judicial Power of the United States, shall be vested in one supreme Court, and in such inferior Courts as the Congress may from time to time ordain and establish. The Judges, both of the supreme and inferior Courts, shall hold their Offices during good Behaviour, and shall, at stated Times, receive for their Services, a Compensation, which shall not be diminished during their Continuance in Office.

Section 2.

Clause 1: The judicial Power shall extend to all Cases, in Law and Equity, arising under this Constitution, the Laws of the United States, and Treaties made, or which shall be made, under their Authority;—to all Cases affecting Ambassadors, other public Ministers and Consuls;—to all Cases of admiralty and maritime Jurisdiction;—to Controversies to which the United States shall be a Party;—to Controversies between two or more States;—between a State and Citizens of another State;—between Citizens of different States, —between Citizens of the same State claiming Lands under Grants of different States, and between a State, or the Citizens thereof, and foreign States, Citizens or Subjects.

Clause 2: In all Cases affecting Ambassadors, other public Ministers and Consuls, and those in which a State shall be Party, the supreme Court shall have original Jurisdiction. In all the other Cases before mentioned, the supreme Court shall have appellate Jurisdiction, both as to Law and Fact, with such Exceptions, and under such Regulations as the Congress shall make.

Clause 3: The Trial of all Crimes, except in Cases of Impeachment, shall be by Jury; and such Trial shall be held in the State where the said Crimes shall have been committed; but when not committed within any State, the Trial shall be at such Place or Places as the Congress may by Law have directed.

Section 3.

Clause 1: Treason against the United States, shall consist only in levying War against them, or in adhering to their Enemies, giving them Aid and Comfort. No Person shall be convicted of Treason unless on the Testimony of two Witnesses to the same overt Act, or on Confession in open Court.

Clause 2: The Congress shall have Power to declare the Punishment of Treason, but no Attainder of Treason shall work Corruption of Blood, or Forfeiture except during the Life of the Person attainted.

Article IV

Section 1.

Full Faith and Credit shall be given in each State to the public Acts, Records, and judicial Proceedings of every other State. And the Congress may by general Laws prescribe the Manner in which such Acts, Records and Proceedings shall be proved, and the Effect thereof.

Section 2.

Clause 1: The Citizens of each State shall be entitled to all Privileges and Immunities of Citizens in the several States.

Clause 2: A Person charged in any State with Treason, Felony, or other Crime, who shall flee from Justice, and be found in another State, shall on Demand of the executive Authority of the State from which he fled, be delivered up, to be removed to the State having Jurisdiction of the Crime.

Clause 3: No Person held to Service or Labour in one State, under the Laws thereof, escaping into another, shall, in Consequence of any Law or Regulation therein, be discharged from such Service or Labour, but shall be delivered up on Claim of the Party to whom such Service or Labour may be due.

Section 3.

Clause 1: New States may be admitted by the Congress into this Union; but no new State shall be formed or erected within the Jurisdiction of any other State; nor any State be formed by the Junction of two or more States, or Parts of States, without the Consent of the Legislatures of the States concerned as well as of the Congress.

Clause 2: The Congress shall have Power to dispose of and make all needful Rules and Regulations respecting the Territory or other Property belonging to the United States; and nothing in this Constitution shall be so construed as to Prejudice any Claims of the United States, or of any particular State.

Section 4.

The United States shall guarantee to every State in this Union a Republican Form of Government, and shall protect each of them against Invasion; and on Application of the Legislature, or of the Executive (when the Legislature cannot be convened) against domestic Violence.

Article V.

The Congress, whenever two thirds of both Houses shall deem it necessary, shall propose Amendments to this Constitution, or, on the Application of the Legislatures of two thirds of the several States, shall call a Convention for proposing Amendments, which, in either Case, shall be valid to all Intents and Purposes, as Part of this Constitution, when ratified by the Legislatures of three fourths of the several States, or by Conventions in three fourths thereof, as the one or the other Mode of Ratification may be proposed by the Congress; Provided that no Amendment which may be made prior to the Year One thousand eight hundred and eight shall in any Manner affect the first and fourth Clauses in the Ninth Section of the first Article; and that no State, without its Consent, shall be deprived of its equal Suffrage in the Senate.

Article VI

Clause 1: All Debts contracted and Engagements entered into, before the Adoption of this Constitution, shall be as valid against the United States under this Constitution, as under the Confederation.

Clause 2: This Constitution, and the Laws of the United States which shall be made in Pursuance thereof; and all Treaties made, or which shall be made, under the Authority of the United States, shall be the supreme Law of the Land; and the Judges in every State shall be bound thereby, any Thing in the Constitution or Laws of any State to the Contrary notwithstanding.

Clause 3: The Senators and Representatives before mentioned, and the Members of the several State Legislatures, and all executive and judicial Officers, both of the United States and of the several States, shall be bound by Oath or Affirmation, to support this Constitution; but no religious Test shall ever be required as a Qualification to any Office or public Trust under the United States.

Article VII.

The Ratification of the Conventions of nine States, shall be sufficient for the Establishment of this Constitution between the States so ratifying the Same.

Amendments to the Constitution

Amendment I [1791]

Congress shall make no law respecting an establishment of religion, or prohibiting the free exercise thereof; or abridging the freedom of speech, or of the press; or the right of the people peaceably to assemble, and to petition the Government for a redress of grievances.

Amendment II [1791]

A well regulated Militia, being necessary to the security of a free State, the right of the people to keep and bear Arms, shall not be infringed.

Amendment III [1791]

No Soldier shall, in time of peace be quartered in any house, without the consent of the Owner, nor in time of war, but in a manner to be prescribed by law.

Amendment IV [1791]

The right of the people to be secure in their persons, houses, papers, and effects, against unreasonable searches and seizures, shall not be violated, and no Warrants shall issue, but upon probable cause, supported by Oath or affirmation, and particularly describing the place to be searched, and the persons or things to be seized.

Amendment V [1791]

No person shall be held to answer for a capital, or otherwise infamous crime, unless on a presentment or indictment of a Grand Jury, except in cases arising in the land or naval forces, or in the Militia, when in actual service in time of War or public danger; nor shall any person be subject for the same offence to be twice put in jeopardy of life or limb; nor shall be compelled in any criminal case to be a witness against himself, nor be deprived of life, liberty, or property, without due process of law; nor shall private property be taken for public use, without just compensation.

Amendment VI [1791]

In all criminal prosecutions, the accused shall enjoy the right to a speedy and public trial, by an impartial jury of the State and district wherein the crime shall have been committed, which district shall have been previously ascertained by law, and to be informed of the nature and cause of the accusation; to be confronted with the witnesses against him; to have compulsory process for obtaining witnesses in his favor, and to have the Assistance of Counsel for his defence.

Amendment VII [1791]

In Suits at common law, where the value in controversy shall exceed twenty dollars, the right of trial by jury shall be preserved, and no fact tried by a jury, shall be otherwise re-examined in any Court of the United States, than according to the rules of the common law.

Amendment VIII [1791]

Excessive bail shall not be required, nor excessive fines imposed, nor cruel and unusual punishments inflicted.

Amendment IX [1791]

The enumeration in the Constitution, of certain rights, shall not be construed to deny or disparage others retained by the people.

Amendment X [1791]

The powers not delegated to the United States by the Constitution, nor prohibited by it to the States, are reserved to the States respectively, or to the people.

Amendment XI [1798]

The Judicial power of the United States shall not be construed to extend to any suit in law or equity, commenced or prosecuted against one of the United States by Citizens of another State, or by Citizens or Subjects of any Foreign State.

Amendment XII [1804]

The Electors shall meet in their respective states, and vote by ballot for President and Vice-President, one of whom, at least, shall not be an inhabitant of the same state with themselves; they shall name in their ballots the person voted for as President, and in distinct ballots the person voted for as Vice-President, and they shall make distinct lists of all persons voted for as President, and of all persons voted for as Vice-President, and of the number of votes for each, which lists they shall sign and certify, and transmit sealed to the seat of the government of the United States, directed to the President of the Senate;—The President of the Senate shall, in the presence of the Senate and House of Representatives, open all the certificates and the votes shall then be counted;—The person having the greatest number of votes for President, shall be the President, if such number be a majority of the whole number of Electors appointed; and if no person have such majority, then from the persons having the highest numbers not exceeding three on the list of those voted for as President, the House of Representatives shall choose immediately, by ballot, the President. But in choosing the President, the votes shall be taken by states, the representation from each state having one vote; a quorum for this purpose shall consist of a member or members from two-thirds of the states, and a majority of all the states shall be necessary to a choice. And if the House of Representatives shall not choose a President whenever the right of choice shall devolve upon them, before the fourth day of March next following, then the Vice-President shall act as President, as in the case of the death or other constitutional disability of the President. The person having the greatest number of votes as Vice-President, shall be the Vice-President, if such number be a majority of the whole number of Electors appointed, and if no person have a majority, then from the two highest numbers on the list, the Senate shall choose the Vice-President; a quorum for the purpose shall consist of two-thirds of the whole number of Senators, and a majority of the whole number shall be necessary to a choice. But no person constitutionally ineligible to the office of President shall be eligible to that of Vice-President of the United States.

Amendment XIII [1865]

Section 1. Neither slavery nor involuntary servitude, except as a punishment for crime whereof the party shall have been duly convicted, shall exist within the United States, or any place subject to their jurisdiction.

Section 2. Congress shall have power to enforce this article by appropriate legislation.

Amendment XIV [1868]

Section 1. All persons born or naturalized in the United States, and subject to the jurisdiction thereof, are citizens of the United States and of the State wherein they reside. No State shall make or enforce any law which shall abridge the privileges or immunities of citizens of the United States; nor shall any State deprive any person of life, liberty, or property, without due process of law; nor deny to any person within its jurisdiction the equal protection of the laws.

Section 2. Representatives shall be apportioned among the several States according to their respective numbers, counting the whole number of persons in each State, excluding Indians not taxed. But when the right to vote at any election for the choice of electors for President and Vice President of the United States, Representatives in Congress, the Executive and Judicial officers of a State, or the members of the Legislature thereof, is denied to any of the male inhabitants of such State, being twenty-one years of age, and citizens of the United States, or in any way abridged, except for participation in rebellion, or other crime, the basis of representation therein shall be reduced in the proportion which the number of such male citizens shall bear to the whole number of male citizens twenty-one years of age in such State.

Section 3. No person shall be a Senator or Representative in Congress, or elector of President and Vice President, or hold any office, civil or military, under the United States, or under any State, who, having previously taken an oath, as a member of Congress, or as an officer of the United States, or as a member of any State legislature, or as an executive or judicial officer of any State, to support the Constitution of the United States, shall have engaged in insurrection or rebellion against the same, or given aid or comfort to the enemies thereof. But Congress may by a vote of two-thirds of each House, remove such disability.

Section 4. The validity of the public debt of the United States, authorized by law, including debts incurred for payment of pensions and bounties for services in suppressing insurrection or rebellion, shall not be questioned. But neither the United States nor any State shall assume or pay any debt or obligation incurred in aid of insurrection or rebellion against the United States, or any claim for the loss or emancipation of any slave; but all such debts, obligations and claims shall be held illegal and void.

Section 5. The Congress shall have power to enforce, by appropriate legislation, the provisions of this article.

Amendment XV [1870]

Section 1. The right of citizens of the United States to vote shall not be denied or abridged by the United States or by any State on account of race, color, or previous condition of servitude.

Section 2. The Congress shall have power to enforce this article by appropriate legislation.

Amendment XVI [1913]

The Congress shall have power to lay and collect taxes on incomes, from whatever source derived, without apportionment among the several States, and without regard to any census or enumeration.

Amendment XVII [1913]

The Senate of the United States shall be composed of two Senators from each State, elected by the people thereof, for six years; and each Senator shall have one vote. The electors in each State shall have the qualifications requisite for electors of the most numerous branch of the State legislatures.

When vacancies happen in the representation of any State in the Senate, the executive authority of such State shall issue writs of election to fill such vacancies: Provided, That the legislature of any State may empower the executive thereof to make temporary appointments until the people fill the vacancies by election as the legislature may direct.

This amendment shall not be so construed as to affect the election or term of any Senator chosen before it becomes valid as part of the Constitution.

Amendment XVIII [1919]

Section 1. After one year from the ratification of this article the manufacture, sale, or transportation of intoxicating liquors within, the importation thereof into, or the exportation thereof from the United States and all territory subject to the jurisdiction thereof for beverage purposes is hereby prohibited.

Section. 2. The Congress and the several States shall have concurrent power to enforce this article by appropriate legislation.

Section. 3. This article shall be inoperative unless it shall have been ratified as an amendment to the Constitution by the legislatures of the several States, as provided in the Constitution, within seven years from the date of the submission hereof to the States by the Congress.

Amendment XIX [1920]

The right of citizens of the United States to vote shall not be denied or abridged by the United States or by any State on account of sex.

Congress shall have power to enforce this article by appropriate legislation.

Amendment XX [1933]

Section 1. The terms of the President and Vice President shall end at noon on the 20th day of January, and the terms of Senators and Representatives at noon on the 3d day of January, of the years in which such terms would have ended if this article had not been ratified; and the terms of their successors shall then begin.

Section. 2. The Congress shall assemble at least once in every year, and such meeting shall begin at noon on the 3d day of January, unless they shall by law appoint a different day.

Section. 3. If, at the time fixed for the beginning of the term of the President, the President elect shall have died, the Vice President elect shall become President. If a President shall not have been chosen before the time fixed for the beginning of his term, or if the President elect shall have failed to qualify, then the Vice President elect shall act as President until a President shall have qualified; and the Congress may by law provide for the case wherein neither a President elect nor a Vice President elect shall have qualified, declaring who shall then act as President, or the manner in which one who is to act shall be selected, and such person shall act accordingly until a President or Vice President shall have qualified.

Section. 4. The Congress may by law provide for the case of the death of any of the persons from whom the House of Representatives may choose a President whenever the right of choice shall have devolved upon them, and for the case of the death of any of the persons from whom the Senate may choose a Vice President whenever the right of choice shall have devolved upon them.

Section. 5. Sections 1 and 2 shall take effect on the 15th day of October following the ratification of this article.

Section. 6. This article shall be inoperative unless it shall have been ratified as an amendment to the Constitution by the legislatures of three-fourths of the several States within seven years from the date of its submission.

Amendment XXI [1933]

Section 1. The eighteenth article of amendment to the Constitution of the United States is hereby repealed.

Section 2. The transportation or importation into any State, Territory, or possession of the United States for delivery or use therein of intoxicating liquors, in violation of the laws thereof, is hereby prohibited.

Section 3. This article shall be inoperative unless it shall have been ratified as an amendment to the Constitution by conventions in the several States, as provided in the Constitution, within seven years from the date of the submission hereof to the States by the Congress.

Amendment XXII [1951]

Section 1. No person shall be elected to the office of the President more than twice, and no person who has held the office of President, or acted as President, for more than two years of a term to which some other person was elected President shall be elected to the office of the President more than once. But this article shall not apply to any person holding the office of President when this article was proposed by the Congress, and shall not prevent any person who may be holding the office of President, or acting as President, during the term within which this article becomes operative from holding the office of President or acting as President during the remainder of such term.

Section 2. This article shall be inoperative unless it shall have been ratified as an amendment to the Constitution by the legislatures of three-fourths of the several states within seven years from the date of its submission to the states by the Congress.

Amendment XXIII [1961]

Section 1. The District constituting the seat of government of the United States shall appoint in such manner as the Congress may direct:

A number of electors of President and Vice President equal to the whole number of Senators and Representatives in Congress to which the District would be entitled if it were a state, but in no event more than the least populous state; they shall be in addition to those appointed by the states, but they shall be considered, for the purposes of the election of President and Vice President, to be electors appointed by a state; and they shall meet in the District and perform such duties as provided by the twelfth article of amendment.

Section 2. The Congress shall have power to enforce this article by appropriate legislation.

Amendment XXIV [1964]

Section 1. The right of citizens of the United States to vote in any primary or other election for President or Vice President, for electors for President or Vice President, or for Senator or Representative in Congress, shall not be denied or abridged by the United States or any state by reason of failure to pay any poll tax or other tax.

Section 2. The Congress shall have power to enforce this article by appropriate legislation.

Amendment XXV [1967]

Section 1. In case of the removal of the President from office or of his death or resignation, the Vice President shall become President.

Section 2. Whenever there is a vacancy in the office of the Vice President, the President shall nominate a Vice President who shall take office upon confirmation by a majority vote of both Houses of Congress.

Section 3. Whenever the President transmits to the President pro tempore of the Senate and the Speaker of the House of Representatives his written declaration that he is unable to discharge the powers and duties of his office, and until he transmits to them a written declaration to the contrary, such powers and duties shall be discharged by the Vice President as Acting President.

Section 4. Whenever the Vice President and a majority of either the principal officers of the executive departments or of such other body as Congress may by law provide, transmit to the President pro tempore of the Senate and the Speaker of the House of Representatives their written declaration that the President is unable to discharge the powers and duties of his office, the Vice President shall immediately assume the powers and duties of the office as Acting President.

Thereafter, when the President transmits to the President pro tempore of the Senate and the Speaker of the House of Representatives his written declaration that no inability exists, he shall resume the powers and duties of his office unless the Vice President and a majority of either the principal officers of the executive department or of such other body as Congress may by law provide, transmit within four days to the President pro tempore of the Senate and the Speaker of the House of Representatives their written declaration that the President is unable to discharge the powers and duties of his office. Thereupon Congress shall decide the issue, assembling within forty-eight hours for that purpose if not in session. If the Congress, within twenty-one days after receipt of the latter written declaration, or, if Congress is not in session, within twenty-one days after Congress is required to assemble, determines by two-thirds vote of both Houses that the President is unable to discharge the powers and duties of his office, the Vice President shall continue to discharge the same as Acting President; otherwise, the President shall resume the powers and duties of his office.

Amendment XXVI [1971]

Section 1. The right of citizens of the United States, who are 18 years of age or older, to vote, shall not be denied or abridged by the United States or any state on account of age.

Section 2. The Congress shall have the power to enforce this article by appropriate legislation.

Amendment XXVII [1992]

No law varying the compensation for the services of the Senators and Representatives shall take effect until an election of Representatives shall have intervened.

Appendix C

NAS ASRS Form 277B

DO NOT REPORT AIRCRAFT ACCIDENTS AND CRIMINAL ACTIVITIES ON THIS FORM.
ACCIDENTS AND CRIMINAL ACTIVITIES ARE NOT INCLUDED IN THE ASRS PROGRAM AND SHOULD NOT BE SUBMITTED TO NASA.
ALL IDENTITIES CONTAINED IN THIS REPORT WILL BE REMOVED TO ASSURE COMPLETE REPORTER ANONYMITY.

(SPACE BELOW RESERVED FOR ASRS DATE/TIME STAMP)

IDENTIFICATION STRIP: *Please fill in all blanks to ensure return of strip.*
NO RECORD WILL BE KEPT OF YOUR IDENTITY. This section will be returned to you.

TELEPHONE NUMBERS where we may reach you for further details of this occurrence:

HOME   Area _____   No. _____   Hours _____
WORK   Area _____   No. _____   Hours _____

NAME _____   TYPE OF EVENT/SITUATION _____
ADDRESS/PO BOX _____   _____
_____   DATE OF OCCURRENCE _____
CITY _____ STATE ____ ZIP _____ LOCAL TIME (24 hr. clock) _____

PLEASE FILL IN APPROPRIATE SPACES AND CHECK ALL ITEMS WHICH APPLY TO THIS EVENT OR SITUATION.

REPORTER	FLYING TIME	CERTIFICATES/RATINGS		ATC EXPERIENCE	
o Captain	total _____ hrs.	o student	o private	o FPL	o Developmental
First Officer		o commercial	o ATP	radar _____ yrs.	
o pilot flying	last 90 days _____ hrs.	o instrument	o CFI	non-radar _____ yrs.	
o pilot not flying		o multiengine	o F/E	supervisory _____ yrs.	
o Other Crewmember	time in type _____ hrs.	o _____		military _____ yrs.	
o _____					

AIRSPACE		WEATHER		LIGHT/VISIBILITY		ATC/ADVISORY SERV.	
o Class A (PCA)	o Special Use Airspace	o VMC	o ice	o daylight	o night	o local	o center
o Class B (TCA)	o airway/route _____	o IMC	o snow	o dawn	o dusk	o ground	o FSS
o Class C (ARSA)	o unknown/other _____	o mixed	o turbulence	ceiling _____ feet		o apch	o UNICOM
o Class D (Control Zone/ATA)	_____	o marginal	o tstorm	visibility _____ miles		o dep	o CTAF
o Class E (General Controlled)	_____	o rain	o windshear	RVR _____ feet		Name of ATC Facility:	
o Class G (Uncontrolled)		o fog	o _____				

	AIRCRAFT 1			AIRCRAFT 2		
Type of Aircraft (Make/Model)	(Your Aircraft) _____	o EFIS		(Other Aircraft) _____	o EFIS	
		o FMS/FMC			o FMS/FMC	
Operator	o air carrier	o military	o corporate	o air carrier	o military	o corporate
	o commuter	o private	o other _____	o commuter	o private	o other _____
Mission	o passenger	o training	o business	o passenger	o training	o business
	o cargo	o pleasure	o unk/other _____	o cargo	o pleasure	o unk/other _____
Flight plan	o VFR	o SVFR	o none	o VFR	o SVFR	o none
	o IFR	o DVFR	o unknown	o IFR	o DVFR	o unknown
Flight phases at time of occurrence	o taxi	o cruise	o landing	o taxi	o cruise	o landing
	o takeoff	o descent	o missed apch/GAR	o takeoff	o descent	o missed apch/GAR
	o climb	o approach	o other _____	o climb	o approach	o other _____
Control status	o visual apch	o on vector	o on SID/STAR	o visual apch	o on vector	o on SID/STAR
	o controlled	o none	o unknown	o controlled	o none	o unknown
	o no radio	o radar advisories		o no radio	o radar advisories	

If more than two aircraft were involved, please describe the additional aircraft in the "Describe Event/Situation" section.

LOCATION	CONFLICTS		
Altitude _____ o MSL o AGL	Estimated miss distance in feet: horiz _____ vert _____		
Distance and radial from airport, NAVAID, or other fix _____	Was evasive action taken?	o Yes	o No
_____	Was TCAS a factor? TA o RA		o No
Nearest City/State _____	Did GPWS activate?	o Yes	o No

NASA ARC 277B (January 1994) **GENERAL FORM**

NAS ASRS Form 277B | 323

NATIONAL AERONAUTICS AND SPACE ADMINISTRATION

NASA has established an Aviation Safety Reporting System (ASRS) to identify issues in the aviation system which need to be addressed. The program of which this system is a part is described in detail in FAA Advisory Circular 00-46D. Your assistance in informing us about such issues is essential to the success of the program. Please fill out this form as completely as possible, enclose in an sealed envelope, affix proper postage, and and send it directly to us.

The information you provide on the identity strip will be used only if NASA determines that it is necessary to contact you for further information. THIS IDENTITY STRIP WILL BE RETURNED DIRECTLY TO YOU. The return of the identity strip assures your anonymity.

AVIATION SAFETY REPORTING SYSTEM

Section 91.25 of the Federal Aviation Regulations (14 CFR 91.25) prohibits reports filed with NASA from being used for FAA enforcement purposes. This report will not be made available to the FAA for civil penalty or certificate actions for violations of the Federal Air Regulations. Your identity strip, stamped by NASA, is proof that you have submitted a report to the Aviation Safety Reporting System. We can only return the strip to you, however, if you have provided a mailing address. Equally important, we can often obtain additional useful information if our safety analysts can talk with you directly by telephone. For this reason, we have requested telephone numbers where we may reach you.

Thank you for your contribution to aviation safety.

NOTE: AIRCRAFT ACCIDENTS SHOULD NOT BE REPORTED ON THIS FORM. SUCH EVENTS SHOULD BE FILED WITH THE NATIONAL TRANSPORTATION SAFETY BOARD AS REQUIRED BY NTSB Regulation 830.5 (49CFR830.5).

Please fold both pages (and additional pages if required), enclose in a sealed, stamped envelope, and mail to:

NASA AVIATION SAFETY REPORTING SYSTEM
POST OFFICE BOX 189
MOFFETT FIELD, CALIFORNIA 94035-0189

DESCRIBE EVENT/SITUATION

Keeping in mind the topics shown below, discuss those which you feel are relevant and anything else you think is important. Include what you believe really caused the problem, and what can be done to prevent a recurrence, or correct the situation. (USE ADDITIONAL PAPER IF NEEDED)

CHAIN OF EVENTS
- How the problem arose
- Contributing factors
- How it was discovered
- Corrective actions

HUMAN PERFORMANCE CONSIDERATIONS
- Perceptions, judgments, decisions
- Actions or inactions
- Factors affecting the quality of human performance

Selected Bibliography

Anthony, Robert A. *Symposium on the 50th Anniversary of the APA: The Supreme Court and the APA: Sometimes They Just Don't Get It*, 10 Admin. L. J. Am. U. 1 (Spring 1996).

CCH Editorial Staff. *Aviation Law Reporter.* 4 vols. Chicago: CCH Incorporated, November 2001.

Elder, Bill, Comment: *Free Flight: The Future of Air Transportation Entering the Twenty-First Century*, 62 J. Air L. & Com. 871 (February/March 1997).

Holmes, Eric Mills. *Holmes's Appleman on Insurance*, vol. 1, 2d ed. Minneapolis, MN: West Publishing Co., 1996.

Krause, Charles F., and Krause, Kent C. *Aviation Tort and Regulatory Law*, 2d ed. Eagan, MN: West Group, 2002.

Kuchta, Joseph D. *Federal Aviation Decisions*, 5 vols. New York: Clark Boardman Callaghan, 1999.

McKay, Jennifer. *Notes: The Refinement of the Warsaw System: Why the 1999 Montreal Convention Represents the Best Hope for Uniformity*, 34 Case W. Res. J. Int'l L. 73 (Fall 2002).

Nelson, Deborah L., and Howicz, Jennifer L. *Williston on Sales*, vol. 1, 5th ed. New York: Clark Boardman Callaghan, 2000.

Pike & Fischer, Inc. *Uniform Commercial Code Case Digest.* New York: Clark Boardman Callaghan, 1995.

NTSB Bar Association, Select Comm. On Aviation Policy. *Aviation Professionals and the Threat of Criminal Liability—How Do We Maximize Aviation Safety?* 67 J. Air L. & Com. 3 (Summer 2002).

Rollo, Vera Foster. *Aviation Law: An Introduction.* Lanham, MD: Maryland Historical Press, 1985.

Rollo, Vera Foster. *Aviation Insurance.* Lanham, MD: Maryland Historical Press, 1986.

Speiser, Stuart M., and Krause, Charles F. *Aviation Tort Law.* Eagan, MN: West Group, 2000.

Yodice, John S. *Aviation Lawyer's Manual: Representing the Pilot in FAA Enforcement Actions.* Lanham, MD: Maryland Historical Press, 1986.

Index

Absolute liability crimes, 69
Acceptance (of contract), 161
Accidents:
   definition of, 284, 287–289
   NTSB investigation of, 123
Accused, rights and protections
   for, 73
ACs (Advisory Circulars), 125
Acts, criminal, 69
Actual cause test, 100
Actus reas, 69
ADA (Americans with Disabilities
   Act), 265
ADAP (Airport Development Aid
   Program), 223
Adaptability of legal system, 10
ADEA (Age Discrimination in
   Employment Act), 265
Adjudication, 123, 131–133
Administrative actions (FAA), 137
Administrative agencies, 119–153
   adjudication by, 131–133
   *Air Transport Association of America
      v. DOT and FAA*, 143–145
   authority for creation of, 5
   checks on, 138–142
   delegation of authority to, 41
   Department of Transportation, 122
   enforcement by, 126–131
   Federal Aviation Administration,
      122–123, 133–138
   functions of, 121, 123–124
   Garvey v. NTSB and Merrell,
      145–149
   interdependence of functions
      within, 123–124
   National Transportation Safety
      Board, 123
   and non-instrument-rated pilot
      problem, 149–153
   NTSB Identification: NYC05FA001,
      150–153

Administrative agencies (*Cont.*):
   power of, 5
   reasons for establishing, 120–121
   rulemaking by, 124–125,
      142–145
   and *stare decisis* doctrine, 4
   state, 120
   and "validly adopted" regulations,
      145–149
Administrative law judges (ALJs),
   130, 132
Administrative Procedure Act (APA)
   of 1946, 124–125
Administrator v. Dailey, 132
Administrator v. Parker, 132
Administrator v. Ramaprakash, 129
Adverse possession, 220
Advisory Circulars (ACs), 125
*Aeronautical Information Manual
   (AIM)*, 125
Aeronautics Branch, Department of
   Commerce, 122
Age 60 rule, 43, 272–276
Age Discrimination in Employment
   Act (ADEA), 265
Agency law, 254–261
   contract liability, 258–259
   creating/terminating relationships,
      255–256
   duties of parties, 257–258
   tort liability, 259–261
   and types of agency relationship, 255
Agreements for sale, 182–185
*AIM (Aeronautical Information
   Manual)*, 125
AIP (Airport Improvement
   Program), 223
Air Commerce Act of 1926, 122
Air France v. Saks, 288–289
Air Transport Association, 142
*Air Transport Association of America
   v. DOT and FAA*, 142–145

Aircraft:
   leasing, 228–229
   multiple ownership of, 229–230
   registering, 225, 227
   storing, 225, 228
   transferring ownership of, 224–225
Aircraft noise regulation, 56–58
Aircraft Owners and Pilots
   Association, 142
Airman/pilot certificates, 83–85,
   133–135
Airport Development Aid Program
   (ADAP), 223
Airport Improvement Program
   (AIP), 223
Airport proprietor exception, 58–59
Airports, public use, 223–224
ALI (*see* American Law Institute)
ALJs (*see* Administrative law judges)
Alternative dispute resolution, 25–26
Altseimer v. Bell, 112–114
Amendments to the Constitution, 49,
   312–318
American Law Institute (ALI), 69,
   93, 160
Americans with Disabilities Act
   (ADA), 265
Answer, 18–20
APA (*see* Administrative Procedure
   Act of 1946)
Appeals, 23–24
   of administrative agency actions,
      138–139
   courts of, 11–13 (*See also* Federal
      Court of appeals)
   in criminal cases, 71
   of FAA orders, 131, 132
"Arbitrary or capricious" test, 139
Arbitration, 25, 262
Arraignment, 71
Arrests, 70
Articles of Confederation, 44

Index

Artificial person, 202
Artisans' liens, 177
ASAP (Aviation Safety Action Program), 137
ASRP (*see* Aviation Safety Reporting Program)
Assault, 93–94
Assets, ownership of, 196
Assumption of risk, 104, 106
Attachment jurisdiction, 16
Authority, distribution of, 41–49
  and commerce clause, 44–48
  and federal preemption, 42–44
  under federalism system, 41
  and full faith and credit clause, 48–58
  to U.S. Congress, 41–42
Availability of legal system, 10
Aviation Safety Action Program (ASAP), 137
Aviation Safety Reporting Program (ASRP), 135–136
Aviation Safety Reports, 136, 319–321

Bailments, 228
Bankruptcy, 177–179, 185–187
Battery, 94
Bermuda I agreement, 283
Bermuda II agreement, 283
Betts v. Brady, 73
Beyond rights, skies as, 283
Bill of Rights, 49
Blackstone, Sir William, 71
The Board (*see* National Transportation Safety Board)
Boards of directors, 204
Borrowing (government functions), 40–41
Braniff Airways, Inc. v. Nebraska State Board of Equalization and Assessments et al., 59–62
Breach of contract, 168–170
Breach of duty, 92, 98–99
British Airways Board v. Port Authority of New York, 58, 59
Brockelsby v. United States of America and Jeppesen, 108–112
Brown v. Board of Education of Topeka, 52

Burden of proof, 92
Bureau of Air Commerce, 122
Burke v. Pan World Airways, Inc., 97–98
Business entities, 191–212
  and corporate opportunity doctrine, 208–211
  corporations, 202–207
  Klinicki v. Lundgren, 208–211
  limited liability companies, 200–202
  limited partnerships, 197–200
  and line between business and personal matters, 211–212
  Nelsen v. Morris, 211–212
  partnerships, 194–197
  sole proprietorships, 192–194
Business insurance, 235
Businesses, basic rights of, 49–52
But for rule, 100

CAA (*see* Civil Aeronautics Administration)
CAB (*see* Civil Aeronautics Board)
California v. Ciraolo, 85–88
Capacity (contracts), 162
Cape Town Convention, 286
Capital punishment, 73
Care, duty of, 97–98
Careless operations laws, 76
Case briefs, 299–302
Case law, 4, 26–27
Causal connection (property law), 243–247
Causation in fact test, 100
Certificates, 83
  FAA inspection of, 127
  revocation of, 133–135
Cessna Aircraft Company, 114
CFR (Code of Federal Regulations), 124
Chattel, 96
Checks and balances, 39–40, 138–140
Chicago Convention of 1944, 282–283, 287
Chiron v. NTSB, 153
City of Burbank et al. v. Lockheed Air Terminal, Inc. et al., 43, 56–58
Civil Aeronautics Act of 1938, 122

Civil Aeronautics Administration (CAA), 122–123
Civil Aeronautics Authority, 122
Civil Aeronautics Board (CAB), 122, 123
Civil law, 7, 8
Civil penalty laws, 142
Civil procedure, 17
Civil Rights Act of 1964, 264
Clinton v. City of New York, 41
Code (statutory) law, 4–5
Code of Federal Regulations (CFR), 124
Collective bargaining, 261–263
Comity, doctrine of, 49
Commentaries on the Laws of England (Sir William Blackstone), 71
Commerce clause, 44–48
Commercial law, 159–187
  and agreements for sale, 182–185
  and bankruptcy, 185–187
  contracts, 160–170
  Dallas Aerospace, Inc. v. CIS Air Corporation, 180–182
  debtor-creditor issues, 174–179
  Edward Miles, Richard W. Keenan and Kenneth L. "Dusty" Burrow, Appellants, v. John F. Kavanaugh, Appellee, 183–185
  In re: UAL Corporation, et al., Debtors. Chapter 11, Case No. 02-B-48191 (Jointly Administered), 185–187
  sales law, 170–174
  warranties, 179–182
Commercial speech, 52
Common (case) law, 4, 5, 26–27
Community property, 220
Comparative negligence (fault, responsibility), 103–104, 106–107
Compensatory damages, 169
Complaint, 17–18
Complete Auto Transit, Inc. v. Brady, 47–48, 62
Conciliation, 25–26
Concurrent jurisdiction, 14
Concurrent ownership, 219

Index

Congress of the United States:
  authority of, 41–42
  commerce regulation by, 44–46
  control of administrative agencies by, 140
  delegation of authority by, 41
  express preemption by, 43
Congressional Review Act of 1996, 140
Consequences, foreseeable, 100–101
Consequential damages, 169
Consideration (contracts), 161–162
The Constitution of the United States of America, 37–62, 303–318
  and aircraft noise regulation, 56–58
  amendments to, 312–318
  Articles of, 39, 303–312
  authority for, 39
  basic rights of individuals and businesses, 49–52
  Braniff Airways, Inc. v. Nebraska State Board of Equalization and Assessments et al., 59–62
  checks and balances under, 39–40
  City of Burbank et al. v. Lockheed Air Terminal, Inc. et al., 56–58
  commerce clause, 42–44
  court system under, 11–12
  distribution of authority between federal and state governments, 41–49
  full faith and credit clause, 48–49
  and long-arm statutes, 16
  and priority of laws, 5
  protections for defendants, 71–73
  as source of law, 3
  structure and organization of the federal government, 39–41
  supremacy clause, 42–43
  and "taking" of property, 52–56
  and taxation, 59–62
  United States v. Causby et ux., 53–56
Constitutionality of laws, 39
Constitutions, state, 3–5, 38
Consumer debt adjustment (Chapter 13) bankruptcy, 179, 180
Contract liability, 258–259

Contracts, 160–170
  capacity requirement for, 162
  consideration requirement for, 161–162
  defenses to, 164–166
  defined, 160
  insurance, 230, 232–234
  lawfulness requirement for, 162–164
  mutual agreement requirement for, 160–161
  of partnerships, 195
  remedies for breach of, 168–170
  rights and duties of, 167–168
  sales, 171
  written vs. oral, 166
Contributory negligence, 103, 106
Conventions, 280 (*See also specific conventions*)
Conversion of personal property, 96–97
Coownership, 194
Corporate opportunity doctrine, 208–211
Corporations, 202–207
  duties in, 204–205
  formation of, 203
  liability issues for, 205–206
  operating, 204
  taxation of, 206–207
Counterclaims, 19
Courts, 11–13
  appellate court authority, 4
  federal, 11–13 (*See also* United States Supreme Court)
  state, 12–13
  (*See also specific cases*)
Covenants not to compete, 163
Credit, 175–177
Criminal law, 7–8, 67–88
  acts and intents in, 69
  affecting aviation activities, 74–76
  and airman/pilot certificates, 83–85
  California v. Ciraolo, 85–88
  classifications of, 68–69
  constitutional protections for defendants, 71–73
  criminal procedure under, 69–71
  and evidence obtained by flying over private property, 85–88
  federal, 74–76

Criminal law (*Cont.*):
  and levels of aviation regulation, 76–79
  and regulatory vs. criminal violations, 79–83
  Robert David Ward v. State of Maryland, 76–79
  state, 76
  tort law vs., 92
  United States of America v. SabreTech, 80–83
  United States v. Evinger, 83–85
Criminal procedure, 69–71
Criminal violations, 79–83
Cruel or unusual punishments, 73
Custom (international law), 281

Dallas Aerospace, Inc. v. CIS Air Corporation, 180–182
Damages:
  monetary (contracts), 169
  redress for (*see* Tort law)
David E. Hollins v. Delta Airlines, 269–272
De novo review, 139
Debtor-creditor issues, 174–179
Defamation, 52, 95
Defendants, protections for, 71–73
Delta Airlines, 179
Department of Commerce, 122
Department of Labor (DOL), 120
Department of Transportation (DOT), 120–122
  FAA under, 123
  NTSB independence from, 123
  open skies initiative of, 283
  rulemaking by, 123
Department of Transportation (DOT) Act, 122, 123
Depositions, 21
Direct conflict, preemption by, 43
Disability insurance, 234
Disclaimers, 107
Disclosure of information, 127
Discovery, 21
Discrimination:
  employment, 263–269
  against interstate commerce, 46
  racial, 269–272
  in "undue burden" cases, 47–48

Dispute resolution, 11–26
  alternative methods for, 25–26
  in court system, 11–13
  jurisdictional matters, 13–17
  litigation process, 17–25
Disputes, 3
District courts, federal, 11, 27
Documents:
  FAA inspection of, 127
  falsifying, 74
DOL (Department of Labor), 120
DOT (*see* Department of Transportation)
Double jeopardy, 72
Drug testing, 267
Dry leases, 228, 229
Due process, 50–51
Duress:
  as defense to contract, 166
Duties, contract, 167–168
Duty, breach of, 92, 98–99
Duty of care, 97–98, 205

EAJA (Equal Access to Justice Act), 140
Easements, 221
Eastern Airlines, 75, 179
Edward Miles, Richard W. Keenan and Kenneth L. "Dusty" Burrow, Appellants, v. John F. Kavanaugh, Appellee, 183–185
Eighteenth Amendment, 315–316
Eighth Amendment, 73, 313
EIRs (enforcement investigative reports), 130
El Al Israel Airlines, Ltd. v. Tsui Yuan Tseng, 289–293
Eleventh Amendment, 313
Emergency orders (FAA), 133–135
Eminent domain, 50, 51
Emotional distress:
  intentional infliction of, 94–95
  negligent infliction of, 102
Employee protection laws, 263–269
  Age Discrimination in Employment Act, 265
  Americans with Disabilities Act, 265
  Equal Pay Act, 264–265
  Fair Employment Practices Act, 264

Employee protection laws (*Cont.*):
  Family and Medical Leave Act of 1993, 168–169
  Occupational Safety and Health Act of 1970, 266
  privacy, 266–267
  Social Security, 268
  unemployment insurance, 268
  workers' compensation, 267–268
Employees:
  and labor unions, 261–263
  laws protecting, 263–269
Employer-employee relationships, 254, 255, 259, 260
Employment law, 253–276
  and Age 60 rule, 272–276
  agency law, 254–261
  David E. Hollins v. Delta Airlines, 269–272
  employee protection, 263–269
  and labor unions, 261–263
  Professional Pilots Federation, et al. v. FAA, 273–276
  and racial discrimination, 269–272
Enforcement, 123
  flow of, 134
  of judgments, 24–15
Enforcement investigative reports (EIRs), 130
Enron, 210
Entities, business (*see* Business entities)
Environmental Protection Agency (EPA), 120
Equal Access to Justice Act (EAJA), 140
Equal Pay Act, 264–265
Equal protection, 51–52
Equitable damages, 170
Escrow, 239–240
Ethics, law and morals vs., 8, 9
Evidence obtained by flying over private property, 85–88
Exclusionary rule, 72
Executive branch (U.S. government), 39, 139
Express preemption, 43, 62
Express warranties, 172

FAA (*see* Federal Aviation Administration)

FAAP (Federal Airport Aid Program), 223
Fair Employment Practices Act, 264
Fairness of legal system, 10
False imprisonment, 94
Falsifying documents, 74
Family and Medical Leave Act (FMLA) of 1993, 168–169
FARs (*see* Federal Aviation Regulations)
Fault, standards of, 69, 70
Federal Airport Aid Program (FAAP), 223
Federal Arbitration Act, 25
Federal Aviation Act of 1958, 122–123
Federal Aviation Administration (FAA), 122–123, 133–138
  and adjudication of cases, 131–133
  administrative actions by, 137
  aircraft leasing requirements of, 229
  aircraft registry, 224–225
  airport improvement programs, 223–224
  Aviation Safety Action Program of, 137
  civil penalties from, 142
  emergency actions by, 133–135
  enforcement investigative reports of, 130
  impact of, 5, 120
  interpretations of rules by, 125
  investigations by, 127–131, 134
  letters of investigation from, 127–128
  NASA ASRP, 135–136
  Notice of Proposed Certificate Action from, 128–130
  and reexamination for compliance, 138
  regulation by, 120
  regulations of, 27, 124
  rule-making power of, 41, 123
Federal Aviation Agency, 123
Federal Aviation Regulations (FARs), 124, 127, 224
Federal Communications Commission, 142
Federal (U.S.) Court of Appeals:
  appeals of administrative agency actions, 138

Federal (U.S.) Court of Appeals (*Cont.*):
   FAA filings with, 133
   reporters for, 26
Federal court system, 11–13, 39–40
Federal criminal law, 74–76
Federal government:
   delegation of authority to, 41–42
   distribution of authority between state governments and, 41–49
   limits on authority of, 49–52
   overlapping function within, 40–41
   structure and organization of, 39–41
Federal jurisdiction, 28–31
Federal law, 7
Federal preemption, 42–44
Federal statutes, priority of laws and, 5
Federal Trade Commission (FTC), 120, 142
Fee simple absolute, 219
Felonies, 8, 68
Fiduciary duty, 204
Fiduciary relationship, 254
Fifteenth Amendment, 315
Fifth Amendment, 49–51, 72, 312–313
First Amendment, 52, 312
First Conference on Private Air Law (Paris), 284
FMLA (*see* Family and Medical Leave Act of 1993)
FOIA (*see* Freedom of Information Act)
Foreseeability test, 100–101
Form 277B (NASA ASRS), 136, 319–321
Forms of business (*see* Business entities)
Fort Gratiot Sanitary Landfill, Inc. v. Michigan Department of Natural Resources, 46
Fourteenth Amendment, 49–52, 314–315
Fourth Amendment, 71–72, 312
Fractional ownership, 229–230
Fraud:
   as defense to contract, 165
   mail and wire, 74–75
Fraudulent misrepresentation, 165
Freedom of Information Act (FOIA), 141, 142

Freedom of press, 52
Freedom of speech, 52
FTC (*see* Federal Trade Commission)
FTC v. Ruberoid Co., 120
Full faith and credit clause, 48–49
Functions of law, 2–3

GARA (*see* General Aviation Revitalization Act)
Garnishments, 24
Garvey v. NTSB and Merrell, 145–149
General Aviation Revitalization Act (GARA), 112, 114
General law, 281
General partnerships (*see* Partnerships)
Gideon v. Wainwright, 73
Government, protecting process of, 3
Grand jury indictments, 70
Griggs v. Allegheny County, 56
Guarantees, 177
Gun Free School Zone Act of 1990, 45

Hague Protocol of 1955, 295
Hamlet (William Shakespeare), 174
Havana Convention of 1928, 281, 282
Hazardous Materials Transportation Act, 75
Health insurance, 234
Helicopteros Nacionales De Colombia, S.A. v. Hall et al., 31–33
Hold harmless clauses, 163–164
Hull insurance, 231
Hunt v. Washington State Apple Advertising Commission, 46

ICAO (*see* International Civil Aviation Organization)
Implied assumption of risk, 104
Implied preemption, 43
Implied warranties, 172–173
In personam jurisdiction, 15–16
In re: UAL Corporation, et al., Debtors. Chapter 11, Case No. 02-B-48191 (Jointly Administered), 185–187

In rem jurisdiction, 16
Independent contractors, 254–255, 260–261
Independent Safety Board Act of 1975, 123
Indictments, 70
Individuals, basic rights of, 49–52
In-flight (-motion) insurance, 235
Information, disclosure of, 127
Injunctions, 170
Injury:
   as element of negligence, 102
   redress for (*see* Tort law)
Inside directors, 204
Inspections, 127
Insurable interest in aircraft, 232, 239, 241–243
Insurance, 230–235, 267–268
Intangible property, 230
Intent, criminal, 69
Intentional infliction of emotional distress, 95
Intentional torts, 93–97
   against persons, 93–95
   against property, 95–97
Intermediate scrutiny test (equal protection), 51–52
International aviation law, 279–295
   Air France v. Saks, 288–289
   and definition of "accidents," 287–289
   El Al Israel Airlines, Ltd. v. Tsui Yuan Tseng, 289–293
   and forum for filing claims, 293–295
   and International Civil Aviation Organization, 287
   origins of, 280–281
   private, 283–286
   public, 281–283
   Resham Jeet Singh, Gursharan Jeet Kaur, Individually and as Guardians of Gurpreet Kaur v. Tarom Romanian Air Transport, 293–295
   and suits in actions based on local law, 289–293
International Civil Aviation Organization (ICAO), 282, 287
International Shoe Co v. Washington, 33

International Standards and Practices, 287
International Technical Committee of Aerial Legal Experts, 284
Interpretations of regulations/mandates, 125, 139
Interrogatories, 21
Interstate commerce, 45–48
Intervening cause, 101
Inverse condemnation, 50
Investigation:
    of accidents (NTSB), 123
    in enforcing statutes/regulations, 127–131

James Bowman v. American Home Assurance Company, 241–243
Johnson, Lyndon, 122
Joint tenancy, 219
Journal of Air Law and Commerce (Southern Methodist University), 27
Judges, 11, 12, 39
Judgment on pleadings, motion for, 20
Judgment-proof debtors, 175
Judicial branch (U.S. government), 39, 138–139
Judicial circuits, 11
Jurisdiction, 13–17
    concurrent, 14
    defined, 13
    federal, 28–31
    in federal courts, 11–12
    over parties (personal), 14–17
    personal, 31–33
    state, 15
    in state courts, 12–13
    subject matter, 13–15
Jury trials, 22, 26

Kassell v. Consolidated Freightways, Corp., 47
Kelo v. City of New London, Connecticut, 50
Key-person life insurance, 235
Klinicki v. Lundgren, 208–211

Labor unions, 261–263
Land, trespass to, 95–96

Law(s):
    code (statutory), 4–5
    common (case), 4, 5
    definitions of, 2
    functions of, 2–3
    priority of, 5, 6
    researching, 27
    sources of, 3–6
    (*See also specific laws; specific types of law, e.g.:* Criminal law)
Law review articles, 27
Lawyer's Edition, 26
Lear Romec, 114
Leasing:
    aircraft, 228–229
    real property, 222–223
Legal brief, 24
Legal precedents, 4
Legal research, 26–27
Legal systems, effectiveness of, 9–10 (*See also* United States legal system)
Legislative branch (U.S. government), 39, 140
Legislative enactments, 5
Legislative rules, 124 (*See also* Regulation(s))
Letter of Correction (FAA), 137
Letter of investigation (LOI), 127–128
Liability:
    in agency relationships, 258–261
    for corporations, 205–206
    dollar limits to, 285
    for limited partnerships, 198–199
    of LLC members, 201, 202
    for partnerships, 195–196, 198
    for sole proprietors, 193
    under Warsaw Convention, 284, 285
Liability insurance, 231–232
Libel, 95
Liens, 177
Life insurance, 235
Limited liability companies (LLCs), 200–202
Limited partnerships, 197–200
Line-Item Veto Act, 41
Liquidated damages, 169–170
Liquidation (Chapter 7) bankruptcy, 179

Litigation process, 17–25
    answer, 18–20
    appeals, 23–24
    civil procedure, 17
    complaint, 17–18
    enforcing judgments, 24–15
    pre-trial, 19–22
    summons, 18–19
    trial, 22–23
LLCs (*see* Limited liability companies)
Logbooks, FAA inspection of, 127
LOI (*see* Letter of investigation)
"Long-arm statutes," 16
Loss, risk of, 174
Lucia v. Teledyne, 28–31

Machado v. Administrator, 137
Mail fraud, 74–75
Malum in se, 68
Marbury v. Madison, 39
Martin v. OSHRC, 149
Material interference, 43
Mechanics' liens, 177
Mediation, 21, 25–26, 262
Medical certificates, 83
Mens reas, 69
Minitrials, 26
Miranda rights, 72
Misdemeanors, 8, 69
Mistakes, contracts and, 164–165
Model Penal Code (American Law Institute), 69, 70
Modified comparative negligence, 103–104, 107
Montreal Convention of 1999, 284–286, 295
Morals, law and ethics vs., 8, 9
Motion for judgment on pleadings, 20
Motion for summary judgment, 22
Mutual mistakes, 164–165

NASA (*see* National Aeronautic and Space Administration)
NASA ASRS Form 277B, 136, 319–321
National Aeronautic and Space Administration (NASA), 135–136

Index

National Labor Relations Act (NLRA), 261
National Labor Relations Board (NLRB), 120
National Mediation Board (NMB), 120, 262
National Transportation Safety Board (NTSB), 123
   accident investigation function of, 123
   adjudication of cases by, 40, 123, 131–133
   administrative law judges, 130
   EAJA rules by, 140
   and FAA emergency orders, 133–135
   and FAA reexamination of airmen, 138
   impact of, 120
   NYC05FA001 Accident Report, 150–153
   public meetings of, 142
   safety recommendations from, 120
   and substantial evidence test, 139
   web site catalogue of cases, 27
Negligence, 97–104
   and agency, 259–260
   breach of duty, 98–99
   comparative, 103–104, 106–107
   contributory, 103, 106
   defenses to, 102–104
   duty of care, 97–98
   elements of, 97
   injury, 102
   proximate cause, 99–102
Negligence per se, 98–99
Nelsen v. Morris, 211–212
New York Times Co. v. Sullivan, 52
"Nexus" test, 48
Nineteenth Amendment, 316
Ninth Amendment, 313
NLRA (National Labor Relations Act), 261
NLRB (National Labor Relations Board), 120
NMB (*see* National Mediation Board)
Noise regulation, 56–58
Nolo contendere, 71
Non-instrument-rated pilots, 149–153
Norris-LaGuardia Act of 1932, 261
Northwest Airlines, 179

Notice of Proposed Certificate Action (NPCA), 128–130
Notices (tort liability), 107
Not-in-flight (-motion) insurance, 231
NPCA (*see* Notice of Proposed Certificate Action)
NTSB (*see* National Transportation Safety Board)
NTSB Identification: NYC05FA001, 150–153

Occupational Safety and Health Act of 1970, 266
Occupational Safety and Health Administration (OSHA), 127, 266
Offer, defined, 160–161
Open skies agreements, 283
Operating agreement (LLCs), 200–201
Opinion and Order (NTSB), 132–133
Oral Initial Decisions, 132
Orr v. Orr, 52
OSHA (*see* Occupational Safety and Health Administration)
Ownership:
   of aircraft, 224–225, 229–230
   of assets, 196
   fractional, 229–230
   of real property, 219–220

Palsgraf v. Long Island Railroad, 100–101
Pan Am, 179
Paris Convention of 1919, 281, 282
Parties, jurisdiction over (personal jurisdiction), 14–17, 31–33
Partnerships, 194–197
Patents, 230
People v. DeFore, 71
Perfection of security interest, 176
Personal jurisdiction (*see* Parties, jurisdiction over)
Personal property, 224–230
   conversion of, 96–97
   intangible, 230
   tangible, 224–230
   trespass to, 96
Persons, intentional torts against, 93–95

Philko Aviation, Inc, v. Shacket et ux., 236–240
Pilots, non-instrument-rated, 149–153
Plaintiff misconduct, 106
Plaintiffs, 92
Plea bargains, 71
Precedents, 4
Preemption, federal, 42–44, 62
Presentment clause, 41
President, agency control by, 139
Presidential Emergency Board, 262–263
Press, freedom of, 52
Pretrial conferences, 21–22
Pretrial process, 19–22
Privacy Act of 1974, 141
Private international aviation law, 283–286
Private pilot certificates, 83
Private property:
   evidence obtained by flying over, 85–88
   preservation of, 3
   rights to, 3
Privity, 107
Procedural rules, 125
Product misuse, 106
Professional Pilots Federation, et al. v. FAA, 273–276
Proof:
   burden of, 92
   standard of, 7, 8
Property:
   and eminent domain, 50, 51
   and full faith and credit clause, 48–49
   intangible, 230
   intentional torts against, 95–97
   jurisdiction over, 16
   real, 218–223
   seizure of, 24
   "taking" of, 52–56
   takings clause, 49–50
   tangible, 224–230
   (*See also* Personal property)
Property law, 217–247
   and causal connection, 243–247
   and insurable interest in aircraft, 239, 241–243
   and insurance, 230–235
   James Bowman v. American Home Assurance Company, 241–243

Property law (*Cont.*):
  personal property, 224–230
  Philko Aviation, Inc, v. Shacket et ux., 236–240
  private property rights, 3
  public use airports, 223–224
  real property, 218–223
  South Carolina Insurance Company v. Lois S. Collins, 245–247
  Western Food Products Company, Inc. v. United States Fire Insurance Company, 243–245
Protections:
  for accused, 73
  for employees, 263–269
Proximate cause, 99–102
Public hearings, 142
Public international aviation law, 281–283
Public use airports, 223–224
Punitive damages, limits on, 107–108
Pure comparative negligence, 103, 106–107

Quasi in rem jurisdiction, 16

Racial discrimination, 269–272
Railway Labor Act (RLA), 261, 262
Rational relationship test, 51
RCRA (Resource Conservation and Recovery Act), 75
Real property, 218–223
  easements, 221
  leasing, 222–223
  ownership of, 219–220
  test for, 218
  transfers of, 220–221
  zoning issues with, 221–222
Reasonable doubt, 68
Reasonable person test, 98–99
Reasonableness test, 163
Reckless operations laws, 76
Reformation, 170
Registering aircraft, 225, 227
Regulation(s):
  of aviation, levels of, 76–79
  under commerce clause, 44–48
  creation of, 124–125

Regulation(s) (*Cont.*):
  interpretations of, 125
  noise, 56–58
  origin of, 122
  researching, 27
  state, 5
  "validly adopted," 145–149
Rental agreements, 222–223
Reorganization (Chapter 11) bankruptcy, 179, 180
Reporters (case law), 6
Repose, statute of, 107, 112–114
Res ipsa loquitur, 99
Rescission of contract, 168–169
Resham Jeet Singh, Gursharan Jeet Kaur, Individually and as Guardians of Gurpreet Kaur v. Tarom Romanian Air Transport, 293–295
Resource Conservation and Recovery Act (RCRA), 75
Respondeat superior, 260
Restatement of Torts (American Law Institute), 93
Rights:
  for accused, 73
  assignment of, 167
  of individuals/businesses, 49–52
Risk:
  assumption of, 104, 106
  of insurers, 230
  of loss, 174
RLA (*see* Railway Labor Act)
Robert David Ward v. State of Maryland, 76–79
Rulemaking, 123
  by administrative agencies, 124–125, 142–145
  APA requirements for, 124–125
  by FAA, 123

S corporations, 206–207
Safety:
  documentary evidence for, 74
  employee, 265–269
Sales:
  of aircraft, 224–226
  of real property, 220–221
Sales law, 170–174

Santa Monica Airport Association, et al. v. City of Santa Monica, 59
Sarbanes-Oxley Act of 2002, 210–211
Search and seizure, 71–72
SEC (*see* Securities and Exchange Commission)
Second Amendment, 312
Secured transactions, 175–177
Securities and Exchange Commission (SEC), 120, 142
Self-incrimination, 72
Separation of powers, doctrine of, 41
Seventeenth Amendment, 315
Seventh Amendment, 313
Shakespeare, William, 174
Shares, corporate, 203
Simpson, O. J., 92
Sixteenth Amendment, 315
Sixth Amendment, 73, 313
Skidmore v. Swift & Co., 125
Slander, 95
Smooth limits insurance, 231–232
Social Security, 268
Social Security Administration (SSA), 40
Sole proprietorships, 192–194, 201
Sources of law, 3–6
South Carolina Insurance Company v. Lois S. Collins, 245–247
Southern Methodist University, 27
Special (special jurisdiction) courts, 12
Special skills/knowledge, negligence and, 99
Specific performance, 170
Speech, freedom of, 52
Speedy trial, 73
SSA (Social Security Administration), 40
Stability of legal system, 9
Standard of proof:
  in civil cases, 7
  in criminal cases, 8
Stare decisis, doctrine of, 4
State administrative agencies, 120
State constitutions, 3–5, 38
State court systems, 12–13, 15–17
State criminal law, 76
State governments:
  distribution of authority between federal government and, 41–49
  limits on authority of, 49–52

State jurisdiction, 15
State laws/statutes, 7
    and priority of laws, 5
    reporters for, 27
    taxation, 62
State regulations, priority of laws and, 5
States' rights, 3
Statute of Frauds, 166, 255
Statute of limitations, 286
Statute of repose, 107, 112–114
Statutes, researching, 27
Statutory law (*see* Code law)
Storing aircraft, 225, 228
Strict liability crimes, 69
Strict product liability, 104–112
    Brockelsby v. United States of America and Jeppesen, 108–112
    defenses to, 105–108
    elements of, 104–105
Strict scrutiny test (due process), 51
Strict scrutiny test (equal protection), 52
Strikes, labor, 263
Subject matter, classifying law by, 6–7
Subject matter jurisdiction, 13–15
Sublimited insurance coverage, 231
Substantial evidence test, 139
Substantive due process, 51
Summary judgment, motion for, 22
Summary jury trials, 26
Summons, 18–19
Sunshine Act, 142
Superseding cause, 101
Supremacy clause, 42–43
Supreme Court Reporter, 26
Supreme courts:
    federal (*see* United States Supreme Court)
    state, 13
Surety arrangements, 177
Surrounding circumstances (in negligence), 98

"Taking" of property, 52–56
Takings clause (Fifth Amendment), 49–50
Tangible property, 224–230

Taxation, 59–62
    and agency relationships, 255
    of corporations, 206–207
    of limited partnerships, 200
    of LLC members, 202
    of partnerships, 196–197
    of sole proprietorships, 193–194
    in "undue burden" cases, 47–48
Teledyne Continental Motors, 114
Tenancy:
    in common, 219
    by the entirety, 219–220
Tenth Amendment, 114, 313
Terrorism, federal law on, 75
Third Amendment, 312
Third-party beneficiaries, 168
Thirteenth Amendment, 314
Title, transfer of, 173–174
Tort law, 91–114
    Altseimer v. Bell, 112–114
    Brockelsby v. United States of America and Jeppesen, 108–112
    criminal law vs., 92
    and GARA statute of repose, 112–114
    intentional torts, 93–97
    negligence, 97–104
    objectives of, 92
    strict product liability, 104–112
    wrongful death, 104
Tort liability, 259–261
Tort reform, 108
Trade names, 193
Trademarks, 230
Transfer of title, 173–174
Transferred intent, doctrine of, 94
Treaties, 4
    and priority of laws, 5
    as source of international law, 280
Trespass:
    to land, 95–96
    to personal property, 96
Trials:
    criminal, 71, 73
    NTSB, 132
    in U.S. legal system, 22–23
    under Warsaw Convention, 285–286
TWA, 179
Twelfth Amendment, 313–314
Twentieth Amendment, 316

Twenty-fifth Amendment, 318
Twenty-first Amendment, 317
Twenty-fourth Amendment, 317
Twenty-second Amendment, 317
Twenty-seventh Amendment, 318
Twenty-sixth Amendment, 318
Twenty-third Amendment, 317
Tyco, 210

U.C.C. (*see* Uniform Commercial Code)
"Undue burden" regulations, 46–48
Undue influence, 165–166
Unemployment insurance, 268
Uniform Commercial Code (U.C.C.), 166, 171, 174–176
Unilateral mistakes, 164
Unions, 261–263
United Airlines, 185–187
United Nations, 287
United States Coast Guard, actions by, 123
United States Code (U.S.C.), 74
United States Constitution (*see* The Constitution of the United States of America)
United States Court of Appeals (*see* Federal Court of Appeals)
United States legal system, 1–33
    alternative dispute resolution methods, 25–26
    classifications of law, 6–8
    court system, 11–13
    and distinctions between law, ethics, and morals, 8, 9
    and effectiveness of legal systems, 9–10
    federal jurisdiction case, 28–31
    functions of law, 2–3
    Helicopteros Nacionales De Colombia, S.A. v. Hall et al., 31–33
    jurisdictional matters, 13–17
    legal research, 26–27
    litigation process, 17–25
    Lucia v. Teledyne, 28–31
    personal jurisdiction case, 31–33
    sources of law, 3–6
United States of America v. SabreTech, 75, 80–83

United States Reports, 26
United States Supreme Court, 12
   and commerce clause, 44
   power of judicial review, 39–40
   reporters for, 26
   (See also specific cases)
United States v. Causby et ux., 53–56
United States v. Evinger, 83–85
United States v. Lopez, 45
United States v. Morrison, 45
United States v. State of New York, 59
United States v. Virginia, 52
Unsecured credit, 175
Unwarranted by the facts standard, 139

U.S.C. (*see* United States Code)
"Validly-adopted" interpretations, 139
"Validly-adopted" regulations, 145–149
Venue, 16–17
Violations (criminal), 69
Voir dire, 22

Wackenhut Corporation and Delta Airlines, Inc. v. Lippert Supreme Court of Florida, 300–301
Wagner Act, 261
Warning Notice (FAA), 137
Warranties, 171–173, 179–182
Warrants, arrest, 70

Warsaw Convention, 284–286
Western Food Products Company, Inc. v. United States Fire Insurance Company, 243–245
Wet leases, 228, 229
Wickard v. Filburn, 44–45
Wire fraud, 74–75
Workers' compensation, 267–268
WorldCom, 210
Writ of certiori, 12, 79
Wrongful death, 104

Zone of danger, 101
Zoning, 221–222